AF581867

Advanced Strategies to Preserve Quality and Extend Shelf Life of

Advanced Strategies to Preserve Quality and Extend Shelf Life of Foods

Editors

Matteo Alessandro Del Nobile
Amalia Conte

MDPI • Basel • Beijing • Wuhan • Barcelona • Belgrade • Manchester • Tokyo • Cluj • Tianjin

Editors
Matteo Alessandro Del Nobile
University of Foggia
Italy

Amalia Conte
University of Foggia
Italy

Editorial Office
MDPI
St. Alban-Anlage 66
4052 Basel, Switzerland

This is a reprint of articles from the Special Issue published online in the open access journal *Foods* (ISSN 2304-8158) (available at: https://www.mdpi.com/journal/foods/special_issues/Preserve_Quality).

For citation purposes, cite each article independently as indicated on the article page online and as indicated below:

LastName, A.A.; LastName, B.B.; LastName, C.C. Article Title. *Journal Name* **Year**, *Volume Number*, Page Range.

ISBN 978-3-0365-3913-3 (Hbk)
ISBN 978-3-0365-3914-0 (PDF)

Contents

About the Editors

Matteo Alessandro Del Nobile received his Ph.D. Degree in Material Science from the University of Naples "Federico II". Since 2001, he has worked at the University of Foggia where he is full professor. Most of his early research efforts were put towards determining the relationships that exist between the physical and chemical structure of polymers and their mass transport properties. This topic has been approached both from a theoretical and from an experimental point of view. The knowledge he gained studying polymers has more recently been applied to food science. He has published over 250 papers on topics related to food science with particular attention on food packaging, food preservation and functional food. He has been Principal Investigator of many national projects and is the coordinator of the Ph.D. in "Translational Medicine and Food: Innovation, Safety and Management" at the University of Foggia.

Amalia Conte She graduated at the University of Foggia in Food Science and Technology and since 2008 she has been a researcher. Her research interests include food packaging, functional food, shelf-life prolongation and novel technologies. In the last 10 years, she has been involved in successful national research projects. To date, she has co-authored more than 200 scientific publications in international peer-reviewed journals, 20 book chapters, 1 book and 1 edited book, as well as serving as guest editor for four Special Issues. She is a peer-reviewer for various scientific journals in the food science sector. She is the CEO of Spin Off and is co-inventor of five patents.

 MDPI

Editorial

Introduction to the Special Issue: Advanced Strategies to Preserve Quality and Extend Shelf Life of Foods

Amalia Conte * and Matteo A. Del Nobile *

Department of Agricultural Sciences, Food and Environment, University of Foggia, Via Napoli 25, 71122 Foggia, Italy
* Correspondence: amalia.conte@unifg.it (A.C.); matteo.delnobile@unifg.it (M.A.D.N.)

Citation: Conte, A.; Del Nobile, M.A. Introduction to the Special Issue: Advanced Strategies to Preserve Quality and Extend Shelf Life of Foods. *Foods* **2022**, *11*, 1052. https://doi.org/10.3390/foods11071052

Received: 15 March 2022
Accepted: 2 April 2022
Published: 6 April 2022

We are pleased to present this Special Issue, which includes 13 papers that highlight the most important research activities in the field of food quality assurance and shelf-life extension. The goal of this Special Issue was to broaden the current knowledge of advanced approaches to guarantee the maintenance of the properties of packaged products during storage. The most consolidated strategies in the literature concern the use of heat and modified atmospheres. However, knowledge gained in the sector has broadened the perspective and found valid and effective alternative in the use of bioactive compounds, industrial food by-products, adoption of active packaging solutions or the application of novel mild treatments, such as pulsed light, ultrasounds, high-pressure processing and cold plasma.

The 11 research articles/communication/ and 2 reviews that comprise this Special Issue highlight the most recent research and investigations into this exciting area, covering the following topics: (i) vacuum packaging; (ii) cork closures; (iii) innovative active packaging; (iv) emerging technologies; (v) the reuse of by-products; and (vi) secondary shelf life.

Interesting results that bring to light the issues concerning the effects of vacuum packaging on surface color and lipid oxidation of beef steaks were presented by Reyes et al. [1]. The results from this study suggest that the use of vacuum packaging for beef steaks is plausible for maintaining quality characteristics during extended display periods.

The study of Amaro et al. [2] aimed at investigating the impact of different technical cork stoppers on the quality preservation and shelf life of sparkling wines. The volatile compositions of two Italian sparkling wines sealed with a sparkling cork with two natural cork discs (2D) and a micro-agglomerated (MA) cork were determined during bottle aging (12 to 42 months) after disgorging. The results unveiled that the type of closure has a greater impact on the volatile composition of sparkling wines at longer post-bottling periods, and 2D stoppers preserve the fruity and sweet aromas of sparkling wines better after 42 months of bottle storage.

The next four papers dealt specifically with the effects of active packaging on food shelf life. In particular, Ambrosio et al. [3] proved the positive effects of an active polypropylene-based packaging functionalized with the antimicrobial peptide 1018K6 on microbial growth, physicochemical properties and sensory attributes of raw salmon fillets and hamburgers of Sarda sarda during storage. Roy et al. [4] developed a pullulan/chitosan-based multifunctional edible composite film by reinforcing mushroom-mediated zinc oxide nanoparticles (ZnONPs) and propolis. The system was advantageously used for wrapping pork belly. Gutiérrez-Jara et al. [5] coated sweet cherries by electro-spraying with an edible nano-emulsion of alginate and soybean oil, with or without a $CaCl_2$ cross-linker to reduce fruit cracking. It was interesting to observe that the use of the nano-emulsion + $CaCl_2$ coating on sweet cherries helps to reduce cracking and maintain fruit quality at 4 °C for about 1 month. Finally, Socaciu et al. [6] studied the effects of a whey-protein-isolate-based film incorporated with tarragon essential oil on the quality and shelf-life of refrigerated brook trout. The selected essential oil conferred antioxidant and antimicrobial properties to the film. Thus, the developed active packaging system could be a promising material for fresh fish packaging.

As regards the adoption of active compounds of natural origin, two papers have been published, one dealing with shelf-life extension of chilled pork by optimal ultrasonicated ceylon spinach (*Basella alba*) extracts [7] and another one on the potential of algae extracts for extending the shelf life of rainbow trout (*Oncorhynchus mykiss*) fillets [8]. In the first study, Phimolsiripol et al. [7] found that fresh pork treated with the ultrasonicated extracts at 100 and 120 mg/mL had lower values of thiobarbituric acid reactive substances (TBARS) than the control (without dipping). From the food safety standpoint, as measured by the total microbial count, the fresh pork dipped with 100–120 mg/mL spinach extract could be kept at 0 °C for 7 days, 2 to 3 days longer than control meat at 0 and 4 °C, respectively. The results of Saez et al. [8] on the shelf life of rainbow trout demonstrated that algae extracts are also naturally effective agents for preserving fish.

In the context of natural compounds used for shelf-life extension, another two studies have been also published in the current Special Issue. This is the case of one article and one review dealing with fruit and vegetable by-products, whose valorization is considered a hot topic. In the article of Panza et al. [9], olive paste, a by-product from olive oil production, was valorized as breading for fresh fish sticks stored for 15 days at 4 °C. The results proved that the enrichment with olive paste increased the total phenols, the flavonoids and the antioxidant activity of the breaded fish samples compared to the control, without compromising the sensory parameters. The overview of Nardella et al. [10] collects the recent applications of fruit and vegetable by-products as valid components to prolong food shelf life. This review provides a detailed picture of the state-of-art of the literature on the topic in the last 10 years. The review highlights the potential of by-products and the clear advances in terms of food sustainability, even though the current situation still limits by-product diffusion. The authors also underlined that for future perspectives of by-products recycling, multidisciplinary research is of striking importance, as it is able to promote the scale-up of by-products and encourage their adoption at the industrial level.

As regards the emerging technologies and food shelf life, one article and one review were published in the current Special Issue. In particular, the article deals with the effects of gaseous ozone on microbiological quality of Andean blackberries (*Rubus glaucus* benth) [11]. Andean blackberries are highly perishable. Ozone was applied prior to storage at 0.4, 0.5, 0.6 and 0.7 ppm for 3 min, and this treatment was found effective in maintaining the quality of blackberries throughout refrigerated storage. The authors suggest that higher doses could be advisable to enhance its antimicrobial activity. The review of Tavares et al. [12] deals with emergent preservation techniques (chilling and super-chilling) as a complement, or even replacement of conventional preservation methodologies (refrigeration and freezing), to assure fish safety and extend shelf life without compromising food safety. In addition, the use of novel food packaging methodologies (edible films and coatings) was also presented and discussed, along with a new storage methodology, hyperbaric storage, that uses storage pressure control as a hurdle microbial development and slows down organoleptic decay at subzero, refrigeration and room temperatures.

One paper dealing with secondary shelf life (SSL) is also included in this Special Issue. SSL represents the time after package opening during which the food product retains a required level of quality. The study of Nicosia et al. [13] suggests the possibility to significantly extend or even omit the SSL indications for industrial pesto sauces because the product remained acceptable for a time longer than that reported on the label. This study could have practical outcomes at the domestic level in terms of food waste reduction and at industrial level in terms of reduced household stock turnover and consequent cost savings.

Taken together, these studies are clear evidence of how the achievement of food shelf-life extension is still a complex and multifaceted process. Food manufacturers have to meet consumer demands for freshness and convenience without compromising the safety of foods, and the food industry is thus continuously challenged and seeking sustainable and practical methods to ensure the safety of products and guarantee the maximum level of security for consumers. The innovative and exciting research included in this Special Issue highlights the interest and potential of this emerging area, addressing some of the most pressing global issues.

In summary, all the papers published in this Special Issue highlighted a large portion of the research activities in the field of advanced application of novel processing, antimicrobial/antioxidant substances as well as from by-products and active packaging. The development of these topics and the exploration of their combined use will remain a very active research area in the coming decades.

We sincerely hope that the readers will find this Special Issue interesting and informative.

Author Contributions: A.C. and M.A.D.N. equally contributed to organizing the Special Issue, to editorial work and to writing this editorial. All authors have read and agreed to the published version of the manuscript.

Funding: This research received no external funding.

Acknowledgments: We thank and congratulate all authors for submitting the manuscripts of high quality. We are very grateful to the authors who have shared their scientific knowledge. We also thank the reviewers' willingness for their careful evaluations to improve the papers.

Conflicts of Interest: The authors declare no conflict of interest.

References

1. Reyes, T.M.; Wagoner, M.P.; Zorn, V.E.; Coursen, M.M.; Wilborn, B.S.; Bonner, T.; Brandebourg, T.D.; Rodning, S.P.; Sawyer, J.T. Vacuum Packaging Can Extend Fresh Color Characteristics of Beef Steaks during Simulated Display Conditions. *Foods* **2022**, *11*, 520. [CrossRef] [PubMed]
2. Amaro, F.; Almeida, J.; Oliveira, A.S.; Furtado, I.; Bastos, M.L.; Guedes de Pinho, P.; Pinto, J. Impact of Cork Closures on the Volatile Profile of Sparkling Wines during Bottle Aging. *Foods* **2022**, *11*, 293. [CrossRef] [PubMed]
3. Ambrosio, R.L.; Gogliettino, M.; Agrillo, B.; Proroga, Y.T.R.; Balestrieri, M.; Gratino, L.; Cristiano, D.; Palmieri, G.; Anastasio, A. An Active Peptide-Based Packaging System to Improve the Freshness and Safety of Fish Products: A Case Study. *Foods* **2022**, *11*, 338. [CrossRef] [PubMed]
4. Roy, S.; Priyadarshi, R.; Rhim, J.W. Development of Multifunctional Pullulan/Chitosan-Based Composite Films Reinforced with ZnO Nanoparticles and Propolis for Meat Packaging Applications. *Foods* **2021**, *10*, 2789. [CrossRef] [PubMed]
5. Gutiérrez-Jara, C.; Bilbao-Sainz, C.; McHugh, T.; Chiou, B.S.; Williams, T.; Villalobos-Carvajal, R. Effect of Cross-Linked Alginate/Oil Nanoemulsion Coating on Cracking and Quality Parameters of Sweet Cherries. *Foods* **2021**, *10*, 449. [CrossRef] [PubMed]
6. Socaciu, M.I.; Fogarasi, M.; Simon, E.L.; Semeniuc, C.A.; Socaci, S.A.; Podar, A.S.; Vodnar, D.C. Effects of Whey Protein Isolate-Based Film Incorporated with Tarragon Essential Oil on the Quality and Shelf-Life of Refrigerated Brook Trout. *Foods* **2021**, *10*, 401. [CrossRef] [PubMed]
7. Phimolsiripol, Y.; Buadoktoom, S.; Leelapornpisid, P.; Jantanasakulwong, K.; Seesuriyachan, P.; Chaiyaso, T.; Leksawasdi, N.; Rachtanapun, P.; Chaiwong, N.; Sommano, S.R.; et al. Shelf Life Extension of Chilled Pork by Optimal Ultrasonicated Ceylon Spinach (*Basella alba*) Extracts: Physicochemical and Microbial Properties. *Foods* **2021**, *10*, 1241. [CrossRef] [PubMed]
8. Sáez, M.I.; Suárez, M.D.; Alarcón, F.J.; Martínez, T.F. Assessing the Potential of Algae Extracts for Extending the Shelf Life of Rainbow Trout (*Oncorhynchus mykiss*) Fillets. *Foods* **2021**, *10*, 910. [CrossRef] [PubMed]
9. Panza, O.; Lacivita, V.; Palermo, C.; Conte, A.; Del Nobile, M.A. Food By-Products to Extend Shelf Life: The Case of Cod Sticks Breaded with Dried Olive Paste. *Foods* **2020**, *9*, 1902. [CrossRef] [PubMed]
10. Nardella, S.; Conte, A.; Del Nobile, M.A. State-of-Art on the Recycling of By-Products from Fruits and Vegetables of Mediterranean Countries to Prolong Food Shelf Life. *Foods* **2022**, *11*, 665. [CrossRef] [PubMed]
11. Horvitz, S.; Arancibia, M.; Arroqui, C.; Chonata, E.; Virseda, P. Effects of Gaseous Ozone on Microbiological Quality of Andean Blackberries (*Rubus glaucus* Benth). *Foods* **2021**, *10*, 2039. [CrossRef] [PubMed]
12. Tavares, J.; Martins, A.; Fidalgo, L.G.; Lima, V.; Amaral, R.A.; Pinto, C.A.; Silva, A.M.; Saraiva, J.A. Fresh Fish Degradation and Advances in Preservation Using Physical Emerging Technologies. *Foods* **2021**, *10*, 780. [CrossRef] [PubMed]
13. Nicosia, C.; Fava, P.; Pulvirenti, A.; Antonelli, A.; Licciardello, F. Domestic Use Simulation and Secondary Shelf Life Assessment of Industrial *Pesto alla Genovese*. *Foods* **2021**, *10*, 1948. [CrossRef] [PubMed]

Article

Vacuum Packaging Can Extend Fresh Color Characteristics of Beef Steaks during Simulated Display Conditions

Tristan M. Reyes [1], Madison P. Wagoner [1], Virginia E. Zorn [1], Madison M. Coursen [1], Barney S. Wilborn [1], Tom Bonner [2], Terry D. Brandebourg [1], Soren P. Rodning [1] and Jason T. Sawyer [1,*]

[1] Department of Animal Sciences, Auburn University, Auburn, AL 36849, USA; tzr0039@auburn.edu (T.M.R.); mpw0035@auburn.edu (M.P.W.); vez0001@auburn.edu (V.E.Z.); madison.coursen@auburn.edu (M.M.C.); wilbobs@auburn.edu (B.S.W.); tdb0006@auburn.edu (T.D.B.); rodnisp@auburn.edu (S.P.R.)

[2] Winpak Ltd., 100 Saulteaux Crescent, Winnipeg, MB R3J 3T3, Canada; tom.bonner@winpak.com

* Correspondence: jts0109@auburn.edu; Tel.: +1-334-844-1517

Abstract: Packaging technology is evolving, and the objectives of this study were to evaluate instrumental surface color, expert color evaluation, and lipid oxidation (TBARS) on beef *longissimus lumborum* steaks packaged in vacuum-ready packaging (VRF) or polyvinyl chloride (PVC) overwrap packaging. Paired strip loins (Institutional Meat Purchasing Specifications # 180) were cut into 2.54-cm-thick steaks and assigned randomly to one of two packaging treatments, VRF or PVC. Steaks packaged in VRF were lighter in color ($p < 0.05$) as the display period increased, whereas steaks packaged in PVC became darker ($p < 0.05$). Redness (a*) values were greater ($p < 0.05$) for PVC steaks until day 5, whereas VRF steaks had a greater ($p < 0.05$) surface redness from day 10 to 35 of the display period. Calculated spectral values of red to brown were greater ($p < 0.05$) for steaks in VRF than PVC. In addition, expert color evaluators confirmed VRF steaks were less brown and less discolored ($p < 0.05$) from day 5 to 35 of the display. Nonetheless, lipid oxidation was greater ($p < 0.05$) for PVC steaks from day 10 through day 35 of the display. Results from this study suggest that the use of vacuum packaging for beef steaks is plausible for maintaining surface color characteristics during extended display periods.

Keywords: instrumental color; overwrapped packaging; simulated retail display; TBARS; vacuum packaging

Citation: Reyes, T.M.; Wagoner, M.P.; Zorn, V.E.; Coursen, M.M.; Wilborn, B.S.; Bonner, T.; Brandebourg, T.D.; Rodning, S.P.; Sawyer, J.T. Vacuum Packaging Can Extend Fresh Color Characteristics of Beef Steaks during Simulated Display Conditions. *Foods* **2022**, *11*, 520. https://doi.org/10.3390/foods11040520

Academic Editors: Matteo Alessandro Del Nobile and Amalia Conte

Received: 22 December 2021
Accepted: 9 February 2022
Published: 11 February 2022

Publisher's Note: MDPI stays neutral with regard to jurisdictional claims in published maps and institutional affiliations.

1. Introduction

Vacuum packaging using form-and-fill technology is a packaging method that is becoming one of the most prominent packaging systems in use within the retail meat industry [1]. Unfortunately, previous research focused on form-and-fill vacuum packaging for use with fresh meat storage in a retail setting is limited. Previous efforts in vacuum packaging uses for fresh meat have focused on using bag or skin technologies [2]. Form-and-fill packaging systems use one film to construct a pouch with time, pressure, and heat. After forming the pouch, meat products are placed into the pouch and a second film is overlayed and sealed within the vacuum chamber. Furthermore, vacuum packaging has accounted for 40% of packaging types within meat cases, with most products packaged using a roll-stock machine [1]. It has been noted that PVC overwrapped packaged beef has decreased in use by 46% from 2018 to 2021 [1].

While the meat surface color is still regarded as one of the greatest determining factors consumers utilize when purchasing fresh beef in the retail setting [3,4], packaging technologies are pivotal in maintaining the surface color of fresh meat. PVC is a packaging method used with fresh meat that allows oxygen and other gasses to permeate through the film in large quantities allowing oxygen to bind with myoglobin. The oxymyoglobin state of beef is often correlated with a fresher and more wholesome product by consumers due to a bright cherry red color [5]. Creating a shift from the current industry's primary packaging methods of PVC to vacuum packaging is unclear; however, many advantages such as the extension of shelf life and color stability may exist with the use of vacuum packaging in fresh meat applications. Vacuum packaging allows meat products to remain more color stable over extended periods of time within retail coolers [6]. Reportedly, vacuum packaging has been known to extend the storage period of fresh meat products by reducing the amount of residual oxygen within the package [7,8].

With the ability to extend fresh meat storage through the use of vacuum packaging, it is a packaging system that is quickly becoming an essential part of the solution for meeting sustainability programs and reducing food waste for the meat industry. Food waste has been characterized as edible food that is not consumed and often discarded by consumers or retailers [9]. It has also been reported that meat, poultry, and fish were the top food groups contributing to an estimated food loss approaching $48 billion in 2010 [9]. In addition, approximately 43 billion pounds of food at the retail level and 90 billion pounds at the consumer level have not been consumed [9]. Aside from food loss and food waste issues that still reside within the meat and food industry, there exist excessive food packaging materials entering the waste management system destined for landfills. In 2017, there were approximately 26.3 billion pounds of beef, 25.6 billion pounds of pork, and 42.2 billion pounds of chicken that American meat companies processed [10].

The packaging of fresh meat products is a necessity for the purpose of maintaining a fresh and wholesome product during retail display for consumer purchases. With the volume of packaging necessary to address the meat industry's demand of packaged meat products, it is essential that a packaging option be investigated for extending storage times of meat products. New packaging technologies could assist in reducing the volume of markdowns and throwaways that occur at the retail counter. A large percentage of fresh meat has been packaged with a form-and-fill roll stock machine, which utilizes multi-layered packaging films [11]. Multi-layered vacuum packaging is constructed with a wide variety of materials that can include amorphous polyethelene terephthalate, polyolefines, ethylene vinyle alcohol, polyvinylidene di-chloride, and nylon [10–13]. Currently, the ability to recycle multi-layered films lacks economic viability due to the nature of the film layering [14].

Nonetheless, multi-layered vacuum packaging films are growing in popularity for vacuum packaging platforms; unfortunately, these packaging films are often constructed without sustainable or recycle-ready materials. Limitations in recycle-ready packaging materials can create difficulties downstream from the consumer with sustainable meat packaging due to challenges in the delamination process of multi-layered films [10,15]. Nevertheless, an investigation into using multi-layered films is a necessity to extend the fresh-meat shelf life. With a need for greater storage periods of fresh meat by retailers, customers, and consumers, the agriculture industry could focus its efforts on becoming more sustainable through innovative developments of packaging materials for meat and meat products. Therefore, the objectives of the current study were to investigate the feasibility of using VRF vacuum packaging film in place of PVC overwrapping on beef strip loin steaks and the subsequent impacts on surface color characteristics during a simulated retail display period.

2. Materials and Methods

2.1. Raw Materials

Cattle (n = 7) were harvested under simulated commercial conditions according to USDA humane slaughter standards at the Auburn University Lambert Powell Meat Laboratory after a 12 h rest period. After harvest, carcasses were chilled for 48 h at 2 °C. Following carcass chilling, beef carcasses were subsequently fabricated into left- and right-side paired (IMPS # 180) boneless beef strip loins, vacuum packaged (3 mil, Clarity Vacuum Pouches, Kansas City, MO, USA), and stored in the absence of light for 10 days to simulate boxed beef fabrication and logistics. After aging, beef strip loins were cut into 2.54-cm-thick steaks (N = 112 steaks/packaging treatment) using a BIRO bandsaw (Model 3334, BIRO

Manufacturing Company, Marblehead, Ohio, USA). At the time of steak cutting, steaks from each loin were allocated randomly to one of two packaging treatments, VRF or PVC. The allocated steaks were placed onto a plastic tray and allowed to bloom for 30 min prior to packaging.

2.2. Packaging and Simulated Display Conditions

After steak portioning, steaks allocated to vacuum packaging (VRF) were packaged using a Reiser form-and-fill vacuum packaging machine (Optimus OL0924, Variovac, Zarrentin, Germany) and sealed. Steaks were packaged in VRF packaging films (O_2 transmission rate = 0.8 cc/sq. m^2/24 h/atm). Steaks allocated to traditional overwrapping (PVC) were placed onto a foam tray (2s, Genpak, Charlotte, NC, USA) with an absorbent moisture pad (DRI-LOC AC-50, Novipax, Oak Brook, IL, USA) and wrapped by hand with a polyvinyl chloride film (O_2 transmission rate = 14,000 cc O_2/m/24 h/atm).

Packaged steaks were placed onto lighted shelves within a refrigerated retail display case (Model TOM- labels 60DXB-N, Turbo Air Inc., Long Beach, CA, USA). Packages of steaks were displayed for 35 days at 3 °C ± 1.2 °C, and the case temperature throughout the display period was monitored with temperature data recorders (Model-TD2F, ThermoWorks, American Fork, UT, USA) placed on the center of each display shelf. Packages of steaks were displayed on shelves under continuous LED lighting with an intensity of 2297 lux for each shelf. Lighting intensity was measured (ILT10C, International Light Technologies, Peabody, MA, USA) throughout the duration of the simulated display period. Additionally, packages of steaks were distributed evenly across all shelves and rotated daily from side to side and front to back to simulate consumer movement. Fresh meat characteristics of instrumental color, surface color, lipid oxidation, purge loss, and pH were measured on days 0, 5, 10, 15, 20, 25, 30, and 35 throughout the simulated display period.

2.3. Instrumental Color

Throughout the 35-day simulated retail display period, the instrumental surface color was measured on packaged steaks (n = 28) with a HunterLab MiniScan EZ colorimeter, Model 45/0 LAV (Hunter Associates Laboratory Inc., Reston, WV, USA). Prior to surface color readings, the colorimeter was standardized using a black and white tile. Instrumental color values were determined from the mean of three readings through the surface of each unopened package using illuminant A, an aperture of 31.8 mm, and a 10° observer. Packages of steaks were evaluated for lightness (L*), redness (a*), and yellowness (b*) using the Commission Internationale de l' Eclairage guidelines for surface color [16]. In addition, the hue angle was calculated as $\tan^{-1}$ (b*/a*), with a greater value indicative of the surface color shifting from red to yellow. Chroma (C*) was calculated as $\sqrt{a^{*2} + b^{*2}}$, where a larger value indicates a more vivid color. Lastly, reflectance values within the spectral range of 400 to 700 nm were used to capture the surface color changes from red to brown by calculating the reflectance ratio of 630 nm:580 nm and the relative values of deoxymyoglobin (DMb = {[1.395 − ({A572 − A700}/{A525 − A700})]} × 100), metmyoglobin (MMb = {2.375 × [1 − ({A473 − A700}/{A525 − A700})]} × 100), and oxymyoglobin (OMb = DMb − MMb) according to color guidelines previously described [17].

2.4. Expert Color Evaluation

A five-member, expert color panel was used to evaluate the surface color of packaged beef boneless strip steaks during the simulated retail display period. Color measuring experts used anchors for scoring surface color discoloration previously described and modified from meat color guidelines [12]. At 16:00 h on the day of simulated display, experts rated surface color changes for steaks (n = 28) every 5 days for 35 days of refrigerated storage. Surface color ratings were created for steaks packaged under vacuum (VRF) for the initial beef color (1 = extremely bright purple-red, 2 = bright purple-red, 3 = moderately bright purple-red, 4 = slightly purple-red, 5 = slightly dark purple, 6 = moderately dark purple, 7 = dark purple, 8 = extremely dark purple), whereas packages of PVC overwrapped steaks were rated for the initial beef color (1 = extremely bright cherry-red, 2 = bright cherry-red, 3 = moderately bright, 4 = slightly bright cherry-red, 5 = slightly dark cherry-red, 6 = moderately dark red, 7 = dark red, 8 = extremely dark red). Both VRF- and PVC-packaged steaks were rated for the amount of browning (1 = no evidence of browning, 2 = dull, 3 = grayish, 4 = brownish gray, 5 = brown, and 6 = dark brown) and percent (%) discoloration (1 = no discoloration [0%], 2 = slight discoloration [1–20%], 3 = small discoloration [21–40%], 4 = modest discoloration [41–60%], 5 = moderate discoloration [61–80%], 6 = extensive discoloration [81–100%]).

2.5. *Purge Loss and Fresh Muscle pH*

Prior to conducting lipid oxidation analysis, steaks were removed from their respective packaging materials, blotted dry, and weighed on an analytic balance (PB3002-S, Mettler Toledo, Columbus, OH, USA). Purge loss was calculated as [(packaged steak weight − steak weight) ÷ packaged steak weight × 100)]. After capturing the purge loss for each steak, fresh muscle pH was measured in duplicate with a glass electrode inserted into two random locations within the steak and attached to a pH meter (Model-HI99163, Hanna Instruments, Woonsocket, RI, USA). Prior to measuring, the pH probe was calibrated (pH 4.0 and 7.0) using 2-point standard buffers (Thermo Fisher Scientific, Chelmsford, MA, USA) and again after 10 readings.

2.6. *Lipid Oxidation*

Packaged steaks ($n = 56$) were removed from their packaging material and sampled for 2-thiobarbituric acid reactive substances (TBARS) using a previously described method [18]. Steaks were trimmed of all external fat and connective tissue then minced together to form a uniform sample of the entire steak. Approximately 2 g of minced muscle was homogenized with 8 mL of cold (1 °C) 50 mM phosphate buffer (pH of 7.0 at 4 °C) containing 0.1% EDTA, 0.1% n-propyl gallate, and 2 mL trichloroacetic acid (Sigma-Aldrich, Saint Louis, MO, USA). Homogenized samples were subsequently filtered through Whatmann No. 4 filter paper, and duplicate 2-mL aliquots of the clear filtrate were transferred into 10-mL borosilicate tubes, mixed with 2 mL of 0.02 M 2-thiobarbituric acid reagent (Sigma-Aldrich, Saint Louis, MO, USA) then boiled for 20 min. After boiling, tubes were placed into an ice bath for 15 min. Absorbance was measured at 533 nm with a spectrophotometer (Turner Model–SM110245, Barnstead International, Dubuque, IA, USA) and multiplied using a factor of 12.21 to obtain the TBARS value (mg malonaldehyde/kg of meat).

2.7. *Statistical Analysis*

Data were analyzed with the GLIMMIX procedures of SAS (ver. 9.4; SAS Institute Inc. Cary, NC, USA) with treatment serving as the lone fixed effect and replication serving as the random effect for instrumental color, expert color, lipid oxidation, purge loss, and pH. All data were analyzed in a modified randomized design with steak serving as the experimental unit. For expert surface color rating data, the expert color panelist was included as a random factor, and panelist × day of display was included as a random, repeated factor (with a first-order autoregressive covariance structure). Least-squares means were generated, and when significant ($p \leq 0.05$) F-values were observed, least-squares means were separated using a pair-wise *t*-test (PDIFF option).

3. Results and Discussion

3.1. *Instrumental Beef Color*

The instrumental surface color of packaged steaks was measured throughout a 35-day simulated retail period. An interaction of the packaging method × day of display on steak surface lightness (L*) occurred (Table 1). Steaks packaged in PVC were lightest ($p < 0.05$) on day 0 and became darker as the length of display period increased (Table 1). However, from day 20 through day 35 of the display, steaks packaged using VRF were lighter ($p < 0.05$) than steaks packaged using PVC methods (Table 1). Additionally, surface redness (a*) for beef steaks packaged in PVC were redder ($p < 0.05$) from day 0 through day 15 of the display period (Table 1), whereas steaks packaged in VRF became significantly redder ($p < 0.05$) until the conclusion of the study on day 35 (Table 1). Greater a* values are indicative of a redder fresher color and have a greater consumer appeal at the time of the consumers' purchasing decision. PVC-packaged steaks maintained greater ($p < 0.05$) values for yellowness (b*) throughout the duration of simulated retail display than steaks packaged in VRF (Table 1). The changes in surface color for steaks packaged using VRF indicated surface lightness and redness were more stable throughout the entire simulated retail period than steaks packaged in PVC. As expected during a simulated retail period, fresh steaks packaged in an oxygen-rich permeable method such as PVC will have a brighter surface color initially. Similar findings have reported that ground beef packaged using PVC methods resulted in greater L* values on day 0, along with greater a* and b* through only 50% of the display period [5] when displayed up to 35 days. Moreover, ground beef patties when packaged with PVC materials have recorded similar results, indicating a* values will decline within the first 5 days of the display period [19]. However, a* values for ground beef patties packaged using a vacuum packaging platform have been reported to increase throughout a display period [19]. Furthermore, a study evaluating the surface color of beef steaks indicated a* values were greater for vacuum packaging rather than other packaging types at the conclusion of a 35-day study [20]. It appears the results for b* values of ground beef and steaks are consistent with the current study, resulting in a decline during a 5-day retail storage period when packaged in PVC. Regardless of the fluctuation of yellowness, the current and previous results suggest b* values are less stable regardless of the packaging method [5,19–21].

Table 1. The interactive impact of packaging method × day of display for instrumental surface color values on fresh beef strip loin steaks during a simulated retail display.

	Day								
	0	5	10	15	20	25	30	35	SEM *
PVC									
L* [1]	46.85 [a]	45.42 [abc]	44.26 [cde]	44.37 [bcde]	43.31 [de]	43.08 [def]	42.47 [efg]	40.63 [g]	0.713
a* [1]	29.57 [a]	25.40 [b]	19.33 [d]	15.93 [e]	15.69 [e]	15.79 [e]	15.93 [e]	15.99 [e]	0.704
b* [1]	21.33 [a]	19.97 [b]	17.71 [c]	15.13 [d]	14.29 [de]	14.03 [de]	13.69 [e]	13.44 [e]	0.392
C* [2]	36.47 [a]	32.32 [b]	26.31 [c]	22.07 [efg]	21.41 [g]	21.31 [g]	21.16 [g]	21.01 [gh]	0.672
Hue (°) [3]	35.76 [d]	38.22 [cd]	43.24 [ab]	43.81 [a]	42.87 [ab]	42.16 [ab]	41.08 [abc]	40.20 [bc]	1.185
RTB [4]	5.28 [a]	3.94 [b]	2.57 [de]	1.92 [f]	1.99 [f]	2.03 [f]	2.07 [f]	2.22 [ef]	0.143
MMb [5]	20.40 [ef]	28.08 [d]	40.18 [abc]	43.22 [a]	41.09 [ab]	38.81 [abc]	37.14 [bc]	34.59 [c]	2.081
DMb [5]	4.89 [f]	7.68 [ef]	14.27 [e]	24.25 [d]	33.14 [c]	34.91 [c]	35.74 [c]	38.74 [c]	3.161
OMb [5]	74.71 [a]	64.25 [b]	45.55 [c]	32.53 [d]	25.77 [e]	26.28 [e]	27.12 [e]	26.67 [e]	1.815
VRF									
L* [1]	41.17 [fg]	42.56 [efg]	43.32 [de]	44.86 [abcd]	45.41 [abc]	46.37 [ab]	46.58 [a]	45.84 [abc]	0.713
a* [1]	15.72 [e]	19.54 [d]	19.56 [d]	19.89 [cd]	20.26 [cd]	20.55 [cd]	20.91 [cd]	21.59 [c]	0.704
b* [1]	11.03 [fg]	9.69 [h]	9.85 [h]	10.26 [gh]	10.48 [gh]	11.07 [fg]	11.59 [f]	12.12 [f]	0.392
C* [2]	19.26 [h]	21.82 [fg]	21.91 [fg]	22.39 [efg]	22.82 [efg]	23.36 [def]	23.92 [de]	24.77 [cd]	0.672
Hue (°) [3]	35.13 [d]	26.41 [e]	26.71 [e]	27.28 [e]	27.31 [e]	28.26 [e]	29.00 [e]	29.32 [e]	1.185
RTB [4]	2.12 [f]	3.18 [c]	3.10 [c]	2.90 [cd]	2.85 [cd]	2.70 [d]	2.57 [de]	2.69 [d]	0.143
MMb [5]	34.97 [c]	14.54 [g]	15.20 [fg]	17.20 [fg]	18.62 [efg]	20.62 [ef]	23.56 [de]	24.18 [de]	2.081
DMb [5]	53.28 [b]	86.97 [a]	87.40 [a]	88.35 [a]	88.95 [a]	86.65 [a]	82.89 [a]	83.98 [a]	3.161
OMb [5]	11.75 [f]	2.69 [h]	2.90 [h]	5.67 [gh]	7.57 [fgh]	7.26 [fgh]	6.63 [gh]	8.16 [fg]	1.815

[1] L* Values are a measure of darkness to lightness (larger value indicates a lighter color); a* values are a measure of redness (larger value indicates a redder color); and b* values are a measure of yellowness (larger value indicates a more yellow color). [2] C* (Chroma) is a measure of total color (larger number indicates a more vivid color). [3] Hue (°) angle represents the change from the true red axis (larger number indicates a greater shift from red to yellow). [4] RTB calculated as 630 nm ÷ 580 nm, which represents a change in the color of red to brown (larger value indicates a redder color). [5] Calculated percentages of deoxymyoglobin (DMb), metmyoglobin (MMb), and oxymyoglobin (OMb) using relative spectral values. [a–h] Mean values within a row and a packaging method lacking common superscripts differ ($p \leq 0.05$). * SEM, Standard error of the mean. Bold font, the packaging methods investigated.

There was a packaging method × day of display interaction for surface color chroma (C*) and hue angles (Table 1). The instrumental surface color of steaks packaged in PVC was more vivid ($p < 0.05$) on day 0 but C* values declined as the duration of display increased. However, steaks packaged with VRF became more vivid ($p < 0.05$) from day 25 through 35 of the simulated retail display period (Table 1). In addition, steaks packaged with PVC had greater ($p < 0.05$) hue angles indicative of a surface color shift from red to yellow from day 5 through 35. It appears that the reduction in oxygen exposure for steaks in VRF packages protected the surface color of steaks by sustaining the vividness and reducing the shift from red to yellow. Similar results for fresh packaged beef C* and hue angle values have been reported to decline during the initial 10 days of a simulated display period when using an oxygen-rich packaging method such as PVC [22]. Changes in surface color values for the hue angle and C* can be used as a great indicator for observing meat discoloration in retail display settings [19–24]. Interestingly, C* (vividness) for steaks packaged in VFR in the current study differ from previous C* results that did not differ throughout a 35-day display period [23]. It should be noted that as the percentage of oxygen exposure to the steak surface increases a reduction in the hue angle and C* will likely occur during retail display periods as the surface color shifts from red to brown with the formation of metmyoglobin [22–24].

The interactive influence for packaging method × day of display remained for calculated spectral values of red to brown (630:580 nm) and relative forms of myoglobin (Table 1). Red to brown values were greater ($p < 0.05$) for steaks packaged in PVC until day 5 of the simulated display period. However, from day 10 through 35, PVC-packaged steaks' surface color showed a greater shift from red to brown. Steaks packaged in VRF had less ($p < 0.05$) discoloration from red to brown after day 5 through day 35 (Table 1). Previous studies have [25] reported similar findings indicating a decline in calculated red to brown values throughout 7 days of simulated display for beef packaged in PVC [25]. It is expected that calculated spectral values for the surface color of fresh beef will shift from a brighter red to brown as the duration of a simulated retail display increases. Steaks packaged in VRF had the greatest ($p < 0.05$) amount of calculated metmyoglobin (MMb) on day 0 (Table 1). However, as expected from days 5 to 35, steaks packaged using PVC had greater ($p < 0.05$) calculated relative values for MMb. As expected, steaks packaged in VRF had greater ($p < 0.05$) calculated deoxymyoglobin (DMb) values throughout the entire simulated retail display period (Table 1) because of limited oxygen exposure. Interestingly, calculated relative values of oxymyoglobin (OMb) were greater ($p < 0.05$) for steaks packaged using PVC packaging materials throughout the entire simulated retail display period (Table 1). The results for calculated spectral values reported are likely due to the oxygen permeability of the PVC package resulting in greater exposure of the steak surface to an oxygen-rich atmosphere. Greater formations of MMb in PVC have been associated with greater amounts of lipid oxidation [26,27] and the relationship of oxidation during the transition of myoglobin pigment from OMb to MMb [26–28].

3.2. Expert Color Evaluation

Fresh steaks were evaluated by experts for visual surface color variations during a simulated retail display for up to 35 days. However, the evaluation of steaks packaged in aerobic PVC packaging materials was discontinued after day 20 due to total surface color deterioration. An interaction of the packaging method × day of display occurred for the surface color evaluation (Table 2). Trained expert evaluators noted that values for the initial beef color, amount of browning, and surface discoloration deteriorated ($p < 0.05$) for steaks packaged in PVC from day 5 through day 20 (Table 2). The surface color of steaks packaged in PVC materials became darker, with a greater amount of browning, and a greater percentage of discoloration as the duration of display increased. As a result of significant surface discoloration, PVC-packaged steaks used for expert color evaluation were discarded on day 20 of the display period. The changes in visual surface color are influential in driving consumer purchasing intent and the lack of storage for PVC steaks may contribute to greater throwaway by the retailer. Steaks packaged in VRF had initial beef colors that decreased ($p < 0.05$), and the amount of browning and surface discoloration were less ($p < 0.05$) than steaks packaged in PVC throughout the duration of the study (Table 2). Interestingly, steaks packaged in VRF were darker at day 0, but the visual steak color turned brighter purple red with less browning and surface discoloration throughout a 35-day simulated retail period. Results from the current study agree with previous findings when using vacuum packaging. Beef's surface color tends to remain visually stable throughout the duration of the study, whereas high-oxygen packaging of fresh beef can show rapid color deterioration [29]. The color stability of fresh beef is dependent on controlling countless factors such as pH, temperature, light, lipid oxidation, residual oxygen, MMb-reducing systems, reducing equivalents, and the oxygen consumption rate [30,31]. It is plausible that the transformation from OMb to MMb in PVC-packaged steaks was due to greater amounts of lipid oxidation. Furthermore, limited surface color variation of steaks packaged in VRF may be attributed to a lack of residual oxygen within the packaging, influencing and reducing the amount of oxidation occurring in vacuum-packaged fresh beef products.

Table 2. Interactive influence of packaging method × day of display for expert surface color evaluation on fresh beef strip loin steaks during a simulated retail display.

	Day of Simulated Display								
	0	5	10	15	20	25	30	35	SEM *
PVC [1]									
Initial Beef Color	1.20 ^{h}	3.34 ^{f}	4.48 ^{e}	5.09 ^{c}	6.50 ^{b}	–	–	–	0.154
Amount of Browning	1.00 ^{e}	1.79 ^{c}	3.94 ^{b}	3.75 ^{b}	4.52 ^{a}	–	–	–	0.080
Surface Discoloration	1.09 ^{efg}	1.79 ^{d}	3.12 ^{c}	4.20 ^{b}	4.76 ^{a}	–	–	–	0.073
VRF [2]									
Initial Beef Color	7.34 ^{a}	5.57 ^{c}	4.17 ^{e}	4.15 ^{e}	3.41 ^{f}	3.42 ^{f}	3.52 ^{f}	2.79 ^{g}	0.154
Amount of Browning	1.00 ^{e}	1.17 ^{de}	1.30 ^{d}	1.00 ^{e}	1.23 ^{d}	1.00 ^{e}	1.00 ^{e}	1.20 ^{de}	0.080
Surface Discoloration	1.03 ^{fg}	1.21 ^{ef}	1.01 ^{g}	1.01 ^{g}	1.23 ^{e}	1.00 ^{g}	1.00 ^{g}	1.06 ^{efg}	0.073

[1] PVC color anchors: Initial Beef Color (1 = Extremely bright cherry-red to 8 = Extremely dark red); Amount of Browning (1 = No Evidence of Browning to 6 = Dark Brown); Surface Discoloration (1 = No discoloration (0%) to 6 = Extensive discoloration (81–100%). [2] VRF color anchors: Initial Beef Color (1 = Extremely bright purple red to 8 = extremely dark purple red); Amount of Browning (1 = No Evidence of Browning to 6 = Dark Brown); Surface Discoloration (1 = No discoloration (0%) to 6 = Extensive discoloration (81–100%). $^{a–h}$ Mean values within a row and packaging method lacking common superscripts differ ($p \leq 0.05$). * SEM, Standard error of the mean. Bold font, the packaging methods investigated.

3.3. *Lipid Oxidation*

There was an interactive effect of the packaging method × day of display for lipid oxidation on fresh beef steaks (Figure 1). The packaging method did not alter ($p > 0.05$) lipid oxidation through day 5 of the simulated retail display period. However, from days 10 through 35 of the storage period, lipid oxidation was greater ($p < 0.05$) for steaks packaged using PVC methods. Lipid oxidation of fresh steaks using PVC packaging from the current study agrees with previous simulated retail storage studies measuring an expected storage period in a retail setting of 3 to 7 days [32]. The exposure to greater amounts of oxygen across the packaging material can result in increased catalysis of lipid oxidation [33,34]. Moreover, greater lipid oxidation can be correlated to reduced consumer palatability due to the deterioration of the surface color and accumulation of off flavors [35]. Unfortunately, the evaluation of sensory taste characteristics was not completed during the current study, but future studies on the extended storage of fresh beef influencing lipid oxidation and sensory characteristics would be warranted.

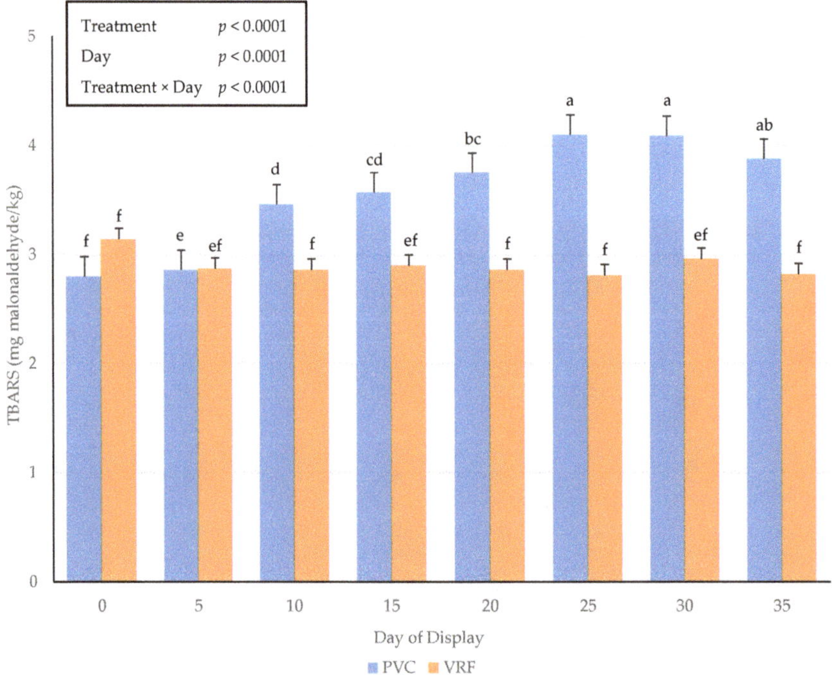

Figure 1. Interactive influence of packaging method × day of display for 2-Thiobarbituric acid reactive substances (TBARS) on beef strip loin steaks during a simulated retail display. Bars lacking common letters differ ($p \leq 0.05$).

3.4. Purge Loss

A packaging method × day of display interaction occurred for the purge loss of fresh beef steaks (Figure 2). The purge loss was greatest ($p < 0.05$) for steaks packaged in PVC materials on day 25 of the simulated display period and the lowest on day 0. The packaging method influenced the purge loss on day 0, with steaks packaged in VRF having a greater ($p < 0.05$) percentage of moisture loss. It is plausible that the method of vacuum packaging using the form-and-fill machine caused more moisture to be pressed out of the steak at the time of package sealing. However, the purge loss in vacuum-packaged meat products can result in an unappealing visual appearance for consumers due to the accumulation of purge in the packaging [36,37]. The results from the current study differ from previous results where values for purge loss using vacuum-packaging platforms were greater than PVC or alternative packaging such as modified atmosphere packaging platforms [38].

3.5. pH

The interactive influence of the packaging method × day of display for fresh muscle pH values is presented in Figure 3. Fresh muscle pH values were recorded within muscle pH values (5.1 to 5.8) throughout the duration of the simulated display period. Values for fresh muscle pH were greatest ($p < 0.05$) on day 10 in steaks packaged using PVC methods. At the time of harvest and before chilling, carcasses were rinsed with an FDA-GRAS (U.S. Food and Drug Administration-Generally Recognized as Safe) organic acid (lactic acid). The combination of vacuum packaging and the organic carcass wash may have contributed to the decline in fresh muscle pH of VRF-packaged steaks, causing a shift in the visual and instrumental surface color variations reported within the current study. Furthermore, it is plausible that pH values for VRF declined due to an increase in lactic acid bacteria that can be present in vacuum-packaged fresh meats. With limited residual oxygen within the vacuum package, favorable conditions for anerobic lactic acid bacteria may have caused fresh muscle pH to decline as lactic acid bacteria populations increased [5,39]. In addition, lactic acid bacteria can be associated with low-pH (<5.8) vacuum-packaged meats due to a lower residual oxygen environment [40].

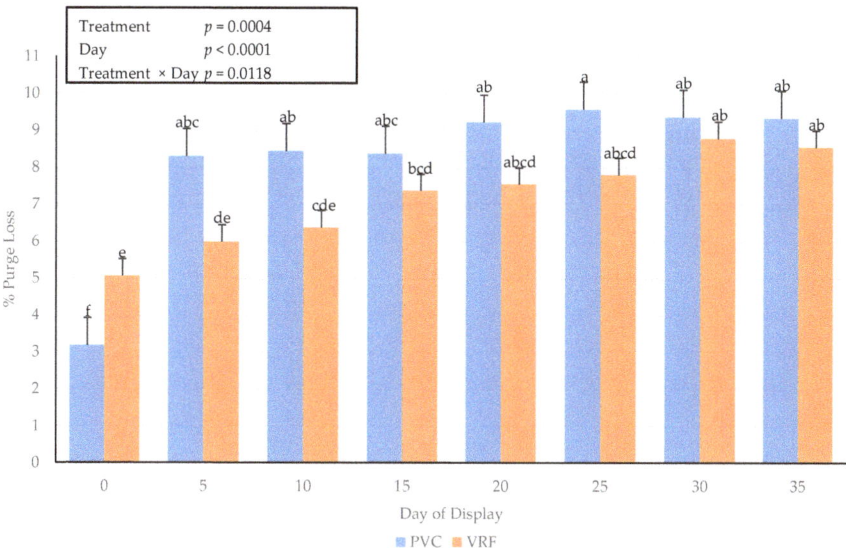

Figure 2. Interactive influence of packaging method × day of display for purge loss (%) on beef strip loin steaks during a simulated retail display. Bars lacking common letters differ ($p \leq 0.05$).

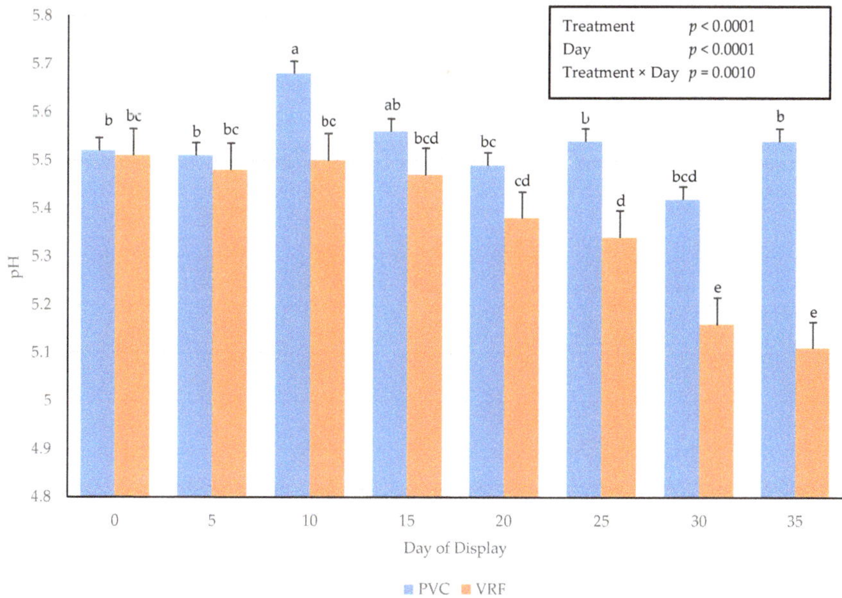

Figure 3. The interactive influence of packaging method × day of display for fresh muscle pH on beef strip loin steaks. Bars lacking common letters differ ($p \leq 0.05$).

4. Conclusions

It is feasible that the storage of beef strip loin steaks using vacuum packaging, VRF, can provide a longer, fresh, refrigerated storage period than steaks packaged in traditional PVC packaging. It is evident that VRF displayed a more color-stable product throughout the duration of simulated retail display. Additionally, VRF maintained less oxidation throughout the display period, whereas steaks packaged in PVC tended to have greater oxidation leading to greater amounts of surface discoloration in beef products. The current results suggest that the vacuum-packaged film used within the current study is an acceptable replacement to traditional packaging methods of PVC for packaging whole-muscle beef steaks for up to 35 days of refrigerated retail storage. However, additional research should be considered to evaluate the sensory taste profiles of vacuum packaging used for extended storage periods and the implications for flavor characteristics of beef steaks.

Author Contributions: Conceptualization, T.M.R., T.B. and J.T.S.; methodology, T.M.R.; validation, M.P.W., V.E.Z. and M.M.C.; formal analysis, J.T.S.; investigation, T.M.R., M.P.W., V.E.Z. and M.M.C.; resources, T.B.; data curation, T.M.R.; writing—original draft preparation, T.M.R.; writing—review and editing, T.M.R., M.P.W., V.E.Z., M.M.C., T.D.B., S.P.R. and J.T.S.; supervision, J.T.S. and B.S.W.; project administration, J.T.S. and B.S.W.; funding acquisition, J.T.S. All authors have read and agreed to the published version of the manuscript.

Funding: This research received no external funding.

Institutional Review Board Statement: Not Applicable.

Informed Consent Statement: Not Applicable.

Data Availability Statement: Not Applicable.

Acknowledgments: The authors would like to extend their upmost appreciation to WINPAK for providing the Variovac thermoforming machine and supplying thermoforming films to complete the study. Additionally, the authors would like to thank the staff of the Lambert Powell Meats Laboratory.

Conflicts of Interest: The authors declare no conflict of interest.

References

1. Beebe, S.; Couch, S. A Snapshot of the Meat Case Today. National Meat Case Study 2021. In Proceedings of the 2021 North American Meat Institute Conference, 31 August 2021.
2. Łopacka, J.; Półtorak, A.; Wierzbicka, A. Effect of MAP, vacuum skin-pack, and combined packaging methods on physicochemical properties of beef steaks stored up to 12 days. *Meat Sci.* **2016**, *119*, 147–153. [CrossRef] [PubMed]
3. Henriott, M.; Herrera, N.; Ribeiro, F.; Hart, K.; Bland, N.; Eskridge, K.; Calkins, C. Impact of myoglobin oxygenation state prior to frozen storage on color stability of thawed beef steaks through retail display. *Meat Sci.* **2020**, *170*, 108232. [CrossRef] [PubMed]
4. Hunt, M.C.; Mancini, R.A.; Hachmeister, K.A.; Kropf, D.H.; Merriman, M.; De Lduca, G.; Milliken, G. Carbon Monoxide in Modified Atmosphere Packaging Affects Color, Shelf Life, and Microorganisms of Beef Steaks and Ground Beef. *J. Food Sci.* **2004**, *69*, FCT45–FCT52. [CrossRef]
5. Lavieri, N.; Williams, S. Effects of packaging systems and fat concentrations on microbiology, sensory and physical properties of ground beef stored at 4 ± 1 °C for 25 days. *Meat Sci.* **2014**, *97*, 534–541. [CrossRef] [PubMed]
6. Suman, S.P.; Hunt, M.C.; Nair, M.N.; Rentfrow, G. Improving beef color stability: Practical strategies and underlying mechanisms. *Meat Sci.* **2014**, *98*, 490–504. [CrossRef]
7. Carreira, L.; Franco, C.; Fente, C.; Cepeda, A. Shelf life extension of beef retail cuts subjected to an advanced vacuum skin packaging system. *Eur. Food Res. Technol.* **2003**, *218*, 118–122. [CrossRef]
8. Strydom, P.E.; Hope-Jones, M. Evaluation of three vacuum packaging methods for retail beef loin cuts. *Meat Sci.* **2014**, *98*, 689–694. [CrossRef]
9. Buzby, J.C.; Farah-Wells, H.; Hyman, J. The Estimated Amount, Value, and Calories of Postharvest Food Losses at the Retail and Consumer Levels in the United States. *USDA-ERS Econ. Inf. Bull.* **2014**, *121*, 39. [CrossRef]
10. North American Meat Institute. The United States Meat Industry at a Glance. Available online: https://www.meatinstitute.org/index.php?ht=d/sp/i/47465/pid/47465 (accessed on 3 December 2021).
11. Pauer, E.; Tacker, M.; Gabriel, V.; Krauter, V. Sustainability of flexible multilayer packaging: Environmental impacts and recyclability of packaging for bacon in block. *Clean. Environ. Syst.* **2020**, *1*, 100001. [CrossRef]
12. Kaiser, K.; Schmid, M.; Schlummer, M. Recycling of Polymer-Based Multilayer Packaging: A Review. *Recycling* **2017**, *3*, 1. [CrossRef]

13. Dixon, J. *Packaging Materials: 9. Multilayer Packaging for Food and Beverages*; ILSI Europe: Brussels, Belgium, 2011; ISBN 9789078637264.
14. Cenci-Goga, B.T.; Iulietto, M.F.; Sechi, P.; Borgogni, E.; Karama, M.; Grispoldi, L. New Trends in Meat Packaging. *Microbiol. Res.* **2020**, *11*, 56–67. [CrossRef]
15. Dilkes-Hoffman, L.S.; Lane, J.L.; Grant, T.; Pratt, S.; Lant, P.; Laycock, B. Environmental impact of biodegradable food packaging when considering food waste. *J. Clean. Prod.* **2018**, *180*, 325–334. [CrossRef]
16. Commission Internationale de l' Eclairage. *Recommendations on Uniform Color Spaces—Color Difference Equations, Psycho-Metric Color Terms*; C.I.E Publication No. 15, Suppl. 2 (E-1.3.1), 1971/(TC-1-3); Bureau Central de la CIE: Paris, France, 1978.
17. American Meat Science Association. *Meat Color Measurement Guidelines*; American Meat Science Association: Champaign, IL, USA, 2012.
18. Buege, J.A.; Aust, S.D. Microsomal lipid peroxidation. *Methods Enzymol.* **1978**, *52*, 302–310. [CrossRef]
19. Mancini, R.; Ramanathan, R.; Suman, S.; Konda, M.; Joseph, P.; Dady, G.; Naveena, B.; López-López, I. Effects of lactate and modified atmospheric packaging on premature browning in cooked ground beef patties. *Meat Sci.* **2010**, *85*, 339–346. [CrossRef] [PubMed]
20. Bağdatli, A.; Kayaardı, S.; Kayaardi, S. Influence of storage period and packaging methods on quality attributes of fresh beef steaks. *CyTA J. Food* **2014**, *13*, 124–133. [CrossRef]
21. VanOverbeke, D.L.; Hilton, G.G.; Green, J.; Hunt, M.; Brooks, C.; Killefer, J.; Streeter, M.N.; Hutcheson, J.P.; Nichols, W.T.; Allen, D.M.; et al. Effect of zilpaterol hydrochloride supplementation of beef steers and calf-fed Holstein steers on the color stability of top sirloin butt steaks1. *J. Anim. Sci.* **2009**, *87*, 3669–3676. [CrossRef] [PubMed]
22. Li, S.; Zamaratskaia, G.; Roos, S.; Båth, K.; Meijer, J.; Borch, E.; Johansson, M. Inter-relationships between the metrics of instrumental meat color and microbial growth during aerobic storage of beef at 4 °C. *Acta Agric. Scand. Sect. A Anim. Sci.* **2015**, *65*, 97–106. [CrossRef]
23. Jeong, J.Y.; Claus, J.R. Color stability of ground beef packaged in a low carbon monoxide atmosphere or vacuum. *Meat Sci.* **2011**, *87*, 1–6. [CrossRef]
24. Yang, X.; Zhang, Y.; Zhu, L.; Han, M.; Gao, S.; Luo, X. Effect of packaging atmospheres on storage quality characteristics of heavily marbled beef longissimus steaks. *Meat Sci.* **2016**, *117*, 50–56. [CrossRef]
25. Stivarius, M.; Pohlman, F.; McElyea, K.; Apple, J. Microbial, instrumental color and sensory color and odor characteristics of ground beef produced from beef trimmings treated with ozone or chlorine dioxide. *Meat Sci.* **2002**, *60*, 299–305. [CrossRef]
26. Turan, E.; Şimşek, A. Effects of lyophilized black mulberry water extract on lipid oxidation, metmyoglobin formation, color stability, microbial quality and sensory properties of beef patties stored under aerobic and vacuum packaging conditions. *Meat Sci.* **2021**, *178*, 108522. [CrossRef] [PubMed]
27. Suman, S.P.; Joseph, P. Myoglobin Chemistry and Meat Color. *Annu. Rev. Food Sci. Technol.* **2013**, *4*, 79–99. [CrossRef] [PubMed]
28. Faustman, C.; Sun, Q.; Mancini, R.; Suman, S.P. Myoglobin and lipid oxidation interactions: Mechanistic bases and control. *Meat Sci.* **2010**, *86*, 86–94. [CrossRef] [PubMed]
29. Rogers, H.; Brooks, J.; Martin, J.; Tittor, A.; Miller, M.; Brashears, M. The impact of packaging system and temperature abuse on the shelf life characteristics of ground beef. *Meat Sci.* **2014**, *97*, 1–10. [CrossRef]
30. Faustman, C.; Cassens, R. The Biochemical Basis for Discoloration in Fresh Meat. A Review. *J. Muscle Foods* **1990**, *1*, 217–243. [CrossRef]
31. Renerre, M.; Labas, R. Biochemical factors influencing metmyoglobin formation in beef muscles. *Meat Sci.* **1987**, *19*, 151–165. [CrossRef]
32. Delmore, R.J. Beef Shelf Life. 2009. Available online: https://www.beefresearch.org/resources/product-quality/fact-sheets/beef-shelf-life (accessed on 3 August 2021).
33. Fernández-López, J.; Sayas-Barberá, E.; Muñoz, T.; Sendra, E.; Navarro, C.; Pérez-Alvarez, J. Effect of packaging conditions on shelf-life of ostrich steaks. *Meat Sci.* **2008**, *78*, 143–152. [CrossRef]
34. Limbo, S.; Torri, L.; Sinelli, N.; Franzetti, L.; Casiraghi, E. Evaluation and predictive modeling of shelf life of minced beef stored in high-oxygen modified atmosphere packaging at different temperatures. *Meat Sci.* **2010**, *84*, 129–136. [CrossRef]
35. Park, S.; Kim, Y.; Lee, H.; Yoo, S.; Shim, J.; Chin, K. Effects of Pork Meat Cut and Packaging Type on Lipid Oxidation and Oxidative Products during Refrigerated Storage (8 °C). *J. Food Sci.* **2008**, *73*, C127–C134. [CrossRef]
36. Lee, K.T. Shelf-life Extension of Fresh and Processed Meat Products by Various Packaging Applications. *Korean J. Packag. Sci. Technol.* **2018**, *24*, 57–64. [CrossRef]
37. Lagerstedt, Å.; Ahnström, M.L.; Lundström, K. Vacuum skin pack of beef—A consumer friendly alternative. *Meat Sci.* **2011**, *88*, 391–396. [CrossRef] [PubMed]
38. Fu, A.-H.; Molins, R.; Sebranek, J. Storage Quality Characteristics of Beef Rib Eye Steaks Packaged in Modified Atmospheres. *J. Food Sci.* **1992**, *57*, 283–287. [CrossRef]
39. Jones, R.J. Observations on the succession dynamics of lactic acid bacteria populations in chill-stored vacuum-packaged beef. *Int. J. Food Microbiol.* **2004**, *90*, 273–282. [CrossRef]
40. Gill, C. Extending the storage life of raw chilled meats. *Meat Sci.* **1996**, *43*, 99–109. [CrossRef]

Article

An Active Peptide-Based Packaging System to Improve the Freshness and Safety of Fish Products: A Case Study

Rosa Luisa Ambrosio [1,†], Marta Gogliettino [2,†], Bruna Agrillo [2,3,4], Yolande T. R. Proroga [5], Marco Balestrieri [2], Lorena Gratino [2], Daniela Cristiano [5], Gianna Palmieri [2,3,*] and Aniello Anastasio [1]

1 Department of Veterinary Medicine and Animal Production, University of Naples Federico II, 80137 Napoli, Italy; rosaluisa.ambrosio@unina.it (R.L.A.); anastasi@unina.it (A.A.)
2 Institute of Biosciences and BioResources, National Research Council (IBBR-CNR), 80131 Napoli, Italy; marta.gogliettino@ibbr.cnr.it (M.G.); bruna.agrillo@ibbr.cnr.it (B.A.); marco.balestrieri@ibbr.cnr.it (M.B.); gratino.lorena@ibbr.cnr.it (L.G.)
3 Materias S.R.L., Corso N. Protopisani 70, 80146 Naples, Italy
4 Department of Biology, University of Naples Federico II di Monte Sant'Angelo, Via Cintia 21, 80126 Naples, Italy
5 Department of Food Microbiology, Istituto Zooprofilattico Sperimentale del Mezzogiorno, 80055 Portici, Italy; proroga.yolande@izsmportici.it (Y.T.R.P.); daniela.cristiano@izsmportici.it (D.C.)
* Correspondence: gianna.palmieri@ibbr.cnr.it; Tel.: +39-081-613-2711
† These authors equally contributed to this work.

Abstract: Fresh fish are highly perishable, owing mainly to their moisture content, high amount of free amino acids and polyunsaturated fatty acids. Microorganisms and chemical reactions cause the spoilage, leading to loss in quality, human health risks and a market value reduction. Therefore, the fishing industry has always been willing to explore new technologies to increase quality and safety of fish products through a decrease of the microbiological and biochemical damage. In this context, antimicrobial active packaging is one such promising solution to meet consumer demands. The main objective of this study was to evaluate the effects of an active polypropylene-based packaging functionalized with the antimicrobial peptide 1018K6 on microbial growth, physicochemical properties and the sensory attributes of raw salmon fillets. The results showed that application of 1018K6-polypropylene strongly inhibited the microbial growth of both pathogenic and specific spoilage organisms (SSOs) on fish fillets after 7 days. Moreover, salmon also kept its freshness as per volatile chemical spoilage indices (CSIs) during storage. Similar results were obtained on hamburgers of *Sarda sarda* performing the same analyses. This work provides further evidence that 1018K6-polymers have good potential as antimicrobial packaging for application in the food market to enhance quality and preserve the sensorial properties of fish products.

Keywords: antimicrobial polymers; antimicrobial peptides; fresh fish; spoilage; fish quality; food safety; food packaging

Citation: Ambrosio, R.L.; Gogliettino, M.; Agrillo, B.; Proroga, Y.T.R.; Balestrieri, M.; Gratino, L.; Cristiano, D.; Palmieri, G.; Anastasio, A. An Active Peptide-Based Packaging System to Improve the Freshness and Safety of Fish Products: A Case Study. *Foods* **2022**, *11*, 338. https://doi.org/10.3390/foods11030338

Academic Editors: Matteo Alessandro Del Nobile and Amalia Conte

Received: 21 December 2021
Accepted: 21 January 2022
Published: 25 January 2022

1. Introduction

Today, health, nutrition and convenience are the major drivers in the global food industry. In this context, fish products have attracted considerable attention as a source of important nutritional components, such as high-quality protein, essential vitamins, minerals and polyunsaturated fatty acids (PUFA) [1,2]. Indeed, fish is considered of key importance for human nutrition all over the world, providing about 17% of the global intake of animal proteins [3]. However, its consumption in many parts of the world is far below the recommended level. As such, high-quality food with an extended shelf-life is essential for both producers and consumers. However, fish is a highly perishable product due to its relevant water activity, nearly neutral pH and specific composition that make it vulnerable to various biochemical, physical and microbial forms of deterioration throughout the production chain, thus causing rejection by consumers. Indeed, spoilage starts quickly after fish are caught, and rigor mortis is responsible for changes in fish after death.

Specifically, the degradation of various components and the formation of new products are accountable for the alterations in odour, flavour, colour and texture that happen during the spoilage process, so that deterioration occurs very rapidly due to mechanisms triggered by the microbial community, endogenous enzymatic activity (autolysis), and the chemical oxidation of lipids [4–8]. Due to all these changes, the shelf-life and quality of fresh fish are very limited, resulting in health risks as well as in enormous economic loss. Therefore, the fish industry is focused on preventing and controlling foodborne illness and microbial growth, which can lead to food spoilage, the major cause of fish loss, spoil estimated at millions of tons per year and accounting for 10% of the total production from capture fisheries and aquaculture [9]. These phenomena are even more evident in lower-income economies, in which spoilage and the quality downgrading of fish products occur due to high ambient temperatures, lack of infrastructure, basic technology and lack of cooling (cold chain) facilities.

Salmon (*Salmo salar*) is one of the most consumed seafood in the world, either fresh or frozen, and a main product of aquaculture, accounting for 93% of total production [10]. Recently, its consumption has increased substantially because it offers several health benefits, mostly due to the presence of essential and vital nutrients for human body, such as omega-3 long-chain fatty acids. As this marine species has become a luxury product in the fishery market, any new strategy developed to enhance its safety and preserve its quality is considered essential for this economic sector. A considerable support in the fight against microbial spoilage may derive from food packaging, which can serve as a carrier of active substances, such as antimicrobials, playing an active role in food quality and shelf life, besides acting as a barrier against moisture, water vapor, gases and solutes. Specifically, antimicrobial packaging is considered a promising form of active packaging based on the immobilization of antimicrobial agents on the surfaces of polymers, thus providing antimicrobial properties to specific materials.

In this context, polypropylene (PP) is an ideal food-safe thermoplastic material for packaging applications because of its low cost and its physical and chemical parameters. Therefore, it would be advantageous to impart antibacterial activity to these polymers, given the growing consumer interest in foods with fewer preservatives. In a previous study, the in-silico-designed antimicrobial peptide 1018K6, extensively characterized both from functional and structural points of view, was covalently bonded to commercial PET (polyethylene terephthalate) films and the ability of the developed antimicrobial packaging to improve the microbial quality and safety of dairy products was clearly demonstrated [11–14]. The applied procedure fulfilled the criteria of an efficient immobilization reaction, such as high yields and remarkable stability of the activated polymers, with no peptide release under different environmental conditions of use even after prolonged incubation times. Surprisingly, the AMP (antimicrobial peptide) was still active and preserved its excellent antimicrobial and antibiofilm abilities against a panel of Gram$^+$ and Gram$^-$ bacterial pathogens upon polymer surface functionalization, along with potent activity against moulds and fungal species, without exhibiting cytotoxic effects on human cells. Specifically, 1018K6 was able to explicate its bactericidal activity against both fungi, such as *Aspergillus brasiliensis*, the Gram$^+$ pathogens *Listeria monocytogenes* and *Staphylococcus aureus* and the Gram$^-$ *Salmonella* Typhimurium and *Escherichia coli* [15]. Furthermore, the optimized technology revealed the possibility of re-using the peptide polymers at least six times, while preserving its antimicrobial properties.

In this paper, 1018K6 was covalently immobilized onto PP surfaces, previously activated by plasma treatment, in order to extend the application field of our AMP-based system. Therefore, the impacts of prepared films on the physicochemical, microbial and sensory properties of fresh salmon fillets throughout storage at 4 °C for 7 days were investigated. A challenge test with *L. monocytogenes* was also performed. In order to validate the antimicrobial effectiveness of 1018K6-PP packaging in controlling the quality decay of fresh fish, the same analyses were performed on a different typology of fish-based foods, burgers of the bonito fish (*Sarda sarda*). Indeed, this typology of product has a softer texture with a lower shear force than other meat products reported in the literature, and it represents a valid solution for meeting consumer preferences for foods that have high nutritional value and are very convenient, being ready-to-cook. Moreover, it is well-known that the production phases of burgers are responsible for higher microbial concentrations than fillets due to the handling and the increased superficial area of the matrix for grinding. All these aspects outline the insidious profile of fish burgers and the challenge of testing the innovative packaging on these products.

2. Materials and Methods

2.1. Atmospheric Plasma Treatments

Openair-Plasma® Technology was used as plasma treatment. The surfaces of commercial PP were cut into square-shaped pieces (4.0 × 4.0 cm dimension), which were cleaned with ethanol prior to use and then were placed on the plate at a distance of 3 cm to the nozzle. In all the treatments, air was used as the processing gas with a power of 440 watts and a speed of 10 mm/s. The effect of plasma on the polymer surfaces was evaluated by using the test Ink (Plasmatreat) in order to assess the wettability of the material.

2.2. Production of 1018K6

The peptide 1018K6 (VRLIVKVRIWRR-NH2) used in this work was purchased from GenScript Biotech (Leiden, Netherlands). It was stored as a lyophilized powder at −20 °C. Analysis by mass spectrometry confirmed the identity of peptide.

2.3. 1018K6 Immobilization on Polymer Surfaces and Release Test

Polymer surfaces pre-activated by atmospheric plasma treatment were incubated into solutions (3 mL) of 1018K6 prepared in distilled water at three concentrations (50, 100 and 200 μM) for about 4 h at 70 °C to fully remove the water. After drying, the functionalized PPs were immersed in a volume of distilled water equal to that evaporated for 16 h at room temperature in agitation, then they were sonicated for 20 min and the recovered solutions analysed by reverse-phase high-performance liquid chromatography (RP-HPLC) to indirectly estimate the immobilization yield. For these analyses, 200 μL of the samples were injected over a μBondapak C18 reverse-phase column (3.9 mm × 300 mm, Waters Corp., Milford, MA, USA) connected to a HPLC system (Shimadzu, Milan, Italy), using a linear gradient of 5–95% 0.1% TFA (Trifluoroacetic acid) in acetonitrile, at a flow rate of 1 mL/min. A reference solution was prepared with the initial peptide concentration used in the coupling reaction and was run in parallel. Therefore, by knowing the added peptide concentration (reference solution), the amount of peptide not covalently attached on the polymer surface was calculated by comparing the peak area and expressed as a percentage. A calibration curve of the C18 column using different 1018K6 concentrations was constructed. All measurements were performed in triplicate on three different preparations.

To determine the stability of 1018K6 on the functionalized polymers, a release assay was performed by RP-HPLC using a linear gradient of 5–95% acetonitrile in 0.1% TFA, at a flow rate of 1 mL/min. A volume of 1 mL of pure water or NaCl 1 M was poured onto the functionalized polymers, which were incubated for 7 days at 4 °C, sonicated for 20 min and then the recovered solutions were loaded on RP-HPLC column. The solutions in contact with the functionalized polymers at time t = 0 were used as control samples and were run in parallel. All measurements were performed in triplicate on three different preparations.

2.4. Sample Preparation

Raw salmon (*Salmo salar*, Linnaeus 1758) from different batches were freshly bought from a local fishery industry (Naples, Italy). To evaluate the antimicrobial effects of the functionalized 1018K6-PPs polymers, two samples were prepared under aseptic conditions from each fillet that was sliced into pieces of approximatively 50 g. As a whole, the samples were separated into two groups: the control group (CTR-PP), including salmon fillets packaged in pre-activated PPs films not-functionalized with 1018K6 and the treated group, including salmon fillets packaged in PPs films functionalized with 1018K6 (1018K6-PP). Both 1018K6-PPs and non-functionalized PPs squares (4 × 4 cm) were placed on Petri dishes (both lid and base) in order to ensure constant contact between the pieces of salmon and the PPs. Subsequently, the packaged samples were refrigerated at 4 ± 1 °C for 7 days. The samples' microbiological, physicochemical properties and quality aspects were analysed at days 0, 4 and 7.

Fish burgers of Atlantic bonito (*Sarda sarda*, Bloch 1793) were purchased from a fishing industry in Naples (Italy). A total of 21 burgers (200 g) were included in the experimental design. The samples were randomly divided into CTR-PP and 1018K6-PP groups and prepared as described above for the salmon samples. Once the fish burgers were packaged, the samples were stored at refrigeration temperature (4 ± 1°C) and sampled at days 0, 3, 5 and 7 to carry out the same analyses as were conducted on the salmon fillets.

The same analyses were performed also on fish samples packaged in PP films that were not subjected to any surface modification.

2.5. *pH and a_w Measurements*

The pH measurements were carried out with a digital pH meter (Crison-Micro TT 2022, Crison Instruments, Barcelona, Spain). Water activity (a_w) was measured with Aqualab 4 TE (Decagon Devices Inc., Northeast Hopkins Court Pullman, Pullman, WA, USA).

2.6. *Microbiological Analyses*

Ten grams of each sample were added to 90 mL (1:10 *w/v*) of sterilized Peptone Water (PW, Oxoid, Madrid, Spain) in a sterile stomacher bag to be homogenized for three minutes at 230 rpm using a peristaltic homogenizer (BagMixer®400 P, Interscience, Saint Nom, France). Ten-fold serial dilutions of each homogenate were prepared. In order to better describe the microbial profile of samples and follow the growth trend of each bacterium responsible for the food alteration, the viable counts of various microorganisms were carried out. Total aerobic bacterial counts (TAB), both mesophilic and psychrophilic, were performed on plate count agar (PCA, Oxoid, Madrid, Spain) incubated at 30 °C for 48/72 h and 7 °C for 10 days, respectively (ISO 4833-1:2013 and ISO 17410:2019); total coliforms on violet red bile lactose agar (VRBL, Oxoid, Madrid, Spain) incubated at 37 °C for 48 h (ISO 4831:2006); Enterobacteriaceae on violet red bile glucose agar (VRBG, Oxoid, Madrid, Spain) incubated at 37 °C for 48 h (ISO 21528-2:2017); lactic acid bacteria (LAB) on MRS agar with Tween 80 (Oxoid, Madrid, Spain), incubated at 30 °C for 72 h (ISO 15214:2015); *Pseudomonas* spp. on pseudomonas agar base with CFC supplement (Oxoid, Madrid, Spain) incubated at 25 °C for 48 h (ISO 13720:2010); β-glucuronidase-positive *Escherichia coli* (ISO 16649-1:2018) on Triptone Bile X-glucoronide Agar (TBX, Oxoid, Madrid, Spain) at 44 °C for 24 h; *Brochothrix thermosphacta* on STAA (streptomycin thallous acetate actidione agar, Oxoid, Madrid, Spain) at 37 °C for 48 h; *Enterococcus faecalis* on KAA (kanamycin aesculin azide, Oxoid, Madrid, Spain) at 37 °C for 48 h; coagulase positive *staphylococci* on Baird-Parker agar (Oxoid, Madrid, Spain) at 37 °C for 24/48 h (ISO 6888-1:1999). After counting, the data were expressed as logarithms of the number of colony-forming units (CFU/g) and means and standard error were calculated.

2.7. *Challenge Test*

Four fillets of approximately 150 g and from different batches were used to evaluate the inter-batch variability. All of them were tested in agreement with the AFNOR-BRD 07/10-04/05-Real Time PCR method in order to evaluate the absence of *L. monocytogenes* contamination. Three strains of *L. monocytogenes* isolated from fish samples were selected, following ISO 11290-1, to perform these analyses and stored in the Zooprophilactic Experimental Institute of Mezzogiorno biobank. All strains were re-suspended in diluent at 0.5 Mcfarland concentration, then a series of 10 times gradient dilution of *L. monocytogenes* was performed until to reach a concentration of 150 CFU/mL. All fillets were contaminated at surface to mimic contamination during the slicing, using one fillet not contaminated as a control. After artificial contamination, the samples were packed between two functionalized films and stored at 5 °C until 96 h. The enumeration of *L. monocytogenes* was performed according to Annex 1 of Reg (CE) 2073/2005 [16] at 24 h, 48 h, 72 h, and 96 h, in agreement with the reference methods EN ISO 11290-2.

2.8. Colour

Colourimetric measurements of the surface appearance of salmon fillets and bonito fish burgers were performed using a Konica Minolta CR 300 colourimeter (Minolta, Osaka, Japan). The data were analysed in the CIELAB colour space, organizing in three orthogonal axes in a Cartesian coordinate system: lightness (L^*), redness (a^*) and yellowness (b^*). Additionally, the angular coordinates of Hue angle [hab = ArcTan(b^*/a^*)], and chroma [Cab = $(a^*_2 + b^*_2)^{1/2}$] were calculated. Total colour difference (ΔE), variation in a^* (Δa^*) and in b^* (Δb^*) were calculated as:

$$\Delta E = \sqrt{(L^*_1 - L^*_2)^2 + (a^*_1 - a^*_2)^2 + (b^*_1 - b^*_2)^2}$$

$$\Delta a^* = a^*_2 - a^*_1$$

$$\Delta b^* = b^*_2 - b^*_1$$

where L^*_2, a^*_2, and b^*_2 are the values recorded at a specific day during the storage; instead, L^*_1, a^*_1, and b^*_1 are values collected at day 0.

ΔE represents the result of changes in lightness (ΔL^*), redness (Δa^*) and yellowness (Δb^*), which do not always change in parallel. For this reason, Δa^* and Δb^* were taken into account. Since the colour may not be homogeneous over the entire surface of fillets and burgers, five superficial measurements were performed for each sample to obtain representative results.

2.9. TBARS, Total Volatile Basic Nitrogen (TVB-N) and Trimethylamine (TMA) Analyses

Lipid oxidation was monitored by determining the thiobarbituric acid ($C_4H_4N_2O_2S$) substances expressed as malondialdehyde ($CH_2(CHO)_2$) concentration (mg/Kg), which represent secondary oxidation products. Measurements were performed according to the method proposed by Ambrosio et al. [17].

The TVB-N and TMA values for all salmon and fish burger samples were quantified according to Conway's micro-diffusion method [18]. The results were expressed in mg of nitrogen per 100 g of sample.

2.10. Sensory Testing

Sensory testing of salmon fillets and bonito fish burgers was undertaken by a panel consisting of five trained panellists. The judge's acceptability study was assessed through a sensory evaluation, taking into account odour, colour and general acceptability. Appropriate attributes have been fixed in order to minimize individual differences and ensure the results' repeatability. Sensory assessments were performed under controlled humidity, light and temperature. A Likert scale (9-point) was used to assess each attribute; in the scale, 9 corresponded to excellent, 8 to very good, 7 to good, 6 to reasonable, 5 to not good (acceptable limit), 4 to disliked, 3 to bad, 2 to very bad, and 1 to completely unacceptable [19]. Coded samples were randomly and simultaneously distributed to each panellist.

2.11. Statistical Analyses

Physicochemical and microbiological data were statistically analysed with generalized linear mixed model of SPSS version 26 (IBM Analytics, Armonk, NY, USA). Analysis of variance was performed to study parameters of salmon fillets and bonito fish burgers at each sampling time, including the fixed effect of packaging used and storage times. An a posteriori contrast was carried out using the Tukey test, considering a p value of <0.05 as statistically significant.

3. Results

3.1. 1018K6 Immobilization on PP Surface

Following confirmation of the excellent antimicrobial properties preserved by 1018K6, even after bonding on different materials, such as PET and nanoparticles [13,20], the peptide was further immobilized on another plastic polymer commonly used in food packaging, polypropylene (PP).

To this aim, commercial PP slides were exposed to plasma treatment to activate the inert polymeric surfaces with reactive -COOH* functional groups that are available to interact with the amine moieties of 1018K6, forming amide bonds [21].

In order to develop an antimicrobial packaging more adequate for food application, the conditions applied in our previous studies to functionalize the polymeric materials were modified. Specifically, the covalent attachment of 1018K6 on the pre-activated PP polymers was executed by a one-step immobilization process involving the immersion of the polymeric surfaces in a water solution containing the peptide at different concentrations. Thereafter, the slides were kept at 70 °C for about 4 h to completely remove the water and to drive the coupling reaction. To validate the success of our immobilization procedure, the test ink was applied, confirming the increase in the surface hydrophilicity of the AMP-functionalized PP slides following the immobilization procedure, due to the introduction of polar groups on the hydrophobic polymer. Moreover, reverse-phase high-precision liquid chromatography (RP-HPLC) analysis was performed to quantify the amount of 1018K6 immobilized on PP surfaces. For this investigation, 1018K6-PP slides, after the coupling reactions, were immersed in distilled water and incubated for 16 h at room temperature under agitation. Then, the polymers were subjected to sonication for 20 min at room temperature and the recovered solutions loaded on an RP-C18 column. By knowing the initial peptide concentrations that were used in the conjugation reaction, the amount of the peptide attached to PP slides was indirectly determined by comparing the peak area in the RP-chromatograms. The data obtained from these analyses showed that the immobilization yield varied from 23%, starting from a peptide concentration of 50 μM, to 5%, when 200 μM was used. The maximum peptide binding (31%) was obtained at 100 μM, which corresponded to a surface coverage of approximately 5.8 nmol/cm^2, confirming that the initial amount of 1018K6 strongly influenced its binding to synthetic slides (Figure 1). Therefore, 100 μM was selected as the peptide concentration for performing all further experiments.

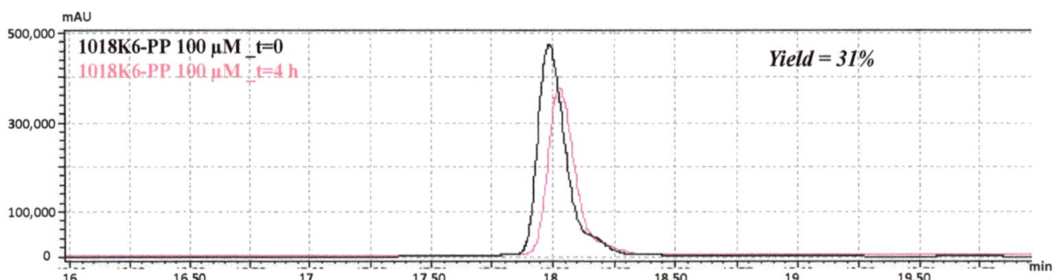

Figure 1. Immobilization yield (%) of 1018K6 on PP surfaces determined by RP-HPLC chromatography on a C18 column after the coupling reaction. PP surfaces pre-activated by plasma treatment, were incubated with a water solution of 1018K6 (100 μM). After the coupling reactions, the supernatants were recovered and analysed by RP-HPLC. The peptide solution (100 μM) at time 0 (t = 0) was used as control. The reported chromatograms are representative of three independent experiments.

Concerning the low binding capacity associated with the highest peptide concentration used, it could be attributed to a steric hindrance effect, which limits polymer–peptide interactions and a phenomenon producing water-soluble microaggregates, which can strongly reduce the availability of bioactive molecules for the immobilization reaction.

One of the most important requirements in applying an antimicrobial packaging in the food industries is the stability of the peptide immobilized on the polymers in the conditions of use, because, in this way, it does not require approval as food additive by EFSA (European Food Safety Authority). For this purpose, the slides functionalized with 100 μM 1018K6 were incubated in pure water or in NaCl 1 M at 4 °C for 7 days and the potential release of the peptide from the polymeric support was monitored by RP-HPLC, using the free peptide as a control. Following these analyses, no peptide-release was observed during 7 days of incubation, confirming the strong attachment via the covalent coupling of the bioactive compound, preventing its release from the surface and highlighting the high stability of the system.

It is worth noting that the projected packaging was reused at least six times in all the subsequent analyses, after washing with EtOH 70% for 1 min, rinsing with water and exposition to UV radiations for 1 h. Surely, this represents an important advantage from the industrial point of view, allowing a substantial decrease in environmental impact based on the concept that "reuse is better than recycling".

3.2. Effects of 1018K6-PP on the Physical Properties and on the Microbiological Quality of Salmon Fillets

It is known that the physicochemical characteristics of raw salmon fillets, such as pH (close to 6) and a high water activity (a_w), make them highly susceptible to microbial growth, which affects the storability of these products [22]. Therefore, the fish industry is actively seeking methods of preservation to improve quality and marketability of this luxury marine food while economizing on costs.

To this aim, the antimicrobial effects of 1018K6-PP on the spoilage microbiota and the intrinsic properties of fresh salmon fillets during refrigerated storage were assessed, using the pre-activated and not-functionalized PP slides as control (CTR). In Figure 2, a representative scheme of the different steps applied for the preparation of salmon fillets employed in the microbiological and physicochemical analyses, is shown.

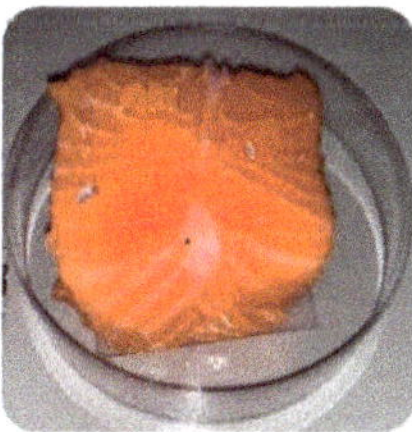

Figure 2. Representative scheme of the experimental preparation of salmon fillets.

As reported in Table 1, the initial pH values (pH > 6) for both samples (CTR and 1018K6-PP) were similar to those reported by other authors [23]. Throughout storage, the pH of salmon fillets in contact with the not-functionalized PP slides (CTR) and 1018K6-PP slightly decreased, recording a significant difference between the two groups on the 4th day. This result could be justified by an increase of acid production due to the homogenous proliferation of lactic acid bacteria occurring in these samples during the experimental analysis [24], although Gonzalez-Rodriguez et al. [25] registered an increase of alkalinity in prepacked salmon slices as a result of the ammonia and amines production by bacteria. As far as the water activity is concerned, no significant differences were observed among the experimental groups, with only a minor increase on the fourth day (Table 1). From the microbiological point of view, the initial concentrations of TAB (total aerobic bacteria) at both 30 °C and 7 °C in raw salmons were somewhat higher compared with values reported in previous works [26,27], probably due to poor handling practices during the processing of fish fillets. However, similar data were reported by Wiernasz et al. [28], which referred to a concentration of 4.3 ± 0.2 Log (CFU/g) for total mesophilic bacteria. Indeed, the performed analyses showed that 1018K6-PP samples did not display significant differences ($p > 0.05$) in the growth kinetics of TAB at 30 °C and 7 °C compared with the control samples at the end of the storage period, indicating that the antimicrobial packaging did not have any effect, either positive or negative, on the microbiota of salmon fillets (Table 1). Albeit the total bacterial count represents a key factor in assessing the microbiological quality and safety of foods, it is well known that *Pseudomonas* spp., Enterobacteriaceae and *Brochothrix thermosphacta* are the main microbial family and genera responsible for the off-flavours and the unpleasant odours typical of deteriorating fish and fish products [29]. As far as the evolution of these bacteria is concerned, the samples stored in 1018K6-PP packaging revealed a significant slowdown in the replication of these microorganisms at the 4th day of conservation. Specifically, the sensitivity of bacteria belonging to the Enterobacteriaceae family to antimicrobial activity of the active packaging could make this product interesting for the food industry and promote its applicability as a potential "controller tool" for *Escherichia coli*. Indeed, the inhibitory effect of the innovative packaging was also evident towards beta-glucuronidase-positive *E. coli*, whose levels in treated samples were always below 1.0 Log (CFU/g) in contrast to the CTR [>2.0 Log (CFU/g)]. This finding becomes more relevant when the microbiological limits recommended for *E. coli* (1.0 and 2.7 Log (CFU/g) for minimum and maximum limit, respectively) by the International Commission on Microbiological Specifications for Foods (ICMSF) for the commercialization of fish and fish products [30], are taking into account. Actually, the innovative packaging makes the salmon fillets hygienically suitable throughout the storage period.

Regarding total coliforms and *Enterococcus faecalis*, the growth curves were very similar for the control and treated groups, but a significant antimicrobial effect of the 1018K6-PP was observed only on the 4th day of storage. Finally, the microbiological results pointed out the ability of the bound peptide to affect the growth of bacteria belonging to *Staphylococcus* genera. Therefore, the antimicrobial coating appears to successfully act on the survival and replicative capacity of this class of microorganisms, showing a significant ($p < 0.01$) difference between the control groups and the treat one on 4th and 7th days. All these findings confirm the results previously obtained with the peptide 1018K6 in a free status [31]. It is worth noting that the same microbiological analyses were performed on salmon fillets packaged in PP films that were not subjected to any surface modification. Interestingly, the obtained results were comparable to those achieved with the pre-activated and not-funzionalized PP films, thus excluding the occurrence of a potential antimicrobial effect determined by the polymeric surfaces activated by plasma alone. It should be pointed out that the microbiological results were not obvious on the basis of two main considerations: (i) the antimicrobial packaging could be unable to kill microbes under conditions of intended use due to the complexity of the fish matrix, which can inactivate the bioactive compound; (ii) 1018K6 could not retain its antimicrobial activity when bound to PP polymers, because the immobilization process could restrict its conformational freedom and influence its orientation, both of which are important features for the peptide activity.

As far as the potential antimicrobial mechanism, two factors could play a dominant role of the bound 1018K6 with respect to its soluble form:

(1) the high local concentration of the peptide tethered to the polymeric surface;
(2) the strong electrostatic interaction between the cationic peptide chains and anionic bacteria cell membranes (instead of membrane insertion), thus leading to an alteration of the potential across the bacterial membrane, which ultimately triggers cellular death.

To sum up, 1018K6-PPs can be considered a promising instrument to positively affect the quality of perishable products such as fresh salmon based on the microbial effects observed. This suggestion is supported by the important antimicrobial data that the new package exerts against specific spoilage microorganisms responsible for spoilage processes in fish and fish products. However, a potential role of 1018K6-PP in the food safety cannot be excluded, taking into account its action against Enterobacteriaceae and *Staphylococcus* spp. In this scenario, the introduction of 1018K6-PP into the food marketplace could guarantee the availability of safe and natural tools capable of limiting damages of bacterial origin.

Table 1. Evaluation of microbiological counts [Log (CFU/g)] in salmon fillets packaged in active PP films functionalized with 1018K6 by storage time.

	Day	0	4	7
		$m \pm sem$	$m \pm sem$	$m \pm sem$
TAB 30 °C	CTR	4.76 ± 0.06^{A}	6.77 ± 0.07^{B}	6.89 ± 0.02^{B}
	1018K6-PP	4.76 ± 0.06^{A}	6.73 ± 0.01^{B}	6.88 ± 0.01^{C}
TAB 7 °C	CTR	2.91 ± 0.04^{A}	$4.24 \pm 0.01^{B,X}$	5.32 ± 0.04^{C}
	1018K6-PP	2.91 ± 0.04^{A}	$4.56 \pm 0.04^{B,Y}$	5.28 ± 0.04^{C}
Coliforms	CTR	1.91 ± 0.05^{A}	$3.91 \pm 0.04^{a,B,X}$	$3.44 \pm 0.19^{b,B}$
	1018K6-PP	1.91 ± 0.05^{A}	$3.56 \pm 0.07^{B,Y}$	3.44 ± 0.18^{B}
Enterobacteriaceae	CTR	0.96 ± 0.01^{A}	$3.86 \pm 0.07^{B,x}$	$3.07 \pm 0.04^{C,X}$
	1018K6-PP	0.96 ± 0.01^{A}	$3.66 \pm 0.04^{B,y}$	$3.96 \pm 0.04^{C,Y}$
Pseudomonas spp.	CTR	4.31 ± 0.09^{A}	$7.32 \pm 0.10^{B,X}$	$7.44 \pm 0.16^{B,X}$
	1018K6-PP	4.31 ± 0.09^{A}	$6.28 \pm 0.05^{B,Y}$	$6.91 \pm 0.04^{C,Y}$
E. coli	CTR	ni^{A}	$2.07 \pm 0.09^{B,X}$	$2.95 \pm 0.03^{C,X}$
	1018K6-PP	*ni*	ni^{Y}	ni^{Y}
Enterococcus faecalis	CTR	3.32 ± 0.08^{A}	$3.96 \pm 0.01^{B,X}$	$5.07 \pm 0.12^{C,x}$
	1018K6-PP	$3.32 \pm 0.08^{a,A}$	$2.96 \pm 0.12^{bA,Y}$	$4.74 \pm 0.06^{B,y}$
B. thermosphacta	CTR	4.98 ± 0.07^{A}	$5.98 \pm 0.03^{B,X}$	$7.36 \pm 0.14^{C,X}$
	1018K6-PP	4.98 ± 0.07^{A}	$6.81 \pm 0.04^{B,Y}$	$5.96 \pm 0.19^{C,Y}$
Staph. coagulase positive	CTR	ni^{A}	$2.26 \pm 0.09^{B,X}$	$3.19 \pm 0.05^{C,X}$
	1018K6-PP	ni^{A}	$ni^{A,Y}$	$1.96 \pm 0.02^{B,Y}$
pH	CTR	6.25 ± 0.01^{A}	$6.18 \pm 0.02^{B,X}$	6.07 ± 0.03^{C}
	1018K6-PP	6.25 ± 0.01^{A}	$6.11 \pm 0.01^{B,Y}$	6.02 ± 0.01^{C}
a_w	CTR	0.973 ± 0.003^{a}	0.981 ± 0.001^{b}	0.982 ± 0.002^{b}
	1018K6-PP	0.973 ± 0.003^{a}	0.979 ± 0.000^{b}	0.980 ± 0.002^{b}

ni: not isolated. In each storage day, three samples by experimental group were analysed. Statistical analysis was performed comparing experimental groups at each sampling time and within each experimental group along the ripening period. All data were presented as mean (m) ± standard error (sem). Different superscript uppercase letters indicate a significant difference at $p < 0.01$. Different superscript lowercase letters indicate a significant difference at $p < 0.05$. $^{a–c}$ In the same row mean values (same group in different days) followed by different letters show significant differences. ^{x,y} In the same column mean values (different groups on the same sampling time) followed by different letters show significant differences.

3.3. Instrumental Colour Analysis of Salmon Fillets

The colour in fish foods is one of the most important qualities influencing consumer decisions to purchase. Therefore, the impact of 1018K6-PPs on the colour of the packaged salmon fillets was investigated for various storage periods. As reported in Table 2, the lightness (L^*) was the only parameter to be influenced significantly by the use of the active packaging, although this phenomenon did not visibly affect the general appearance of the product. Indeed, chroma values are similar in all the samples during the whole experimental period. Our findings are in agreement with Merlo et al. [32], who reported that the use of chitosan film reduces the change in structure of proteins, conferring a darker aspect to treated salmon fillets. This result was justified considering the strong connection between the change in light scattering of the muscle and the variation in lightness. Furthermore, samples packed in 1018K6-PP were found to be slightly more reddish and yellowish (higher values of a^* and b^*, respectively) than the control ones.

According to several authors [32–34] the main value taken into account for this fish family is the redness, which is associated to the consumer's preference and acceptability. Changes in a^* value in salmon are due to the addition of carotenoids, such as astaxanthin and cataxanthines, and related to reddish colour of salmonid fishes. However, the scientific community disagrees, and different opinions are reported in literature. Yeşilayer et al. [35] demonstrated that fillets of farmed Atlantic salmons fed with feed containing carotenoids showed high values of yellowness, demonstrating that the typical red-orange colour is represented by both redness and yellowness values. It is worth underling that a^* and b^* did not differ significantly among all samples by storage time, exhibiting similar ΔE (total colour differences), Δa^*, and Δb^* values. Therefore, 1018K6-PPs did not produce negative effects on colours parameters, potentially preserving this aspect of salmon samples.

Table 2. Changes in colour indices of the salmon fillets packaged in active PP films functionalized with 1018K6 by storage time.

	Day	0	4	7
		$m \pm sem$	$m \pm sem$	$m \pm sem$
L^*	CTR	43.51 ± 1.47^{a}	$45.93 \pm 0.63^{a,X}$	$48.23 \pm 0.67^{b,x}$
	1018K6-PP	43.51 ± 1.47^{A}	$36.13 \pm 1.89^{B,Y}$	$44.48 \pm 1.51^{A,y}$
a^*	CTR	16.78 ± 0.83	19.75 ± 1.41	15.94 ± 1.44
	1018K6-PP	16.78 ± 0.83	20.02 ± 1.25^{a}	16.63 ± 0.59^{b}
b^*	CTR	21.15 ± 1.68	23.55 ± 2.86	15.94 ± 1.99
	1018K6-PP	21.15 ± 1.68	24.40 ± 3.40	17.01 ± 1.62
Chroma	CTR	27.01 ± 1.82	30.76 ± 3.06	22.55 ± 2.43
	1018K6-PP	27.01 ± 1.82	31.63 ± 3.30	23.82 ± 1.52
Hue angle	CTR	51.46 ± 1.01^{A}	49.75 ± 1.64^{a}	$44.80 \pm 0.92^{b,B}$
	1018K6-PP	51.46 ± 1.01^{a}	50.10 ± 2.92	45.42 ± 1.97^{b}
ΔE	CTR		6.31 ± 2.26	7.22 ± 0.88
	1018K6-PP		9.47 ± 1.45	6.45 ± 1.82
Δa^*	CTR		2.97 ± 1.94	-0.83 ± 0.61
	1018K6-PP		3.24 ± 1.53	-0.14 ± 1.21
Δb^*	CTR		2.40 ± 3.14^{a}	-5.21 ± 0.67^{b}
	1018K6-PP		3.24 ± 2.55	-4.15 ± 3.28

On each sampling day, three samples by experimental group were analysed. Statistical analysis was performed comparing experimental groups at each sampling time and within each experimental group along the ripening period. All data were presented as mean (m) ± standard error (sem). Different superscript uppercase letters indicate a significant difference at $p < 0.01$. Different superscript lowercase letters indicate a significant difference at $p < 0.05$. ^{a,b} In the same row mean values (same group in different days) followed by different letters show significant differences. ^{x,y} In the same column mean values (different groups on the same sampling time) followed by different letters show significant differences.

3.4. Effect of 1018K6-PP on Chemical Parameters of Salmon Fillets

It is common to evaluate the "age" of the food through the study of the microbiological community in order to evaluate the presence and the concentration of specific spoilage microorganisms (SSOs). However, the spoilage of fish and fish products is associated with the occurrence of off-odours due to the production of volatile substances as a result of the bacterial metabolism. Changes in the odour affect the acceptability to consumers, who associate the freshness of fish products to typical organoleptic features. Due to perishable foods being sensitive to variations in appearance, some of the characteristic volatile organic compounds (VOCs) produced by bacteria can be used as potential chemical spoilage indices (CSIs) in fish and fish products [36]. In this study, two chemical quality indicators were used to assess the ability of 1018K6-PP to preserve the quality and sensorial properties, such as the total volatile basic-nitrogen (TVB-N) and trimethylamine-nitrogen (TMA-N) (Figure 3).

The choice to detect these two VOCs is dictated by the key role that these chemicals play in the freshness of salmon, being the final products of protein degradation [37].

TVB-N includes the measurement of volatile basic nitrogenous compounds, such as trimethylamine (TMA), dimethylamine (DMA) and other nitrogenic substances, which are produced by bacterial or tissue enzymes from the deamination of amino acids. In the current study, the initial amount of TVB-N in all salmon fillets analysed was 7.89 ± 0.21 mg/100 g (Figure 3A). Significant differences ($p < 0.01$) between salmon samples packaged with CTR-PP and 1018K6-PP slides were observed after 4 days of storage. Specifically, 1018K6-PP appeared, indirectly, to slow down the protein degradation in salmon fillets through the control of microbial growth. Indeed, the great amount of the free amino acids in fish [38,39] are used as substrate by bacteria in their metabolism, with the final production of organic acids, sulphur compounds, ammonia and biogenic amines (BAs) [40,41]. Overall, though 1018K6-PP demonstrated to be efficient in reducing the protein degradation, throughout the entire storage period TVB-N values never reached and overcame the legislative limit of 35 mg/100 g specified by the EU 2019/627 for this fish typology [42].

The TMA-N origins by decomposition of trimethylamine N-oxide (TMAO), used from bacteria as a donor of oxygen molecules in their respiratory metabolism in fish and fish products stored at refrigeration temperature [43–45]. Due to the importance of the initial amount of TMAO in the muscle, the concentration of TMA-N is strongly related to the species of fish, and *Salmo salar* is naturally rich in trimethylamine N-oxide [46]. In Figure 3B, the trends of TMA-N over time are displayed. Specifically, the samples packed with 1018K6-PPs showed the lowest TMA-N values ($p < 0.01$) at both 4 and 7 days of storage, in agreement with the above reported values of TVB-N, thus reinforcing the hypothesis that active slides affect the spoilage microbial communities. In fact, it is well-known that the TMA production is mainly operated by bacteria belonging to the Enterobacteriaceae family, which, the results showed, proved sensitive to the antimicrobial activity of the bound peptide, as already reported in Table 1 [47,48]. Although TMA is considered a good indicator of the deterioration progress, no maximum legislative limits for TMA concentrations were defined and different values were proposed. However, according to Shumilina et al. [49], who reported 4.2 mg/100 g as the acceptability limit for fish, the freshness was preserved only in salmon fillets put in contact with 1018K6-PP (TMA < 5 mg/100 g).

Finally, measurements of thiobarbituric acid reactive substances (TBARS) expressed as malonyldialdehyde (MDA) levels, were performed in order to investigate lipid oxidation, which is a very important event determining the quality of foods, especially of those containing highly unsaturated fats, such as fish [50,51]. As shown in Figure 3C, the TBARS values in control fillets increased significantly during refrigerated storage in contrast to that observed in the packaged fillets with 1018K6-PP slides. Therefore, 1018K6-PP is able to exert antioxidant properties, but this finding is not surprising given the well-known correlation between lipid oxidation and bacterial contamination [52].

Indeed, MDA is the main aldehyde produced as a result of the decomposition of unsaturated fatty acids—also a bacterial operation—thus, remarking on the antimicrobial efficacy of the active films. The chemical analyses performed on the salmon fillets packaged in unmodified PP films demonstrated that the plasma activation by itself was not able to allow the polymers to affect the quality of these fish products.

Overall, the comprehensive analyses of microbiological and chemical parameters pointed out two main aspects: the key roles of TVB-N, TMA and MDA as chemical spoilage indices in perishable food and the effectiveness of 1018K6-PP in preserving salmon fillets. As reported by Prabhakar et al. [53], the assumption of the interconnection among bacterial concentrations and chemical metabolites production is already consolidated, as is the link between TVB-N/TMA levels and quality. Therefore, our findings confirmed this strong link, and the candidate 1018K6-PP as a valuable packaging technology capable of guaranteeing longer durability for highly perishable foods, such as raw salmon.

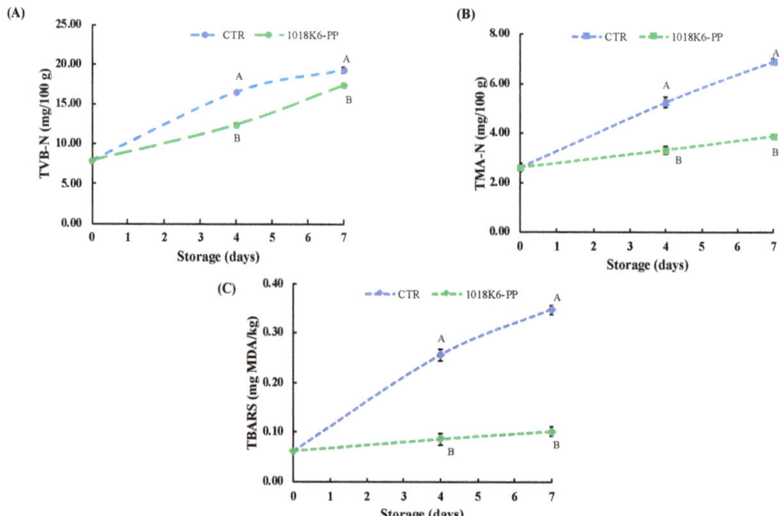

Figure 3. Effects of 1018k6-PP surfaces on the chemical quality of salmon fillets. (**A**) Changes in TVB-N (A), TMA-N (**B**), and TBARS (**C**) of *Salmon salar* fillets packaged in active 1018K6-PP films during storage at 4 °C. CTR (blue lines)—PP films without 1018K6; 1018K6-PP (green lines)—PP films functionalized with 1018K6. Results are means of three independent experiments and error bars represent the standard error (sem). Different letters at each sampling time are used for significantly different samples, according to Tukey test (uppercase letters: $p < 0.01$).

3.5. Panelists' Sensory Evaluation

Sensory perception is the tool through which the consumers choose foods at a store, trusting in their senses and adopting an immediate and easy system for evaluating freshness and quality [54]. In this study, the organoleptic features appeared to be partially influenced by the packaging technology used. As reported in Figure 4, the representation of observed sensory characteristics highlighted an important consequence of the use of antimicrobial slides on the production of off-odours. Despite the initial good quality of all samples, the salmon fillets belonging to the control groups showed signs of spoilage as early as the 4th day of storage at refrigeration temperature. The judges rated the control samples as "poor freshness quality" products, due to the score of odour and of the general appearance obtained by the end of the trial. Contrarily, the treated samples maintained good sensory characteristics over time. In agreement with those reported above for VOCs, the demonstrated antimicrobial activity (Table 1) of 1018K6-PP seems to indirectly control the negative changes in the chemical structure and metabolites production of salmon fillets occurring during storage [55,56]. Furthermore, the scores of overall appearances of treated samples pointed out the absence of negative influences of the novel active packaging on the sensory features, due to the colourless and odourless nature of the 1018K6 molecules.

3.6. Microbial Challenge Testing of *L. monocytogenes* on Salmon Fillets Packaged with 1018K6-PP

Foodborne diseases are a reality affecting thousands of people in industrialized countries every year. Amongst the bacterial pathogens responsible of severe human toxi-infections, *Listeria monocytogenes* is considered one of the most dangerous. Due to their origin and the way in which they are processed, fish products show an increased incidence rate of listeriosis, and then they represent typical food vehicles of high levels of microbiological contamination, taking into account that this bacterium is able to grow also at refrigeration temperatures. Therefore, challenge testing of the food products with

L. monocytogenes is recommended to assess the potential for growth, both qualitatively and quantitatively, in the foods at risk.

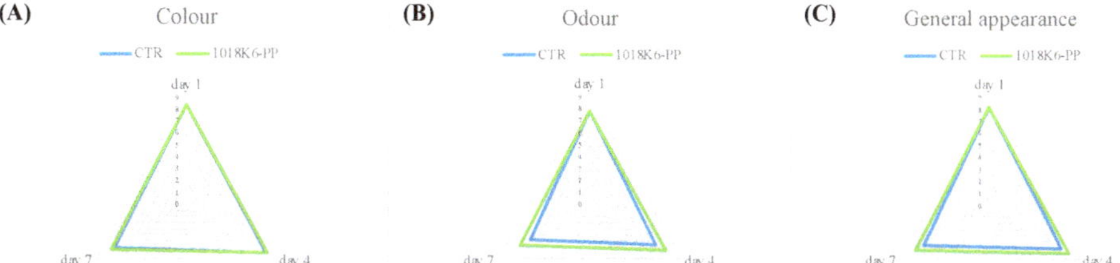

Figure 4. Changes in colour (**A**), odour (**B**), and overall appearance (**C**) of salmon fillets during storage period. CTR (blue lines)—PP films without 1018K6; 1018K6-PP (green lines)—PP films functionalized with 1018K6.

In this context, the anti-listerial efficacy of 1018K6-PP was evaluated in salmon fillets stored at 5 °C for 96 h (Figure 5). In order to confer greater confidence in the assessment of the likelihood of a particular strain to compromise food safety, mixed cultures of three *L. monocytogenes* strains isolated from fish were used at a concentration of ca. 150 CFU/mL. This value of inoculum is representative for the natural contamination of *L. monocytogenes* commonly encountered in fresh foods, taking into account that 100 CFU/mL is the threshold limit considered as low risk for causing listeriosis. In addition, the use of food isolates is recommended because it is likely to represent better the behaviour of naturally contaminating strains.

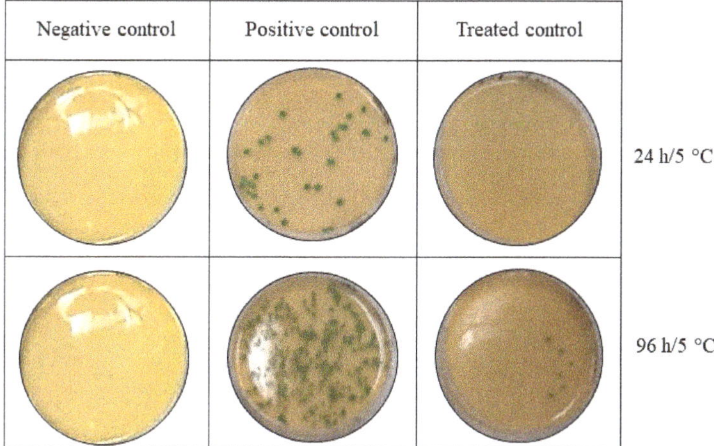

Figure 5. Bactericidal activity of polymer functionalized with 1018K6 against *L. monocytogenes* on salmon fillets. Negative control—untreated salmon fillets; positive control—salmon fillets treated with not-functionalized PP; treated control—salmon fillets treated with 1018K6-PP.

The results of the challenge test performed on salmon fillets indicated that the antimicrobial packaging was effective in inhibiting the growth and survival of the pathogen on the surface of the fresh food during storage in contrast to the untreated control. Indeed, a complete inhibition of *L. monocytogenes* was observed after 72 h incubation, with a slight decrease (95%) at the end of the assay (96 h), thus suggesting that our system could be used to preserve the safety of fish products during storage.

Overall, our results show the positive impact of the effectiveness of 1018K6-PP packaging on food safety when the target microorganism is a foodborne pathogen of great present concern, such as *L. monocytogenes.*

3.7. Evaluation of 1018K6-PPs Slides on the Physicochemical, Microbial and Sensorial Properties of Sarda Sarda Burgers

In order to evaluate the versatility of our active packaging, a different typology of food matrices was included in the experimental design. To this aim, microbiological, physicochemical and sensorial analyses were performed on fish burgers of bonito (*Sarda sarda)* packaged with 1018K6-PP slides. This analysis was aimed at also verifying the effectiveness of 1018K6-PP against minced fish meat, which is notoriously characterized by higher level of microorganisms than fillets because of the shredding process underlying their manufacture [57]. The scheme used to set up the *Sarda sarda* hamburger employed in the experimental trials, is shown in Figure 6.

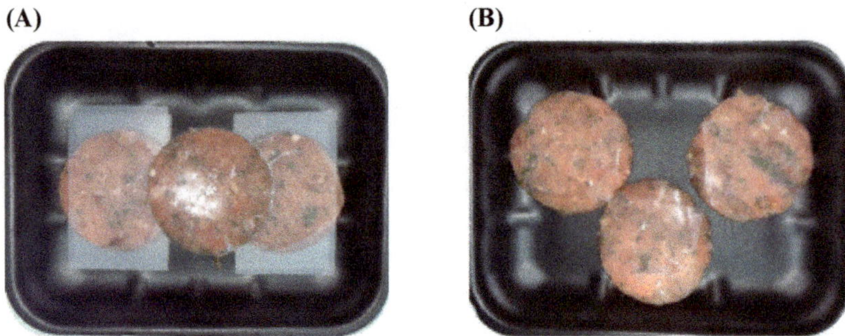

Figure 6. Representative scheme of preparation of *Sarda sarda* burgers employed in the microbiological and physico-chemical analyses. (**A**) Burgers of *Sarda sarda* treated with PPs films not-functionalized with 1018K6 (CTR); (**B**) burgers of *Sarda sarda* treated with PPs films functionalized with 1018K6 (1018K6-PP).

As reported in Table 3, the initial amounts of TAB were significantly different between salmon fillets (Table 1) and fish burgers (Table 3), in which more than 1 Log (CFU/g) of mesophilic bacteria were enumerated. For this reason, fish burgers represent a difficult challenge. Regarding antimicrobial activity, 1018K6-PPs negatively affected the growth of specific microorganisms, including the total bacterial count. During the storage, the mesophilic TAB increased significantly in control samples, until reaching a concentration greater than 8 Log (CFU/g) by the 7th day, in contrast to that observed in the samples packaged with the antimicrobial slides, in which the maximum acceptable limit set by ICMSF for TAB [7 Log (CFU/g)] was never exceeded during 7 days of storage. Furthermore, due to the key role of mesophilic bacteria in the production of metabolites and off-odours, the antimicrobial activity of 1018K6-PP produced a beneficial effect on the overall appearance of fish burgers and their chemical profile. Therefore, our findings not only confirmed the effectiveness of the new package in slowing the growth of the same bacterial communities described for salmon fillets but also the obtained results enhanced the potential role of 1018K6-PP as a tool for monitoring microbiologic contaminations.

Table 3. Evaluation of microbiological counts (Log CFU/g) in *Sarda sarda* burgers packaged in active PP films functionalized with 1018K6 by storage time.

		Day	0	3	5	7
			$m \pm sem$	$m \pm sem$	$m \pm sem$	$m \pm sem$
TAB 30 °C	CTR		$6.25 \pm 0.02^{a,A}$	6.52 ± 0.17^{A}	$6.74 \pm 0.17^{b,A}$	$8.14 \pm 0.08^{X,B}$
	1018K6-PP		6.25 ± 0.02	6.17 ± 0.12	6.22 ± 0.22	6.37 ± 0.34^{Y}
TAB 7 °C	CTR		5.16 ± 0.09^{A}	6.36 ± 0.08^{B}	$6.79 \pm 0.07^{C,X}$	$8.17 \pm 0.04^{D,X}$
	1018K6-PP		5.16 ± 0.09^{A}	6.01 ± 0.16^{B}	$5.94 \pm 0.19^{B,Y}$	$6.50 \pm 0.39^{B,Y}$
Coliforms	CTR		4.61 ± 0.02^{A}	$5.39 \pm 0.09^{a,B,X}$	4.90 ± 0.04^{C}	$5.15 \pm 0.06^{b,B,X}$
	1018K6-PP		4.61 ± 0.02	4.60 ± 0.23^{Y}	4.62 ± 0.16	4.35 ± 0.25^{Y}
Enterobacteriaceae	CTR		3.26 ± 0.17^{A}	$5.96 \pm 0.57^{B,C,X}$	4.98 ± 0.10^{B}	5.26 ± 0.00^{C}
	1018K6-PP		3.26 ± 0.17^{A}	$3.63 \pm 0.46^{a,A,Y}$	$4.97 \pm 0.16^{b,B}$	5.11 ± 0.15^{B}
Pseudomonas spp.	CTR		$5.91 \pm 0.02^{a,A}$	$6.51 \pm 0.20^{bB,x}$	$5.79 \pm 0.10^{A,x}$	$8.59 \pm 0.09^{C,X}$
	1018K6-PP		5.91 ± 0.02^{A}	5.91 ± 0.13^{y}	$5.47 \pm 0.08^{B,y}$	6.44 ± 0.36^{Y}
E. coli	CTR		1.50 ± 0.12^{a}	1.80 ± 0.20^{A}	1.62 ± 0.11^{A}	$1.11 \pm 0.09^{b,B}$
	1018K6-PP		1.50 ± 0.12	1.32 ± 0.18	1.19 ± 0.23	1.28 ± 0.16
Enterococcus faecalis	CTR		4.39 ± 0.13^{A}	$4.41 \pm 0.09^{A,X}$	3.21 ± 0.23^{B}	3.96 ± 0.00^{C}
	1018K6-PP		$4.39 \pm 0.13^{a,A}$	$3.39 \pm 0.13^{B,Y}$	3.81 ± 0.39	3.87 ± 0.21^{b}
B. thermosphacta	CTR		ni^{A}	ni^{A}	$1.98 \pm 0.00^{B,X}$	$1.98 \pm 0.00^{B,X}$
	1018K6-PP		ni	ni	ni^{Y}	ni^{Y}
Staph. coagulase positive	CTR		4.45 ± 0.01^{A}	$5.64 \pm 0.08^{B,X}$	$4.23 \pm 0.14^{A,X}$	$4.37 \pm 0.26^{A,X}$
	1018K6-PP		4.45 ± 0.01^{A}	$4.12 \pm 0.16^{a,A,Y}$	$3.6 \pm 0.08^{b,B,Y}$	$3.19 \pm 0.22^{B,Y}$
pH	CTR		6.20 ± 0.01^{A}	$6.18 \pm 0.00^{A,x}$	6.21 ± 0.01^{B}	$6.39 \pm 0.03^{C,X}$
	1018K6-PP		6.20 ± 0.01^{a}	6.21 ± 0.01^{y}	6.23 ± 0.00^{b}	$6.26 \pm 0.03^{b,Y}$
a_w	CTR		0.976 ± 0.006	0.969 ± 0.001^{a}	0.972 ± 0.001	0.974 ± 0.001^{b}
	1018K6-PP		0.976 ± 0.006	0.963 ± 0.009	0.973 ± 0.000	0.974 ± 0.002

ni—not isolated. In each sampling day, three samples were analysed by experimental group. Statistical analysis was performed comparing experimental groups at each sampling time and within each experimental group along the ripening period. All data were presented as mean (m) ± standard error (sem). Different superscript uppercase letters indicate a significant difference at $p < 0.01$. Different superscript lowercase letters indicate a significant difference at $p < 0.05$. $^{a-d}$ In the same row mean values (same group in different days) followed by different letters show significant differences. ^{x,y} In the same column mean values (different groups on the same sampling time) followed by different letters show significant differences.

It is worth noting that the same analyses were performed on bonito burgers packaged in PP films and not subjected to any surface modification and no discrepancy in the results was observed with respect to those obtained with the pre-activated PP films alone.

As far as the determination of colour values, the changes in this parameter in fish burgers over time overlapped the data collected for salmon fillets (Table 4). Specifically, the samples belonging to the control group appeared less dark than the others, affirming the hypothesis of an increase in proteolysis. Moreover, no differences were highlighted among samples in a^* and b^* values and, consequently, in total colour differences (ΔE), variations in a^* (Δa^*), and in b^* (Δb^*).

The graph reported in Figure 7 showed the positive effect of the active packaging on the redness on 3rd and 7th day, although slightly.

Table 4. Changes in colour indices of *Sarda sarda* burgers packaged in active PP films functionalized with 1018K6 by storage time.

	Day	0	3	5	7
		m ± sem	m ± sem	m ± sem	m ± sem
L^*	CTR	42.68 ± 1.09	41.07 ± 0.51 ^{A}	42.95 ± 1.20	45.04 ± 0.77 ^{B,x}
	1018K6-PP	42.68 ± 1.09 ^{a}	39.76 ± 0.74 ^{b,A}	42.32 ± 0.80 ^{a}	42.79 ± 0.70 ^{B,y}
a^*	CTR	6.12 ± 0.89	4.93 ± 0.20 ^{A}	4.87 ± 0.47 ^{A,x}	7.46 ± 0.27 ^{B,X}
	1018K6-PP	6.79 ± 0.34 ^{a,A}	4.74 ± 0.37 ^{B}	5.95 ± 0.20 ^{b,A,y}	6.12 ± 0.63 ^{Y}
b^*	CTR	12.42 ± 0.38 ^{a,A}	15.61 ± 0.31 ^{B,x}	10.71 ± 0.77 ^{A}	10.93 ± 0.60 ^{b,A}
	1018K6-PP	12.42 ± 0.38 ^{a,A}	14.50 ± 0.37 ^{B,y}	12.29 ± 0.38 ^{a,A}	10.53 ± 0.70 ^{b,A}
Chroma	CTR	13.89 ± 0.44 ^{a,A}	16.37 ± 0.33 ^{B,x}	11.79 ± 0.83 ^{b,A,x}	13.24 ± 0.61 ^{A}
	1018K6-PP	14.15 ± 0.43	15.28 ± 0.35 ^{A,y}	13.66 ± 0.37 ^{B,y}	12.21 ± 0.86 ^{B}
Hue angle	CTR	63.86 ± 3.54 ^{a}	72.47 ± 0.62 ^{b,A}	65.48 ± 1.87 ^{B}	55.51 ± 1.09 ^{b,B,x}
	1018K6-PP	61.34 ± 1.08 ^{A}	71.86 ± 1.45 ^{B}	64.11 ± 0.97 ^{A}	60.06 ± 1.83 ^{A,y}
ΔE	CTR		4.20 ± 0.60 ^{A}	3.61 ± 0.88 ^{B}	4.06 ± 0.91 ^{B}
	1018K6-PP		4.60 ± 0.80 ^{A}	2.79 ± 0.70 ^{a,B}	3.65 ± 0.76 ^{b,B}
Δb^*	CTR		3.10 ± 0.00	−1.71 ± 0.91	−1.49 ± 0.54
	1018K6-PP		2.09 ± 0.39	−0.13 ± 0.37	−1.88 ± 0.75

In each sampling day, three samples were analysed by experimental group. Statistical analysis was performed comparing experimental groups at each sampling time and within each experimental group along the ripening period. All data were presented as mean (m) ± standard error (sem). Different superscript uppercase letters indicate a significant difference at $p < 0.01$. Different superscript lowercase letters indicate a significant difference at $p < 0.05$. ^{a,b} In the same row mean values (same group in different days) followed by different letters show significant differences. ^{x,y} In the same column mean values (different groups on the same sampling time) followed by different letters show significant differences.

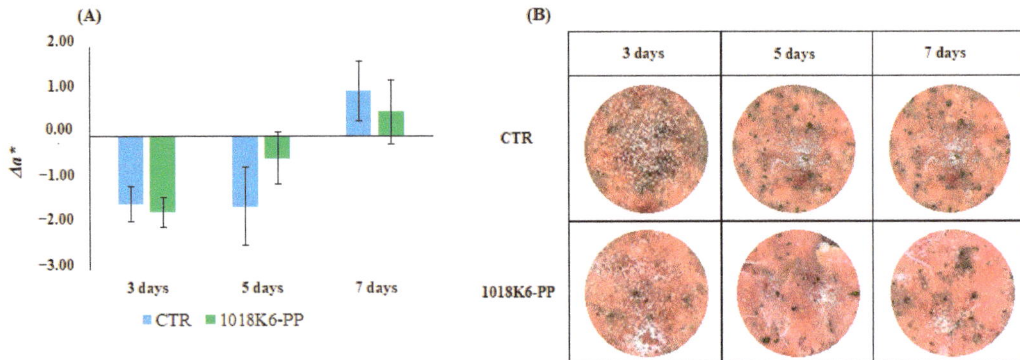

Figure 7. (**A**) Analysis of a^* variation (Δa^*) in bonito fish burgers during the storage period. Results are means of three independent experiments and error bars represent the standard error (sem). CTR (blue)—PP films without 1018K6; 1018K6-PP (green)—PP films functionalized with 1018K6. (**B**) Photos of *Sarda sarda* burgers packaged with the control and functionalized slides at each sampling time.

Moreover, the experimentation on fish burgers marked the important contribution of the antimicrobial molecule in slowing down the protein degradation. Indeed, significant differences were found among samples packed in active films and control ones and the gap recorded between the corresponding TMA-N and TVB-N values proved the concrete beneficial effect of the 1018K6-PP (Figure 8).

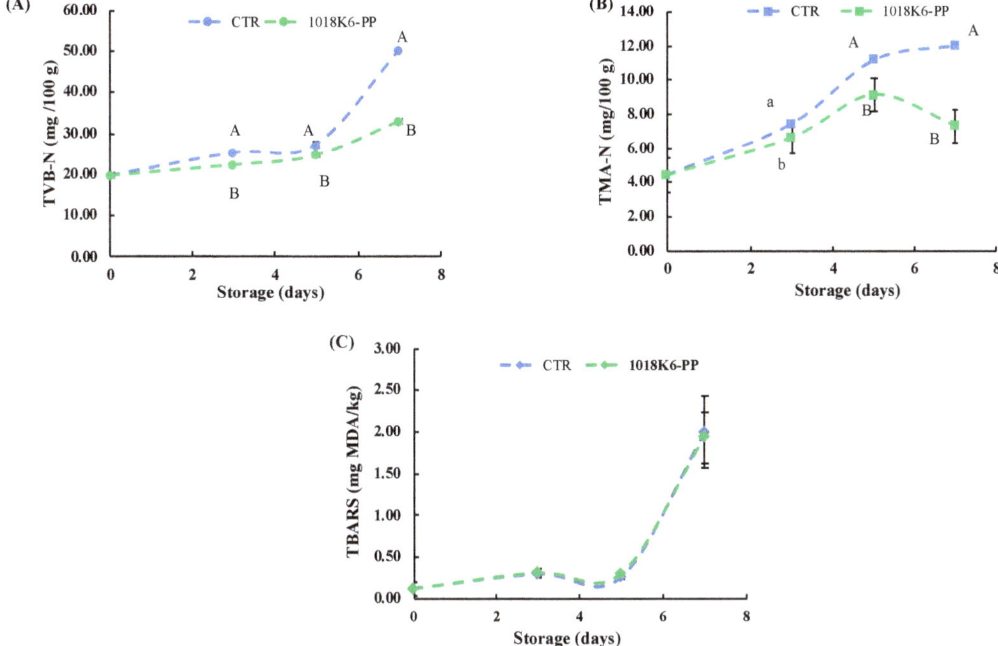

Figure 8. Effects of 1018K6-PP slides on the chemical quality of *Sarda sarda* burgers. Changes in TVB-N (**A**), TMA-N (**B**) and TBARS (**C**) of *Sarda sarda* burgers packaged in active 1018K6-PP films during storage at 4 °C. CTR (blue lines)—PP films without 1018K6; 1018K6-PP (green lines)—PP films functionalized with 1018K6. Results are means of three independent experiments and error bars represent the standard error (sem). Different letters at each sampling time are used for significantly different samples, according to Tukey test (uppercase letters: $p < 0.01$; lowercase letters: $p < 0.05$).

Finally, the off-odours drastically affected the judgments (Figure 9), by which the control samples were labelled as unpleasing foods, probably due to their content in TVB-N and TMA-N. This result was expected considering that the odour weight on the panellists' choices is the most critical sensory characteristic for fish products [58].

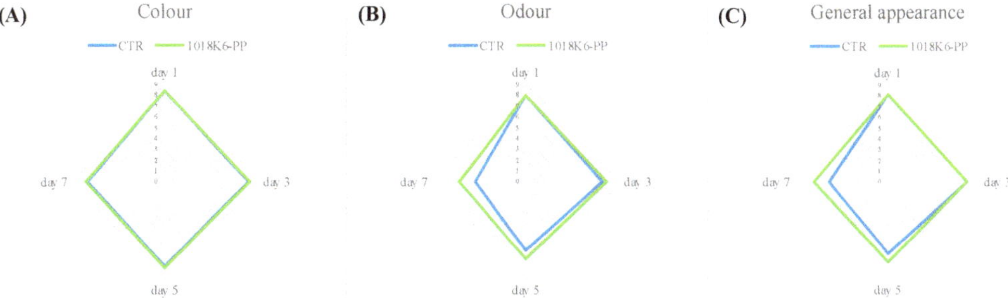

Figure 9. Changes in colour (**A**), odour (**B**), and overall appearance (**C**) of bonito fish burgers during storage period. CTR (blue lines)—PP films without 1018K6; 1018K6-PP (green lines)—PP films functionalized with 1018K6.

Finally, our findings allowed also supposing a positive effect of 1018K6-PP on the quality parameters of bonito burgers, considering the significant differences between the two experimental groups in microorganisms concentrations and CSIs levels, which strongly affected on the sensory appearance of samples. The off-odours and the changes in lightness were demonstrated to be the main visible properties associated to the spoilage processes, so that they could be considered alarm bells for the consumers.

4. Conclusions

As stated, fish is a highly perishable food characterized by a short shelf-life. Refrigeration is probably one of the most used methods for fish preservation, but several deteriorative quality changes occur during storage, particularly in texture, colour and flavour, limiting shelf-life. Therefore, there is an important and urgent need to find alternative strategies to overcome existing challenges that are associated with fish spoilage, which will ultimately benefit both the producers and consumers. In the present study, two different kinds of fish foods, *Salmon salar* fillets and *Sarda sarda* burgers, were used to obtain information about the feasibility of the potential application of 1018K6-PP packaging in the food industry. The results showed that 1018K6-PP helped to maintain the chemical and microbial quality of this kind of product without inducing sensory alterations during refrigerated storage. Therefore, the antimicrobial packaging used in the present study represents an excellent and promising option for the preservation of fish foods due to its antimicrobial, non-toxic and re-usability properties, and thereby reduce the occurrence of foodborne illness.

Author Contributions: Conceptualization, A.A., Y.T.R.P. and G.P.; methodology, R.L.A., L.G., B.A. and D.C.; software, R.L.A., M.B. and D.C.; validation, M.G., B.A. and G.P.; formal analysis, Y.T.R.P., R.L.A., B.A., L.G. and M.B.; investigation, M.G., R.L.A., B.A. and D.C.; resources, A.A., Y.T.R.P. and G.P.; data curation, M.G., R.L.A. and Y.T.R.P.; writing—original draft preparation, M.G., R.L.A. and G.P.; writing—review and editing, M.G. and G.P.; visualization, M.G., G.P. and A.A.; supervision, M.G. and G.P.; project administration, A.A. and G.P.; funding acquisition, G.P., Y.T.R.P. and A.A. All authors have read and agreed to the published version of the manuscript.

Funding: This research was funded by: Ministero della Salute-"Dalle superfici agli alimenti: nuove soluzioni per alimenti più sicuri (NewSan)"-grant number IZSME 06/20 RC, Ricerca Corrente 2020 project; Ministero dello Sviluppo Economico "Sviluppo di piattaforme molecolari e cellulari per l'identificazione di prodotti innovativi ad attività NUTRAceutica da Biotrasformazioni mediante organismi ESTremofili (NUTRABEST)"- grant number F/200050/01-03/X45, Fondo per la Crescita Sostenibile-Sportello "AGRIFOOD" PON I&C 2014-2020 project.

Institutional Review Board Statement: Not applicable.

Informed Consent Statement: Not applicable.

Data Availability Statement: Data is contained within the article.

Acknowledgments: The authors would like to thank Francesca Segreti for technical support and Valentina Brasiello for administrative assistance.

Conflicts of Interest: The authors declare no conflict of interest. The funders had no role in the design of the study; in the collection, analyses, or interpretation of data; in the writing of the manuscript, or in the decision to publish the results.

References

1. Ackman, R.G. Fatty acids. In *Marine Biogenic Lipids, Fats and Oils*; CRC Press: Boca Raton, FL, USA, 1989; pp. 103–137.
2. Ashie, I.N.A.; Smith, J.P.; Simpson, B.K. Spoilage and Shelf Life Extension of Fresh Fish and Shellfish. *Cri. Rev. Food Sci. Nutr.* **1996**, *36*, 87–121. [CrossRef] [PubMed]
3. FAO. *The State of World Fisheries and Aquaculture 2020: Sustainability in Action*; FAO: Rome, Italy, 2020; ISBN 978-92-5-132692-3.
4. Chotphruethipong, L.; Benjakul, S. Use of cashew (*Anacardium occidentale* L.) leaf extract for prevention of lipidoxidation in Mayonnaise enriched with fish oil. *Turkish J. Fish Aquat. Sci.* **2019**, *19*, 825–836.
5. Maqsood, S.; Benjakul, S.; Abushelaibi, A.; Alam, A. Phenolic compounds and plant phenolic extracts as natural antioxidantsin prevention of lipid oxidation in seafood: A detailed review. *Compr. Rev. Food Sci. Food Saf.* **2014**, *13*, 1125–1140. [CrossRef]

6. Hsieh, R.J.; Kinsella, J.E. Oxidation of polyunsaturatedfatty acids: Mechanisms, products, and inhibition with emphasison fish. *Adv. Food Nutr. Res.* **1989**, *33*, 233–341.
7. Olatunde, O.O.; Benjakul, S.; Vongkamjan, K. Comparative study on nitrogen and argon-based modified atmosphere-packaging on microbiological, chemical, and sensory attributes as well as on microbial diversity of Asian sea bass. *Food Packag. Shelf Life* **2019**, *22*, 100404. [CrossRef]
8. Pongsetkul, J.; Benjakul, S.; Sumpavapol, P.; Osako, K.; Faithong, N. Characterization of endogenous protease and the changes in proteolytic activity of *Acetes vulgaris* and *Macrobrachium lanchesteri* during kapi production. *J. Food Biochem.* **2017**, *41*, e12311. [CrossRef]
9. Socaciu, M.I.; Semeniuc, C.A.; Vodnar, D.C. Edible films and coatings for fresh fish packaging: Focus on quality changes and shelf-life extension. *Coatings* **2018**, *8*, 366. [CrossRef]
10. Rollini, M.; Nielsen, T.; Musatti, A.; Limbo, S.; Piergiovanni, L.; Hernandez Munoz, P.; Gavara, R. Antimicrobial performance of two different packaging materials on the microbiological quality of fresh salmon. *Coatings* **2016**, *6*, 6. [CrossRef]
11. Palmieri, G.; Balestrieri, M.; Proroga, Y.T.R.; Falcigno, L.; Facchiano, A.; Riccio, A.; Capuano, F.; Marrone, R.; Neglia, G.; Anastasio, A. New antimicrobial peptides against foodborne pathogens: From in silico design to experimental evidence. *Food Chem.* **2016**, *211*, 546–554. [CrossRef]
12. Palmieri, G.; Balestrieri, M.; Capuano, F.; Proroga, Y.T.R.; Pomilio, F.; Centorame, P.; Riccio, A.; Marrone, R.; Anastasio, A. Bactericidal and antibiofilm activity of bactenecin-derivative peptides against the food-pathogen *Listeria monocytogenes*: New perspectives for food processing industry. *Int. J. Food. Microbiol.* **2018**, *279*, 33–42. [CrossRef]
13. Agrillo, B.; Balestrieri, M.; Gogliettino, M.; Palmieri, G.; Moretta, R.; Proroga, Y.T.; Cornacchia, A.; Capuano, F.; Smaldone, G.; De Stefano, L. Functionalized polymeric materials with bio-derived antimicrobial peptides for "active" packaging. *Int. J. Mol. Sci.* **2019**, *20*, 601. [CrossRef] [PubMed]
14. Festa, R.; Ambrosio, R.L.; Lamas, A.; Gratino, L.; Palmieri, G.; Franco, C.M.; Cepeda, A.; Anastasio, A. A Study on the Antimicrobial and Antibiofilm Peptide 1018K6 as Potential Alternative to Antibiotics against Food-Pathogen *Salmonella enterica*. *Foods* **2021**, *10*, 1372. [CrossRef] [PubMed]
15. Agrillo, B.; Proroga, Y.T.R.; Gogliettino, M.; Balestrieri, M.; Tatè, R.; Nicolais, L.; Palmieri, G. A Safe and Multitasking Antimicrobial Decapeptide: The Road from De Novo Design to Structural and Functional Characterization. *Int. J. Mol. Sci.* **2020**, *21*, 6952. [CrossRef] [PubMed]
16. European Commission. Commission Regulation (EC) No 2073/2005 of 15 November 2005 on microbiological criteria for foodstuffs. *Off. J. Eur. Union* **2005**, *338*, 1–26.
17. Ambrosio, R.L.; Smaldone, G.; Di Paolo, M.; Vollano, L.; Ceruso, M.; Anastasio, A.; Marrone, R. Effects of Different Levels of Inclusion of Apulo-Calabrese Pig Meat on Microbiological, Physicochemical and Rheological Parameters of Salami during Ripening. *Animals* **2021**, *11*, 3060. [CrossRef]
18. Conway, E.J.; Byrne, A. An absorption apparatus for the micro-determination of certain volatile substances: The mi-cro-determination of ammonia. *Biochem. J.* **1933**, *27*, 419.
19. Angiolillo, L.; Conte, A.; Del Nobile, M.A. A new method to bio-preserve sea bass fillets. *Int. J. Food Microbiol.* **2018**, *271*, 60–66. [CrossRef]
20. Palmieri, G.; Tatè, R.; Gogliettino, M.; Balestrieri, M.; Rea, I.; Terracciano, M.; Proroga, Y.T.; Capuano, F.; Anastasio, A.; De Stefano, L. Small synthetic peptides bioconjugated to hybrid gold nanoparticles destroy potentially deadly bacteria at sub-micromolar concentrations. *Bioconjug. Chem.* **2018**, *29*, 3877–3885. [CrossRef]
21. Baniya, H.B.; Shrestha, R.; Guragain, R.P.; Kshetri, M.B.; Pandey, B.P.; Subedi, D.P. Generation and characterization of an atmospheric-pressure plasma jet (APPJ) and its application in the surface modification of polyethylene terephthalate. *Int. J. Polym. Sci.* **2020**, *2020*, 1–7. [CrossRef]
22. Boziaris, I.S.; Stamatiou, A.P.; Nychas, G.J. Microbiological aspects and shelf life of processed seafood products. *J. Sci. Food Agric.* **2013**, *93*, 1184–1190. [CrossRef]
23. Alves, V.L.; Rico, B.P.; Cruz, R.M.; Vicente, A.A.; Khmelinskii, I.; Vieira, M.C. Preparation and characterization of a chitosan film with grape seed extract-carvacrol microcapsules and its effect on the shelf-life of refrigerated Salmon (*Salmo salar*). *LWT* **2018**, *89*, 525–534. [CrossRef]
24. Kitundu, E.; Young, O.; Seale, B.; Owens, A. Lactic fermentation of cooked, comminuted mussel, *Perna canaliculus*. *Food Microbiol.* **2021**, *99*, 103829. [CrossRef] [PubMed]
25. Gonzalez-Rodriguez, M.N.; Sanz, J.J.; Santos, J.A.; Otero, A.; Garcia-Lopez, M.L. Bacteriological quality of aquacultured freshwater fish portions in prepackaged trays stored at 3 C. *J. Food Prot.* **2001**, *64*, 1399–1404. [CrossRef] [PubMed]
26. Lerfall, J.; Jakobsen, A.N.; Skipnes, D.; Waldenstrøm, L.; Hoel, S.; Rotabakk, B.T. Comparative evaluation on the quality and shelf life of Atlantic salmon (*Salmo salar* L.) filets using microwave and conventional pasteurization in combination with novel packaging methods. *J. Food Sci.* **2018**, *83*, 3099–3109. [CrossRef] [PubMed]
27. Mace, S.; Cornet, J.; Chevalier, F.; Cardinal, M.; Pilet, M.F.; Dousset, X.; Joffraud, J.J. Characterization of the spoilage microbiota in raw salmon steaks stored under vacuum or modified atmosphere packaging combining conventional methods and PCR-TTGE. *Food Microbiol.* **2012**, *30*, 164–172. [CrossRef]

28. Wiernasz, N.; Leroi, F.; Chevalier, F.; Cornet, J.; Cardinal, M.; Rohloff, J.; Pilet, M.F. Salmon gravlax biopreservation with lactic acid bacteria: A polyphasic approach to assessing the impact on organoleptic properties, microbial ecosystem and volatilome composition. *Front. Microbiol.* **2020**, *10*, 3103. [CrossRef] [PubMed]
29. Geeroms, N.; Verbeke, W.; Van Kenhove, P. Consumers' health-related motive orientations and ready meal consumption behaviour. *Appetite* **2008**, *51*, 704–712. [CrossRef]
30. Roberts, T.A.; Bryan, F.L.; Christian, J.H.B.; Kilsby, D.; Olson, J.C., Jr.; Siliker, J.H. *ICMSF: International Commission on Microbiological Specifications for Foods. Microorganisms in Foods. 2. Sampling for Microbial Analysis. Principles and Specific Applications*, 2nd ed.; University of Toronto Press: Toronto, ON, Canada, 1986.
31. Colagiorgi, A.; Festa, R.; Di Ciccio, P.A.; Gogliettino, M.; Balestrieri, M.; Palmieri, G.; Anastasio, A.; Ianieri, A. Rapid biofilm eradication of the antimicrobial peptide 1018K6 against *Staphylococcus aureus*: A new potential tool to fight bacterial bio-films. *Food Control* **2020**, *107*, 106815. [CrossRef]
32. Merlo, T.C.; Contreras-Castillo, C.J.; Saldana, E.; Barancelli, G.V.; Dargelio, M.D.B.; Yoshida, C.M.P.; Ribeiro Junior, E.E.; Massarioli, A.; Venturini, A.C. Incorporation of pink pepper residue extract into chitosan film combined with a modified atmosphere packaging: Effects on the shelf life of salmon fillets. *Food Res. Int.* **2019**, *125*, 108633. [CrossRef]
33. Lerfall, J.; Bendiksen, E.A.; Olsen, J.V.; Østerlie, M. A comparative study of organic- versus conventional Atlantic Salmon. II. Fillet color, carotenoid- and fatty acid composition as affected by dry salting, cold smoking and storage. *Aquaculture* **2015**, *451*, 369–376. [CrossRef]
34. Bjerkeng, B. Carotenoid pigmentation of salmonid fishes–recent progress 1. In *Avances en Nutricion Acuicola V Memorias del V. Simposium Internacional de Nutricion Acuicola*; Merida: Yucatan, Mexico, 2000; pp. 71–89.
35. Yeşilayer, N. Comparison of Flesh Colour Assessment Methods for Wild Brown Trout (*Salmo trutta macrostigma*), Farmed Rainbow Trout (*Oncorhynchus mykiss*) and Farmed Atlantic Salmon (*Salmo salar*). *Pak. J. Zool.* **2020**, *3*, 52. [CrossRef]
36. Parlapani, F.F.; Haroutounian, S.A.; Nychas, G.J.E.; Boziaris, I.S. Microbiological spoilage and volatiles production of gutted European sea bass stored under air and commercial modified atmosphere package at 2 C. *Food Microbiol.* **2015**, *50*, 44–53. [CrossRef] [PubMed]
37. Azizi-Lalabadi, M.; Rafiei, L.; Divband, B.; Ehsani, A. Active packaging for Salmon stored at refrigerator with Polypropylene nanocomposites containing 4A zeolite, ZnO nanoparticles, and green tea extract. *Food Sci. Nutr.* **2020**, *8*, 6445–6456. [CrossRef] [PubMed]
38. Bramstedt, F.; Auerbach, M. The Spoilage of Fresh-Water Fish. In *Fish as Food*; Borgstram, G., Ed.; Academic Press Inc.: San Diego, CA, USA, 1961; Volume 1, pp. 613–637.
39. Jacquot, R. Organic Constituents of Fish and Other Aquatic Animal Foods. In *Fish as Food*; Borgstram, G., Ed.; Academic Press Inc.: San Diego, CA, USA, 1961; Volume 1, pp. 145–209.
40. Jorgensen, L.V.; Huss, H.H.; Dalgaard, P. The effect of biogenic amine production by single bacterial cultures and metabiosis on cold-smoked salmon. *J. Appl. Microbiol.* **2000**, *89*, 920–934. [CrossRef]
41. Rodrigues, M.J.; Ho, P.; Lopez-Caballero, M.E.; Vaz-Pires, P.; Nunes, M.L. Characterization and identification of microflora from soaked cod and respective salted raw materials. *Food Microbiol.* **2003**, *20*, 471–481. [CrossRef]
42. European Commission. Commission implementing regulation (EU) 2019/627 of 15 March 2019 laying down uniform practical arrangements for the performance of official controls on products of animal origin intended for human consumption in accordance with Regulation (EU) 2017/625 of the European Parliament and of the Council and amending Commission Regulation (EC) No 2074/2005 as regards official controls. *Off. J. Eur. Union L* **2019**, *131*, 51–100.
43. Dalgaard, P.; Gram, L.; Huss, H.H. Spoilage and shelf life of cod fillets packed in vacuum or modified atmospheres. *Int. J. Food Microbiol.* **1993**, *19*, 283–294. [CrossRef]
44. Boskou, G.; Debevere, J. Reduction of trimethylamine oxide by *Shewanella* spp. under modified atmospheres in vitro. *Food Microbiol.* **1997**, *14*, 543–553. [CrossRef]
45. Giménez, B.; Roncalés, P.; Beltrán, J.A. The effects of natural antioxidants and lighting conditions on the quality of salmon (*Salmo salar*) fillets packaged in modified atmosphere. *J. Sci. Food Agric.* **2005**, *85*, 1033–1040. [CrossRef]
46. Chung, S.W.C.; Chan, B.T.P. Trimethylamine oxide, dimethylamine, trimethylamine and formaldehyde levels in main traded fish species in Hong Kong. *Food Addit. Contam.* **2009**, *2 Pt B*, 44–51. [CrossRef]
47. Dalgaard, P. Qualitative and quantitative characterization of spoilage bacteria from packed fish. *Int. J. Food Microbiol.* **1995**, *26*, 319–333. [CrossRef]
48. Gram, L.; Dalgaard, P. Fish spoilage bacteria—Problems and solutions. *Curr. Opin. Biotechnol.* **2002**, *13*, 262–266. [CrossRef]
49. Shumilina, E.; Slizyte, R.; Mozuraityte, R.; Dykyy, A.; Stein, T.A.; Dikiy, A. Quality changes of salmon by-products during storage: Assessment and quantification by NMR. *Food Chem.* **2016**, *211*, 803–811. [CrossRef] [PubMed]
50. Raeisi, M.; Tajik, H.; Aliakbarlu, J.; Mirhosseini, S.H.; Hosseini, S.M.H. Effect of carboxymethyl cellulose-based coatings in-cor- porated with *Zataria multiflora* Boiss. essential oil and grape seed extract on the shelf life of rainbow trout fillets. *LWT-Food Sci. Technol.* **2015**, *64*, 898–904. [CrossRef]
51. Secci, G.; Parisi, G. From farm to fork: Lipid oxidation in fish products. A review. *Ital. J. Anim. Sci.* **2016**, *15*, 124–136. [CrossRef]
52. Remya, S.; Mohan, C.O.; Bindu, J.; Sivaraman, G.K.; Venkateshwarlu, G.; Ravishankar, C.N. Effect of chitosan based active packaging film on the keeping quality of chilled stored barracuda fish. *J. Food Sci. Technol.* **2016**, *53*, 685–693. [CrossRef]

53. Prabhakar, P.K.; Vatsa, S.; Srivastav, P.P.; Pathak, S.S. A comprehensive review on freshness of fish and assessment: Ana-lytical methods and recent innovations. *Food Res. Int.* **2020**, *133*, 109157. [CrossRef]
54. Rasekh, J.; Kramer, A.; Finch, R. Objective evaluation of canned tuna sensory quality. *J. Food Sci.* **1970**, *35*, 417–423. [CrossRef]
55. Karabagias, I.; Badeka, A.; Kontominas, M.G. Shelf life extension of lamb meat using thyme or orégano essential oils and modified atmosphere packaging. *Meat Sci.* **2011**, *88*, 109–116. [CrossRef]
56. Françoise, L. Occurrence and role of lactic acid bacteria in seafood products. *Food Microbiol.* **2010**, *27*, 698–709. [CrossRef]
57. Roohinejad, S.; Koubaa, M.; Barba, F.J.; Saljoughian, S.; Amid, M.; Greiner, R. Application of seaweeds to develop new food products with enhanced shelf-life, quality and health-related beneficial properties. *Int. Food Res.* **2017**, *99*, 1066–1083. [CrossRef] [PubMed]
58. Wang, T.; Sveinsdóttir, K.; Magnússon, H.; Martinsdóttir, E. Combined Application of Modified Atmosphere Packaging and Superchilled Storage to Extend the Shelf Life of Fresh Cod (*Gadus morhua*) Loins. *J. Food Sci.* **2007**, *73*, S11–S19. [CrossRef] [PubMed]

Communication

Impact of Cork Closures on the Volatile Profile of Sparkling Wines during Bottle Aging

Filipa Amaro [1,2], Joana Almeida [1,2], Ana Sofia Oliveira [2], Isabel Furtado [2], Maria de Lourdes Bastos [1,2], Paula Guedes de Pinho [1,2] and Joana Pinto [1,2,*]

1 Associate Laboratory i4HB—Institute for Health and Bioeconomy, Laboratory of Toxicology, Department of Biological Sciences, Faculty of Pharmacy, University of Porto, Rua Jorge Viterbo Ferreira, 228, 4050-313 Porto, Portugal; famaro@ff.up.pt (F.A.); joana.m.almeida97@gmail.com (J.A.); mlbastos@ff.up.pt (M.d.L.B.); pguedes@ff.up.pt (P.G.d.P.)

2 UCIBIO/REQUIMTE, Laboratory of Toxicology, Department of Biological Sciences, Faculty of Pharmacy, University of Porto, 4050-313 Porto, Portugal; aoliveira@ff.up.pt (A.S.O.); isabel.c.furtado95@gmail.com (I.F.)

* Correspondence: jipinto@ff.up.pt; Tel.: +35-12-2042-8599

Abstract: This study aimed at investigating the impact of different technical cork stoppers on the quality preservation and shelf life of sparkling wines. The volatile compositions of two Italian sparkling wines sealed with a sparkling cork with two natural cork discs (2D) and a microagglomerated (MA) cork were determined during bottle aging (12 to 42 months) after disgorging, by headspace solid-phase microextraction coupled with gas chromatography-mass spectrometry (HS-SPME-GC/MS). The volatile profile of the sparkling wine #1 sealed with 2D stoppers showed the presence of camphor from 12 to 42 months, along with significant alterations in the levels of several alcohols, ketones, and ethyl esters at 24 and 42 months. The impact of closure type was less pronounced for sparkling wine #2 which also showed the presence of camphor from 12 to 42 months in 2D samples, and significantly higher levels of esters at 24 and 42 months for 2D compared with MA. These results unveiled that the type of closure has a greater impact on the volatile composition of sparkling wines at longer post-bottling periods and 2D stoppers preserve the fruity and sweety aromas of sparkling wines better after 42 months of bottle storage.

Keywords: volatile organic compounds; HS-SPME-GC/MS; Italian sparkling wines; cork stoppers; bottle aging

Citation: Amaro, F.; Almeida, J.; Oliveira, A.S.; Furtado, I.; Bastos, M.d.L.; Guedes de Pinho, P.; Pinto, J. Impact of Cork Closures on the Volatile Profile of Sparkling Wines during Bottle Aging. *Foods* **2022**, *11*, 293. https://doi.org/10.3390/foods11030293

Academic Editors: Matteo Alessandro Del Nobile and Amalia Conte

Received: 29 December 2021
Accepted: 20 January 2022
Published: 22 January 2022

Publisher's Note: MDPI stays neutral with regard to jurisdictional claims in published maps and institutional affiliations.

1. Introduction

Over the last two decades, the global wine market has experienced an increase in the demand for sparkling wines due to changes in consumers' preferences [1]. The aroma composition of these wines is a key attribute to consumers' acceptance and an important indicator of quality [2]. Sparkling wines produced by the traditional method (or *Champenoise*) are relatively complex in terms of aroma composition since they undergo a secondary fermentation in the bottle, followed by different contact periods with lees, when yeasts suffer autolysis releasing nitrogen compounds, polysaccharides, volatile compounds (e.g., ethyl esters), and phenolic compounds, among others (e.g., lipids and nucleic acids) [3], which contribute to the aroma complexity of the greatest sparkling wines.

The impact of several winemaking processes in the aroma composition of sparkling wines has been studied, such as grape variety and maturity [4,5], production methods [6], yeast selection [7], and aging period in contact with lees [7,8]. However, it is well known that the type of bottle closure influences the aroma composition of wines during aging, as has been reported for still wines [9,10], but the aroma comparison between sparkling wines bottled with different types of closures has not been reported so far.

Cork stoppers play a pivotal role in preserving the effervescence (carbon dioxide levels) and the aroma attributes of sparkling wines, making them almost irreplaceable for this type of wine. Nowadays, there are several types of sparkling wine cork stoppers available in the market, from microagglomerated to agglomerated corks, and corks made with agglomerated body plus one, two, or three natural cork discs attached to one of the ends [11]. Importantly, the natural cork discs are obtained from high-quality cork planks, allowing a higher contact of the sparkling wines with this type of material. Hence, this study aimed to investigate, for the first time to our knowledge, the impact of two different technical cork stoppers on the volatile composition of two Italian sparkling wines from 12 to 42 months of aging in bottle after disgorging.

2. Materials and Methods

2.1. Chemicals

1,4-Cineole (98%), 1-decanol (99.9%), 1-hexanol (99.9%), 1-octanol (≥99%), 2-heptanone (99%), 2-nonanone (97%), 2-undecanone (97%), 3-hexen-1-ol (98%), 5-methyl-2-furfural, benzaldehyde (≥99.5%), camphor (99%), decanal (95%), diethyl succinate (≥99%), ethyl 2-methylbutanoate (99%), ethyl butanoate (99%), ethyl decanoate (≥98%), ethyl heptanoate (≥98%), ethyl hexanoate (99%), ethyl isobutanoate (≥98%), ethyl isovalerate (98%), ethyl nonanoate (≥98%), ethyl octanoate (99%), eucalyptol (99%), furfural (99%), hexyl acetate (98%), isoamyl acetate (≥99%), isoamyl alcohol (98%), limonene (99%), linalool oxide (97%), nonanal (≥95%), octanal (≥98%), phenylacetaldehyde (90%), phenylethyl acetate (99%), phenylethyl alcohol (99%), tartaric acid (≥99.5%), α-pinene (99%)), β-cyclocitral (90%), β-damascenone (≥98%), α-ionone (85%), and β-linalool (80%) were supplied by Sigma-Aldrich (Madrid, Spain). Ethanol (99.9%) was purchased from ERBA Reagents (Val de Reuil, France).

2.2. Sparkling Wine Samples

The sparkling wines used in this study were a 2011 Classic Brut Vintage (sparkling wine #1) and a 2005 Reserve Brut Vintage (sparkling wine #2), from different producers in the Piemonte region in Italy. The sparkling wines were produced from the Chardonnay and Pinot Noir grape varieties, using the traditional method with secondary fermentation in the bottle, and were both disgorged in 2017, corresponding to approximately 5 years of aging in contact with lees for sparkling wine #1 and 11 years for sparkling wine #2. Two types of commercially available stoppers were used for bottling of the two sparkling wines, namely one sparkling cork (3–7 mm diameter granules) with two natural cork discs glued at one end, termed as 2D throughout the article, and one microagglomerated cork (1–3 mm diameter granules), termed as MA. After bottling, all samples were kept under controlled temperature conditions in the producers' cellars. Samples were then collected at 12, 24, and 42 months (n = 3–5 bottles per sampling point) for analysis of the volatile fraction.

2.3. Analysis of Volatile Composition by HS-SPME-GC/MS

The analyses of volatile compounds in sparkling wine samples were performed in 2018 (12 months), 2019 (24 months) and 2020 (42 months) using a HS-SPME-GC/MS method adapted from Barros et al. [12]. Briefly, each sparkling wine sample (250 µL) was placed in a 20 mL glass vial which was incubated for 5 min at 45 °C, using a Combi-PAL autosampler (Varian Pal Autosampler, Zwingen, Switzerland). The volatile compounds were then extracted by a 50/30 µm divinylbenzene/carboxen/polydimethylsiloxane (DVB/CAR/PDMS) fiber (Supelco Inc., Bellefonte, PA, USA) for 30 min at 45 °C with a stirring speed of 250 rpm. After extraction, the compounds were thermally desorbed into the GC system for 6 min at 250 °C. All samples were randomly injected.

A 436-GC model (Bruker Daltonics, Bremen, Germany) coupled to a SCION single quadrupole (SQ) mass spectrometer (Bruker Daltonics, Bremen, Germany) and a Bruker Daltonics MS workstation (version 8.2.1, Bruker Daltonics, Bremen, Germany) were used for volatile analysis and quantification. The GC system was equipped with a fused silica capillary column (Rxi-5Sil MS, 30 m × 0.25 mm internal diameter × 0.25 µm; Restek Corporation, Bellefonte, PA, USA) and high-purity helium C-60 (Gasin, Leça da Palmeira,

Portugal) was used as carrier gas at a constant flow rate of 1.0 mL/min. The oven temperature was programmed at 40 °C for 1 min, followed by an increase of 5.0 °C/min to 250 °C, where it was held for 5 min, and then increased at 5.0 °C/min to 300 °C. SQ-MS was conducted in the electron ionization (EI) mode at 70 eV and the transfer line, ion source, and manifold temperatures were maintained at 250, 250, 260, and 41 °C, respectively. Data acquisition was performed in full scan mode with a mass range of 40 to 400 *m/z* and a 500 ms scan time.

For the quantification of volatile compounds, standard compounds were dissolved in a wine model solution (12% ethanol, 5 g/L of tartaric acid, pH 3.2) and analyzed under the same conditions by HS-SPME/GC-MS. The calibration curves were achieved by injecting a range of known concentrations of each compound and computed by the respective area of the peak versus concentration.

2.4. Statistical Analyses

Multiple unpaired *t*-tests were applied to evaluate the differences in the levels of volatile compounds in sparkling wines sealed with 2D compared with MA at each post-bottling time. In addition, ordinary one-way analysis of variance (ANOVA) was computed to assess the differences in volatile concentrations between different post-bottling times. The concentration levels were considered significantly different for p-values < 0.05. All statistical analyses were performed using the software GraphPad Prism 9 (version 9.3.0, San Diego, CA, USA).

3. Results

Bottle closures can affect the aroma composition of wines during aging by three main factors: (1) the oxygen ingress through the bottle which can lead to wine oxidation and the development of oxidized aromas; (2) the desorption of volatile compounds from closures into wine which can lead to pleasant (e.g., terpenes) or unpleasant (e.g., pyrazines) aromas; and (3) the scalping of volatile compounds present in wine by closures [9].

From the 39 volatile compounds quantified, significant alterations were found for 8 compounds in sparkling wine #1 sealed with 2D compared with MA (Figure 1, Table 1), and 3 compounds for sparkling wine #2 (Figure 2, Table 2), during bottle storage from 12–42 months. Interestingly, a lower number of altered volatile compounds was found for sparkling wine #2 which aged longer in contact with lees (11 years) in contrast with sparkling wine #1 (5 years). The levels of the remaining quantified compounds in both sparkling wines are present in Tables S1 and S2.

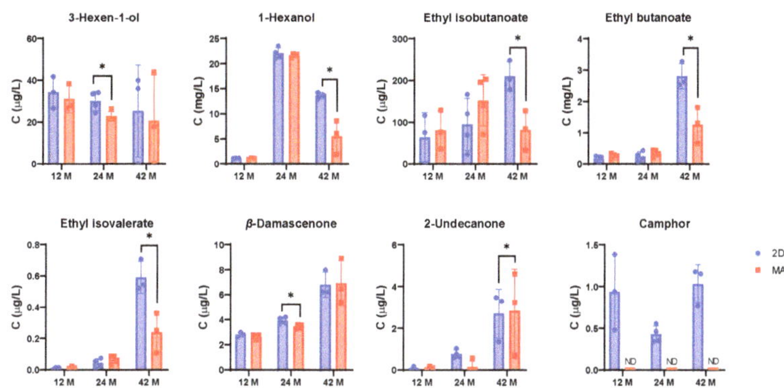

Figure 1. Bar graphs representing the levels of volatile compounds significantly changing in sparkling wine #1 sealed with a sparkling cork with two natural cork discs (2D in blue) and a microagglomerated cork (MA in red) during bottle aging (12 to 42 months). *—$p \leq 0.05$, ND—not detected.

The volatile composition of sparkling wines #1 and #2 was more affected by the type of closure at 24- and 42-months post-bottling, while only a qualitative change in camphor (only present in samples sealed with 2D, Figure 1 and Table 1) was detected in both sparkling wines at 12 months. Camphor is responsible for pleasant aromas—such as herbal, minty, and woody [13]—but the olfactory perception threshold in wine or wine model solution has not been reported so far. At 24 months post-bottling, sparkling wine #1 showed significantly higher levels of 3-hexen-1-ol and β-damascenone in samples sealed with 2D compared with MA and the presence of camphor only in 2D samples (Figure 1, Table 1). 3-hexen-1-ol is characterized by green and leafy odors and was present in concentrations below the olfactory perception threshold (<400 µg/L) reported for wine model solution [14], while β-damascenone is characterized by woody, floral, and herbal odors [14,15], and was present above its olfactory perception threshold (0.05 µg/L) [14]. At 42 months, this sparkling wine showed significantly higher levels of ethyl isobutanoate, ethyl butanoate, and ethyl isovalerate in samples sealed with 2D, as well as significantly lower levels of 2-undecanone and the presence of camphor (Figure 1, Table 1). From these compounds, the three ethyl esters (ethyl isobutanoate, ethyl butanoate, and ethyl isovalerate), and 1-hexanol were present above their olfactory perception thresholds (Table 1) [14], and can contribute with fruity notes [13,15] to the sparkling wine aroma. Interestingly, the levels of 1-hexanol, ethyl butanoate, ethyl isovalerate, 2-undecanone, β-damascenone, and camphor changed significantly with bottle aging, while the levels of 3-hexen-1-ol and ethyl isobutanoate were relatively constant over time (Table 1).

Table 1. Levels of volatile compounds significantly changing in sparkling wine #1 sealed with a sparkling cork with two natural cork discs (2D) and a microagglomerated cork (MA) during bottle aging (12 to 42 months).

Class/Compound	12 Months [1]			24 Months [1]			42 Months [1]			12 vs. 24 vs. 42 Months p	Descriptors [2]	Olfactory Perception Threshold [3]
	2D	MA	p	2D	MA	p	2D	MA	p			
Alcohols												
3-Hexen-1-ol (µg/L)	34.2 ± 6.2	31.0 ± 5.3	ns	30.1 ± 4.0	22.8 ± 2.0	*	25.4 ± 18.0	20.6 ± 18.1	ns	ns	Green, leafy	400 µg/L
1-Hexanol (mg/L)	1.1 ± 0.03	1.0 ± 0.1	ns	22.1 ± 1.0	21.6 ± 0.3	ns	13.7 ± 0.5	5.5 ± 2.7	*	****	Green, fruity	8 mg/L
Ethyl esters												
Ethyl isobutanoate (µg/L)	64.0 ± 48.3	80.7 ± 38.0	ns	95.3 ± 54.0	152.3 ± 53.2	ns	210.1 ± 29.0	81.7 ± 38.7	*	ns	Fruity	15 µg/L
Ethyl butanoate (mg/L)	0.17 ± 0.10	0.25 ± 0.09	ns	0.25 ± 0.14	0.35 ± 0.11	ns	2.81 ± 0.33	1.26 ± 0.47	*	****	Fruity, sweet, apple	0.02 mg/L
Ethyl isovalerate (µg/L)	10.0 ± 0.01	13.6 ± 6.4	ns	42.6 ± 24.9	66.8 ± 22.5	ns	591.7 ± 83.5	240.4 ± 102.9	*	****	Fruity, sweet, spice	3 µg/L
Isoprenoids												
β-Damascenone (µg/L)	2.79 ± 0.15	2.63 ± 0.19	ns	3.93 ± 0.24	3.42 ± 0.15	*	6.81 ± 0.83	6.92 ± 1.47	ns	****	Woody, floral, herbal	0.05 µg/L
Ketones												
2-Undecanone (µg/L)	0.06 ± 0.08	0.09 ± 0.07	ns	0.76 ± 0.17	0.14 ± 0.25	ns	2.71 ± 0.96	2.83 ± 1.64	*	****	Waxy, fruity	NR
Terpenes												
Camphor (µg/L)	0.94 ± 0.37	ND	Q	0.43 ± 0.09	ND	Q	1.03 ± 0.19	BLOQ	Q	*	Herbal, minty, woody	NR

[1] Average concentration and standard deviation of sparkling wine #1 sealed with 2D and MA corks. A $n = 3$ per closure was considered at 12 and 42 months, and a $n = 4$ at 24 months. [2] Descriptors reported in references [13,15]. [3] Olfactory perception thresholds determined in wine model solution as reported in reference [14]. ns—$p > 0.05$, *—$p \leq 0.05$, ****–$p \leq 0.0001$, BLOQ–below limit of quantification, ND—not detected, NR–not reported, Q—qualitative alteration.

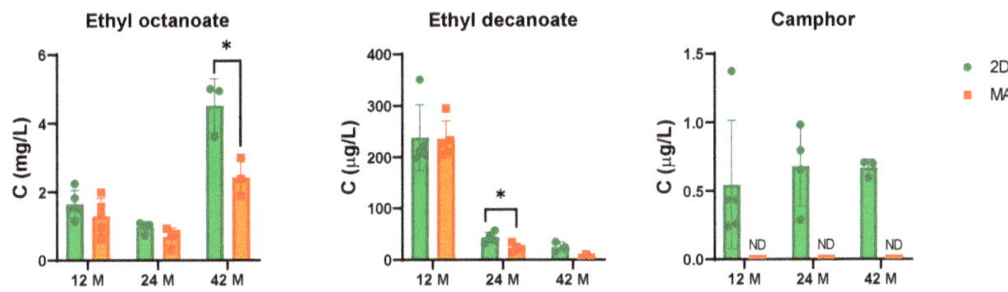

Figure 2. Bar graphs representing the levels of volatile compounds significantly changing in sparkling wine #2 sealed with a sparkling cork with two natural cork discs (2D in green) and a microagglomerated cork (MA in orange) during bottle aging (12 to 42 months). *—$p \leq 0.05$, ND—not detected.

Table 2. Levels of volatile compounds significantly changing in sparkling wine #2 sealed with a sparkling cork with two natural cork discs (2D) and a microagglomerated cork (MA) during bottle aging (12 to 42 months).

Class/Compound	12 Months [1]			24 Months [1]			42 Months [1]			12 vs. 24 vs. 42 Months p	Descriptors [2]	Olfactory Perception Threshold [3]
	2D	MA	*p*	2D	MA	*p*	2D	MA	*p*			
Ethyl esters												
Ethyl octanoate (mg/L)	1.65 ± 0.38	1.30 ± 0.48	ns	0.97 ± 0.14	0.70 ± 0.23	ns	4.52 ± 0.65	2.43 ± 0.45	*	****	Fruity, sweet, waxy	0.6 mg/L
Ethyl decanoate (µg/L)	238.5 ± 57.2	235.2 ± 32.1	ns	43.5 ± 9.3	23.6 ± 7.4	*	24.9 ± 7.7	6.3 ± 3.3	ns	****	Fruity, waxy, sweet apple	200 µg/L
Terpenes												
Camphor (µg/L)	0.55 ± 0.42	ND	Q	0.68 ± 0.26	ND	Q	0.67 ± 0.05	ND	Q	ns	Herbal, minty, woody	NR

[1] Average concentration and standard deviation of sparkling wine #2 sealed with 2D and MA corks. A $n = 5$ per closure was considered at 12 months, a $n = 4$ at 24 months, and a $n = 3$ at 42 months. [2] Descriptors reported in references [13,15]. [3] Olfactory perception thresholds determined in wine model solution as reported in reference [14]. ns—$p > 0.05$, *—$p \leq 0.05$, ****—$p \leq 0.0001$, ND–not detected, NR—not reported, Q—qualitative alteration.

In contrast, at 24 months post-bottling, sparkling wine #2 showed the consistent presence of camphor in samples sealed with 2D, as well as significantly higher levels of ethyl decanoate (Figure 2, Table 2). At 42 months, significantly higher levels of ethyl octanoate and a tendency for higher levels of ethyl decanoate were observed in samples sealed with 2D, along with the presence of camphor (Figure 2, Table 2). The presence of ethyl octanoate in levels above the olfactory perception threshold (>0.6 mg/L) [14] may contribute to the fruity aroma [13,15] of this sparkling wine, while ethyl decanoate may have a lower impact due to its low concentration (<200 µg/L) [14]. Regarding the behavior of these compounds during bottle aging, ethyl octanoate increased significantly, whereas ethyl decanoate showed a significant decrease and camphor levels were constant over time (Table 2).

4. Discussion

Ethyl esters are the main class of aroma compounds released by the autolysis of yeasts in sparkling wines produced by the traditional method and they contribute to the fruity and floral-like aromas of these wines [16]. In our study, the levels of several ethyl esters were significantly higher in both sparkling wines sealed with 2D corks. The preservation of ethyl esters during bottle storage has been a challenge for winemakers as ethyl esters tend to hydrolyze over time due mostly to the low pH of wines [17]. Hence, a stopper able to preserve better the ethyl ester composition of sparkling wines can improve their shelf life and the sensory attributes expected by consumers. Notably, most ethyl esters present in both sparkling wines (Tables 1, 2, S1 and S2)—with exception of ethyl decanoate, hexyl acetate, and phenylethyl acetate—showed a significant increase over time. Despite the behavior of these compounds has been studied during the aging period in contact with lees [7,8], the information about their evolution trends after disgorging is limited [18].

Camphor has been previously identified by our group as only present in wines sealed with natural cork [19], which agrees with the results observed for both sparkling wines sealed with 2D stoppers. The most probable hypothesis is the desorption of camphor from natural cork to wines, which is also corroborated by the detection of this compound in wine model solution extracts of natural cork granules [20]. Based on these facts, camphor seems to be a good marker to discriminate wines bottled with natural cork discs versus other closures.

The levels of two alcohols (3-hexen-1-ol and 1-hexanol), one ketone (2-undecanone) and one isoprenoid (β-damascenone) were also significantly influenced by the type of closure in sparkling wine #1. Alcohols can be substrates for wine oxidation originating their correspondent aldehydes [21]. Thus, the lower levels of 3-hexen-1-ol and 1-hexanol in sparkling wine #1 sealed with MA stoppers may be due to their oxidation in hexanal and 3-hexenal, respectively. Though these aldehydes were not detected in the volatile composition of sparkling wine #1 under our experimental conditions. However, 2-undecanone, a ketone that may be also formed by oxidation [21], was found in higher levels in samples sealed with MA. Finally, β-damascenone is mainly produced from direct degradation of carotenoid molecules during fermentation [22]. Higher levels of this isoprenoid were previously found in a dry white wine sealed with natural cork compared with microagglomerated cork [23], in agreement with the results obtained for sparkling wine #1 at 24 months.

5. Conclusions

These results showed that the type of closure has a greater impact on volatile composition of sparkling wines at longer post-bottling periods (42 months). For both sparkling wines, the sparkling cork with natural cork discs better preserved the fruity and sweety aromas after 42 months of bottle aging, due to the presence of higher amounts of ethyl esters. In addition, the presence of camphor in sparkling wines sealed with a sparkling stopper with natural cork discs seems to be a good marker to discriminate this type of closure versus microagglomerated corks. In general, this work emphasizes the importance of the choice of cork closure for the preservation of the aromatic characteristics of sparkling wines, increasing their shelf life.

Supplementary Materials: The following supporting information can be downloaded at: https://www.mdpi.com/article/10.3390/foods11030293/s1, Table S1: Concentrations of other volatile organic compounds determined in sparkling wine #1, during bottle aging (12 to 42 months), for which no significant change was observed according to the type of closure.; Table S2: Concentrations of other volatile organic compounds determined in sparkling wine #2, during bottle aging (12 to 42 months), for which no significant change was observed according to the type of closure.

Author Contributions: Conceptualization, P.G.d.P. and J.P.; Methodology, P.G.d.P. and J.P.; Software, J.P.; Validation, F.A. and J.A.; Formal analysis, F.A., J.A., A.S.O. and I.F.; Investigation, F.A., J.A., A.S.O. and I.F.; Resources, P.G.d.P. and M.d.L.B.; Data curation, F.A. and J.A.; Writing—original draft preparation, F.A.; Writing—review and editing, J.A., A.S.O., I.F., M.d.L.B., P.G.d.P. and J.P.; Visualization, J.A., A.S.O., I.F., M.d.L.B., P.G.d.P. and J.P.; Supervision, M.d.L.B., P.G.d.P. and J.P.; Project administration, P.G.d.P. and J.P.; Funding acquisition, M.d.L.B. All authors have read and agreed to the published version of the manuscript.

Funding: This research was funded by national funds from FCT-*Fundação para a Ciência e a Tecnologia*, I.P., in the scope of the project UIDP/04378/2020 and UIDB/04378/2020 of the Research Unit on Applied Molecular Biosciences—UCIBIO, and the project LA/P/0140/2020 of the Associate Laboratory Institute for Health and Bioeconomy—i4HB.

Data Availability Statement: Data available upon request.

Conflicts of Interest: The authors declare no conflict of interest. The funders had no role in the design of the study; in the collection, analyses, or interpretation of data; in the writing of the manuscript, or in the decision to publish the results.

References

1. International Organisation of Vine and Wine. OIV Focus the Global Sparkling Wine Market. Available online: https://www.oiv.int/public/medias/7291/oiv-sparkling-focus-2020.pdf (accessed on 15 October 2021).
2. Kemp, B.; Alexandre, H.; Robillard, B.; Marchal, R. Effect of Production Phase on Bottle-Fermented Sparkling Wine Quality. *J. Agric. Food Chem.* **2015**, *63*, 19–38. [CrossRef] [PubMed]
3. Pozo-Bayón, M.Á.; Martínez-Rodríguez, A.; Pueyo, E.; Moreno-Arribas, M.V. Chemical and biochemical features involved in sparkling wine production: From a traditional to an improved winemaking technology. *Trends Food Sci. Technol.* **2009**, *20*, 289–299. [CrossRef]
4. Coelho, E.; Coimbra, M.A.; Nogueira, J.; Rocha, S.M. Quantification approach for assessment of sparkling wine volatiles from different soils, ripening stages, and varieties by stir bar sorptive extraction with liquid desorption. *Anal. Chim. Acta* **2009**, *635*, 214–221. [CrossRef] [PubMed]
5. Esteruelas, M.; González-Royo, E.; Kontoudakis, N.; Orte, A.; Cantos, A.; Canals, J.M.; Zamora, F. Influence of grape maturity on the foaming properties of base wines and sparkling wines (Cava). *J. Sci. Food Agric.* **2015**, *95*, 2071–2080. [CrossRef]
6. Torrens, J.; Riu-Aumatell, M.; Vichi, S.; López-Tamames, E.; Buxaderas, S. Assessment of Volatile and Sensory Profiles between Base and Sparkling Wines. *J. Agric. Food Chem.* **2010**, *58*, 2455–2461. [CrossRef] [PubMed]
7. Martínez-García, R.; Moreno, J.; Bellincontro, A.; Centioni, L.; Puig-Pujol, A.; Peinado, R.A.; Mauricio, J.C.; García-Martínez, T. Using an electronic nose and volatilome analysis to differentiate sparkling wines obtained under different conditions of temperature, ageing time and yeast formats. *Food Chem.* **2021**, *334*, 127574. [CrossRef] [PubMed]
8. Riu-Aumatell, M.; Bosch-Fusté, J.; López-Tamames, E.; Buxaderas, S. Development of volatile compounds of cava (Spanish sparkling wine) during long ageing time in contact with lees. *Food Chem.* **2006**, *95*, 237–242. [CrossRef]
9. Furtado, I.; Lopes, P.; Oliveira, A.S.; Amaro, F.; Bastos, M.D.L.; Cabral, M.; de Pinho, P.G.; Pinto, J. The Impact of Different Closures on the Flavor Composition of Wines during Bottle Aging. *Foods* **2021**, *10*, 2070. [CrossRef] [PubMed]
10. Silva, M.A.; Julien, M.; Jourdes, M.; Teissedre, P.-L. Impact of closures on wine post-bottling development: A review. *Eur. Food Res. Technol.* **2011**, *233*, 905–914. [CrossRef]
11. Pereira, H. Production of cork stoppers and discs. In *Cork*; Elsevier BV: Amsterdam, The Netherlands, 2007; pp. 263–288.
12. Barros, E.P.; Moreira, N.; Pereira, G.E.; Leite, S.G.F.; Rezende, C.M.; de Pinho, P.G. Development and validation of automatic HS-SPME with a gas chromatography-ion trap/mass spectrometry method for analysis of volatiles in wines. *Talanta* **2012**, *101*, 177–186. [CrossRef] [PubMed]
13. The Good Scents Company Information System. Available online: https://www.thegoodscentscompany.com/ (accessed on 2 December 2021).
14. Guth, H. Quantitation and Sensory Studies of Character Impact Odorants of Different White Wine Varieties. *J. Agric. Food Chem.* **1997**, *45*, 3027–3032. [CrossRef]
15. Arn, H.; Acree, T. Flavornet: A database of aroma compounds based on odor potency in natural products. *Developments in Food Science* **1998**, *40*, 27. [CrossRef]
16. Alexandre, H.; Guilloux-Benatier, M. Yeast autolysis in sparkling wine—A review. *Aust. J. Grape Wine Res.* **2006**, *12*, 119–127. [CrossRef]
17. Díaz-Maroto, M.C.; Schneider, R.; Baumes, R. Formation Pathways of Ethyl Esters of Branched Short-Chain Fatty Acids during Wine Aging. *J. Agric. Food Chem.* **2005**, *53*, 3503–3509. [CrossRef] [PubMed]
18. Pérez-Magariño, S.; Ortega-Heras, M.; Bueno-Herrera, M.; Martínez-Lapuente, L.; Guadalupe, Z.; Ayestarán, B. Grape variety, aging on lees and aging in bottle after disgorging influence on volatile composition and foamability of sparkling wines. *LWT* **2015**, *61*, 47–55. [CrossRef]
19. Oliveira, A.S.; Furtado, I.; Bastos, M.D.L.; de Pinho, P.G.; Pinto, J. The influence of different closures on volatile composition of a white wine. *Food Packag. Shelf Life* **2020**, *23*, 100465. [CrossRef]
20. Pinto, J.; Oliveira, A.S.; Lopes, P.; Roseira, I.; Cabral, M.; Bastos, M.D.L.; de Pinho, P.G. Characterization of chemical compounds susceptible to be extracted from cork by the wine using GC-MS and 1H NMR metabolomic approaches. *Food Chem.* **2019**, *271*, 639–649. [CrossRef] [PubMed]
21. Waterhouse, A.L.; Sacks, G.L.; Jeffery, D.W. *Understanding Wine Chemistry*, 1st ed.; Wiley: Hoboken, NJ, USA, 2016. [CrossRef]
22. Daniel, M.A.; Elsey, G.M.; Capone, D.L.; Perkins, M.V.; Sefton, M.A. Fate of Damascenone in Wine: The Role of SO_2. *J. Agric. Food Chem.* **2004**, *52*, 8127–8131. [CrossRef] [PubMed]
23. Moreira, N.; Lopes, P.; Ferreira, H.; Cabral, M.; De Pinho, P.G. Sensory attributes and volatile composition of a dry white wine under different packing configurations. *J. Food Sci. Technol.* **2018**, *55*, 424–430. [CrossRef] [PubMed]

 *foods*

Article

Development of Multifunctional Pullulan/Chitosan-Based Composite Films Reinforced with ZnO Nanoparticles and Propolis for Meat Packaging Applications

Swarup Roy †, Ruchir Priyadarshi † and Jong-Whan Rhim *

Department of Food and Nutrition, BioNanocomposite Research Center, Kyung Hee University, 26 Kyungheedae-ro, Dongdaemun-gu, Seoul 02447, Korea; swaruproy2013@gmail.com (S.R.); ruchirpriyadarshi@gmail.com (R.P.)
* Correspondence: jwrhim@khu.ac.kr
† Both authors contributed equally to this work.

Citation: Roy, S.; Priyadarshi, R.; Rhim, J.-W. Development of Multifunctional Pullulan/Chitosan-Based Composite Films Reinforced with ZnO Nanoparticles and Propolis for Meat Packaging Applications. *Foods* **2021**, *10*, 2789. https://doi.org/10.3390/foods10112789

Academic Editors: Matteo Alessandro Del Nobile and Amalia Conte

Received: 28 October 2021
Accepted: 9 November 2021
Published: 12 November 2021

Abstract: Pullulan/chitosan-based multifunctional edible composite films were fabricated by reinforcing mushroom-mediated zinc oxide nanoparticles (ZnONPs) and propolis. The ZnONPs were synthesized using enoki mushroom extract and characterized using physicochemical methods. The mushroom-mediated ZnONPs showed an irregular shape with an average size of 26.7 ± 8.9 nm. The combined incorporation of ZnONPs and propolis pointedly improved the composite film's UV-blocking property without losing transparency. The reinforcement with ZnONPs and propolis improved the mechanical strength of the pullulan/chitosan-based film by ~25%. Additionally, the water vapor barrier property and hydrophobicity of the film were slightly increased. In addition, the pullulan/chitosan-based biocomposite film exhibited good antioxidant activity due to the propolis and excellent antibacterial activity against foodborne pathogens due to the ZnONPs. The developed edible pullulan/chitosan-based film was used for pork belly packaging, and the peroxide value and total number of aerobic microorganisms were significantly reduced in meat wrapped with the pullulan/chitosan/ZnONPs/propolis film.

Keywords: enoki mushroom; ZnONPs; propolis; pullulan/chitosan; antibacterial; antioxidant activity; meat packaging

1. Introduction

Currently, food spoilage is a growing concern of the global food industry, and food spoilage has a significant impact on food supply and demand. The main cause of food quality degradation due to the oxidation and deterioration of food is the contamination and proliferation of micro-organisms during food storage. Such food spoilage has significantly impacted the quality and safety of perishable foods, especially animal products, due to the high fat content of meat products. In recent years, the intake of processed meat products has been very high, causing a serious spoilage problem in production and consumption. According to reports, ~23% of meat produced is lost or wasted [1]. As consumers demand safe food, there is a growing interest in developing and using edible packaging to preserve meat quality to keep food safe and extend its shelf life [2]. The most important challenge in this situation is to manufacture suitable packaging materials that can extend the shelf life of food. For this purpose, active packaging with antibacterial/antioxidant functions is emerging as one of the tools to control the deterioration of food [3–5].

Active packaging has an advantage over the direct use of antioxidant/antibacterial ingredients in foods because this type of packaging allows for the controlled release of active ingredients from the food packaging surface [6–8]. In this regard, renewable and degradable bio-based polymers of natural origin are suitable candidates to replace synthetic plastics in reducing environmental pollution [9–11]. Currently, various bio-based polymers such as polysaccharides and proteins are being studied. Although they have disadvantages such as low strength and gas barrier properties, they attract great attention as substitutes for petroleum-based plastics [12–14]. Another way to address the limitations of bio-based polymer matrices is to combine biopolymers to create blend films or to create composite materials with functional nanofillers or bioactive compounds [15–19]. For this, the blend of pullulan and chitosan can be an ideal choice as a solid base polymer matrix. Both pullulan and chitosan are edible polysaccharide-based biopolymers known for their excellent film-making potential. Additionally, previous reports have shown that these combinations create a highly compatible film that compensates for the imperfections of each matrix [20–24].

Nanofillers and bioactive natural ingredients are known to enhance the functional properties (antimicrobial and antioxidant) of active packaging films. As one of the commonly used nanofillers, zinc oxide nanoparticles (ZnONPs) are generally recognized as safe (GRAS) to be used in the food sector [25–28]. Adding ZnONPs improves the antimicrobial properties and helps enhance the physical properties, like the mechanical and water vapor barrier properties, of the packaging films [29–31]. As a potential antioxidant filler, the bioactive natural compound propolis plays a useful role due to its flavonoids, essential oils, waxes, and pollen [32–35]. Propolis is a complex mixture mainly obtained from honey bees, and it is a well-known, naturally occurring multifunctional material used since ancient times. The combination of nanofillers (ZnONPs) and natural bioactive compound (propolis) can be a good combination to enhance the physical and functional properties of edible active packaging films. A combination of antibacterial/antioxidant functional fillers in bio-based packaging films can be particularly useful to control the spoilage of meat products.

Therefore, the current work aimed to fabricate multifunctional active packaging films based on a pullulan/chitosan blend film integrated with ZnONPs and propolis. The pullulan/chitosan-based composite film was prepared, the film properties were characterized using analytical methods, and film's applicability to pork loin packaging was tested.

2. Materials and Methods

2.1. Materials

The pullulan was procured from Korea Bio Polymer Co. Ltd. (Bucheon, Gyeonggi-do, Korea). The zinc chloride, potassium hydroxide, and glycerol were acquired from Daejung Chemicals & Metals Co., Ltd. (Siheung, Korea). The chitosan (viscosity: 200–800 cP at 1% acetic acid, MW: 190,000–310,000 based on viscosity, degree of deacetylation: 75–85%), 2,2-diphenyl-1-picrylhydrazyl, 2,2′-azino-bis (3-ethylbenzothiazoline-6-sulfonic acid), and potassium persulfate were purchased from Sigma-Aldrich (St. Louis, MO, USA). The alcoholic extracts of propolis and enoki mushrooms were procured from a local supermarket in Seoul, Korea.

2.2. Preparation of Mushroom-Mediated ZnONP

For the preparation of the ZnONPs, the enoki mushroom extract (ME) was first prepared. For this, 10 g of enoki mushroom was cut into tiny pieces, mixed with 100 mL of DI water, heated at 80 °C for 10 min while being stirred at 500 rpm, and then cooled to room temperature. The undissolved solids were removed by centrifugation at 5000 rpm for 15 min. The collected extract was stored at −20 °C until further use [36,37]. A total of 6.8 g of $ZnCl_2$ was added (0.1 M) to 500 mL of 10% ME, mixed well, and heated until the temperature reached 70 °C. Then, 500 mL of KOH solution (0.2 M) was slowly added to the mixture with vigorous agitation for one hour. The white principate formed was collected by filtration, and the filtrate was washed with DI water until it reached a pH of 7 and then dried at 50 °C overnight to attain powdered ZnONPs.

2.3. Preparation of Pullulan/Chitosan-Based Films

The pullulan/chitosan-based film was prepared using a solution casting method [22], as presented in Scheme 1. For this, 1.5 g of chitosan was dissolved in 100 mL of 1% acetic acid solution with vigorous stirring. A total of 1.5 g of pullulan was dissolved in 100 mL of DI water by heating it at 70 °C with continuous agitation. Then, the chitosan and pullulan solutions were mixed, glycerol 0.9 g (30 wt% of polymer) was added, and then 0.08 g of ZnONPs and 0.2 g of propolis (2 and 5 wt% based on biopolymer, respectively) were added with vigorous stirring. The film-forming solutions were cast on Teflon film-coated glass plates. The dried films were conditioned at 25 °C and 50% RH for 72 h. For evaluation, neat films of pullulan/chitosan, pullulan/chitosan/ZnONPs, and pullulan/chitosan/propolis were also fabricated following an identical process. The prepared films were designated as PLN/CTS, PLN/CTS/ZnO, PLN/CTS/PPS, and PLN/CTS/ZnO/PPS, respectively, depending on the composite materials.

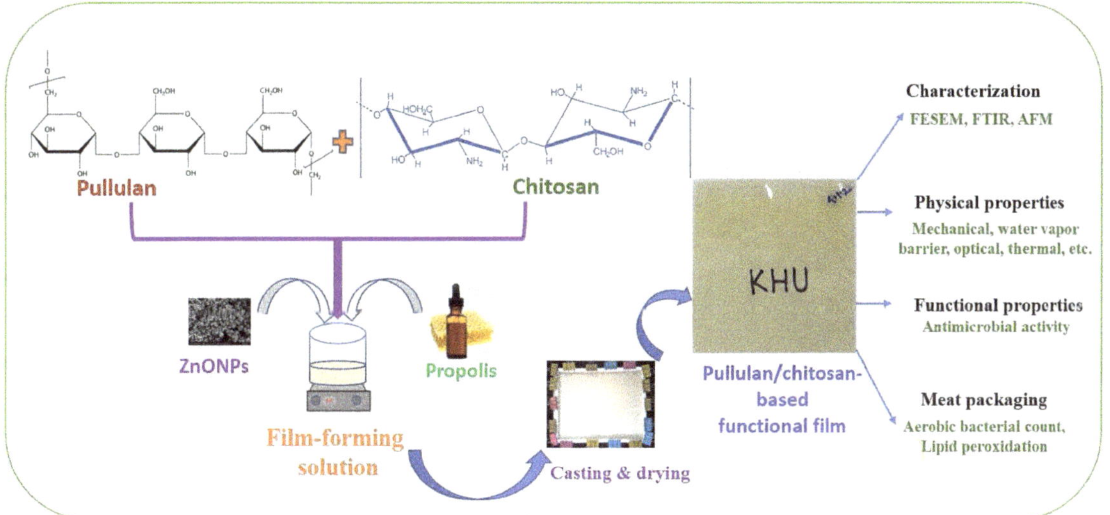

Scheme 1. Schematic presentation for the fabrication of pullulan/chitosan-based biocomposite films.

The details of the characterization methods of mushroom-mediated ZnONPs and the biocomposite films are provided in Supporting Information.

3. Results and Discussion

3.1. Characterization of ZnONPs

The appearance of white residue demonstrated the formation of ZnONPs. The formation of ME-capped ZnONPs was examined by the UV-vis spectra (Figure 1a), and the results showed a clear, typical absorption band around 350 nm due to the ZnONPs. The XRD patterns of the ZnONPs (Figure 1b) showed a distinctive pattern of diffraction peaks at (110), (002), (100), (102), (110), (103), (201), (004), and (202), representing the crystalline ZnONPs. The obtained results are well matched with the earlier reports indicating the formation of ZnONPs [30]. The chemical structure of the ZnONPs capped by ME was studied using FTIR analysis, and the FTIR results of the ZnONPs and ME are shown in Figure 1c. The wideband at 3265 cm^{-1} was owing to the primary amine and O-H stretching, while the peak detected at 2930 cm^{-1} was ascribed to the stretching vibration of the C-H groups [38]. The peak found at 1650 cm^{-1} was owing to the amide peak. Peaks at 1410 and 1035 cm^{-1} were accredited to the O-H bending and stretching vibrations, respectively [39]. The stretching vibration of primary amines and O-H is most likely associated with the bioactive components (proteins, flavonoids, phenolic acids, tannins, and carbohydrates) of the mushroom extract responsible for capping the ZnONPs [40]. The FTIR data established that the ZnONPs were efficaciously capped with several polyphenolic and biomolecules present in the mushroom extract. The observed results are also in line with the formerly available report [41].

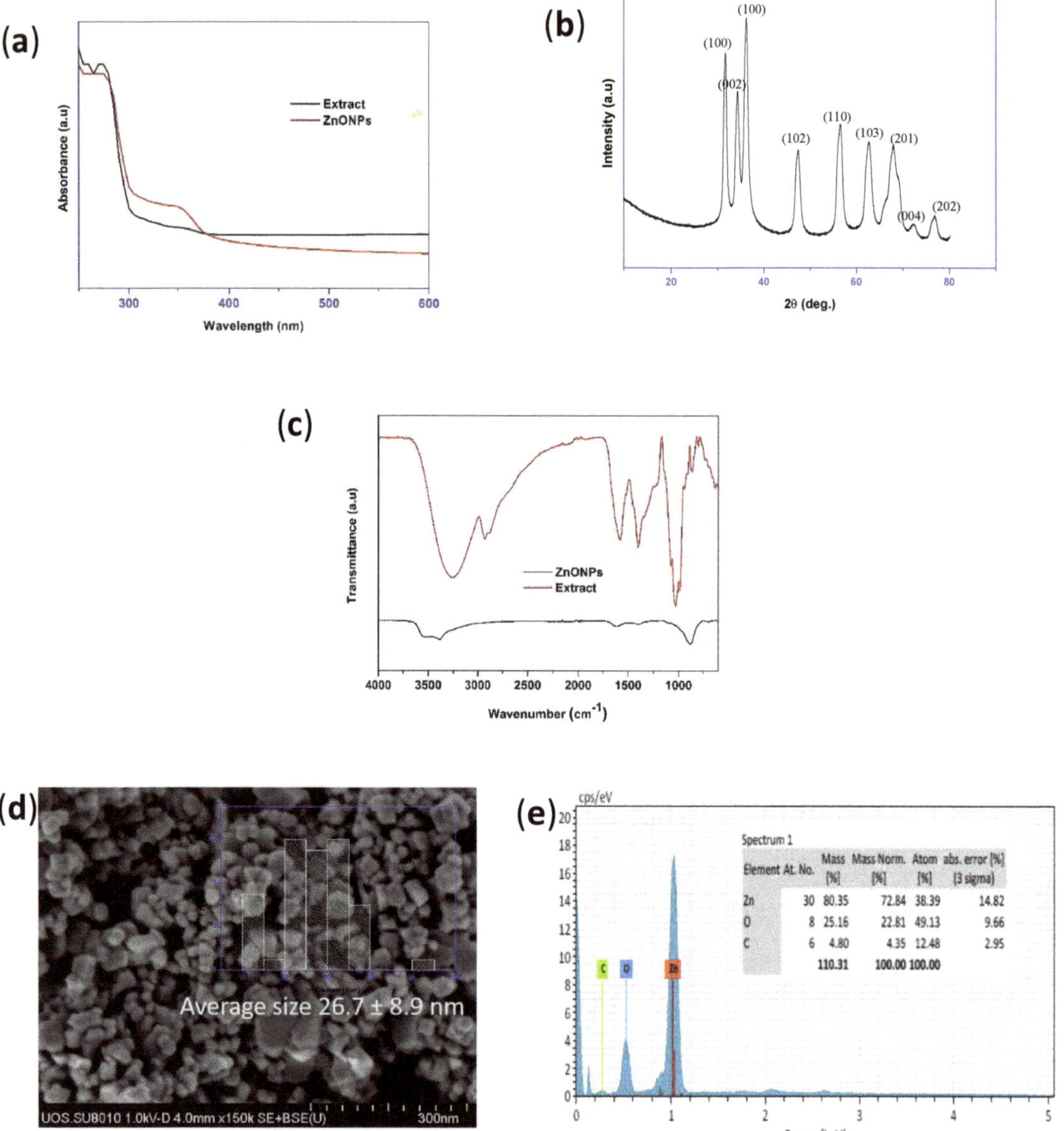

Figure 1. (**a**) UV-vis spectra, (**b**) XRD pattern, (**c**) FTIR spectra, and (**d**,**e**) FESEM and EDX of ZnONPs.

For further characterization of the ZnONPs, the hydrodynamic particle size (Z_{avg}) and polydispersity index (PDI) were analyzed using dynamic light scattering. The Z_{avg} and PDI were 451.3 ± 24.8 nm and 0.36 ± 0.01, respectively (Figure S1, Supporting Information), demonstrating the development of monodispersed nano-sized particles. The stability of the nanoparticles in an aqueous condition was checked by zeta potential, and the obtained data (−25.2 ± 0.5 mV, Figure S1, Supporting Information) revealed the moderate stability of the nanoparticles [42,43]. The stability of the ME-capped ZnONPs solution was very similar to the earlier reported melanin-mediated ZnONPs [30]. For the morphology of the particles, FESEM analysis was performed, and the results are presented in Figure 1d. The microscopic analysis showed irregular-shaped ZnONPs with a size of ~10–40 nm, with an average size of 26.7 ± 8.9 nm. The presence of elemental Zn and O was also further verified from the energy-dispersive X-ray spectroscopy (EDX) (Figure 1e). The elemental analysis was further confirmed by the XPS spectrum (Figure 2). The scan survey spectrum of ME-capped ZnONPs and C1s, N1s, O1s, and Zn2p are shown in Figure 2. The XPS spectra exhibited the binding energies of C1s, N1s, O1s, and Zn2p at about 284, 399, 530, and 1020 eV, respectively. Apart from the peaks of Zn and O, supplementary peaks of C and N were detected, most likely owing to the capping of ME. The results of the binding energy values for Zn2p and O1s are consistent with formerly described data [44]. Thus, the resulting characterization established the formation of ME-capped nano-sized particles.

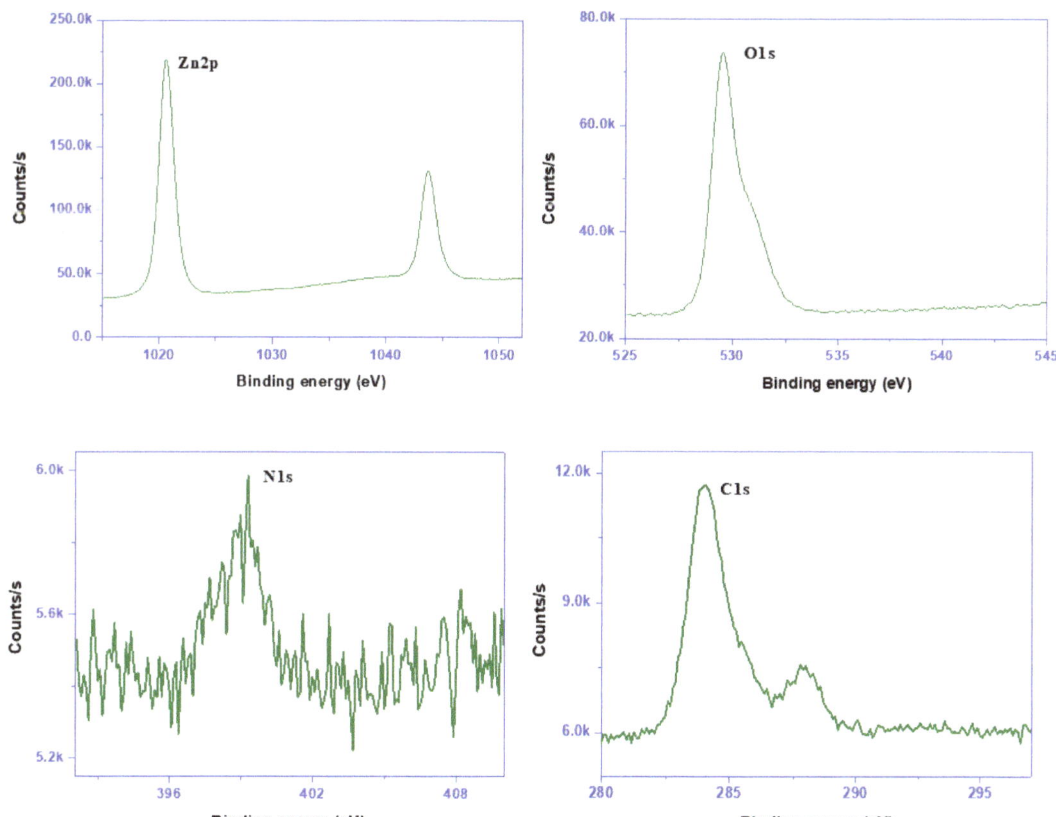

Figure 2. XPS spectra of ZnONPs.

3.2. Properties of the Pullulan/Chitosan-Based Composite Films

3.2.1. Apparent Color and Optical Properties

The appearance and light transmittance spectra of pullulan/chitosan-based binary biocomposite films are presented in Figure 3. The neat pullulan/chitosan (PLN/CTS) film and the ZnONPs-added (PLN/CTS/ZnO) film were slightly yellowish and transparent, while the film with propolis was yellowish due to the color of propolis. The light transmittance spectra of the film showed that all the films were see-through to visible light. The control films also exhibited a peak at 280 nm owing to the UV-light absorption of chitosan. The reinforcement of the film with ZnONPs slightly reduced the UV light transmittance. Nevertheless, the addition of propolis pointedly condensed the UV transmittance of the pullulan/chitosan film. The combined use of ZnONPs and propolis showed a somewhat synergistic effect in the reduction of the UV-light transmittance of the film. The film's UV-light barrier and transparency were assessed, and the results are presented in Table 1. The neat film showed excellent transparency with more than 91% of the T_{660}. The alloying of ZnONPs and propolis alone or in combination in pullulan/chitosan did not reflect any significant reduction of the transparency of the film. Nonetheless, the alloying of ZnONPs slightly reduced the UV-light transmittance (T_{280}), while the mixing of propolis showed a sharp decrease in UV-light transmittance (24.7% to 7.0%) due to the polyphenolic compounds present in the propolis. The ZnONPs and propolis together somewhat synergistically decreased T_{280} to 6.4%. Hence, it can be inferred that adding the fillers makes a highly transparent film and enhances its UV-light barrier properties.

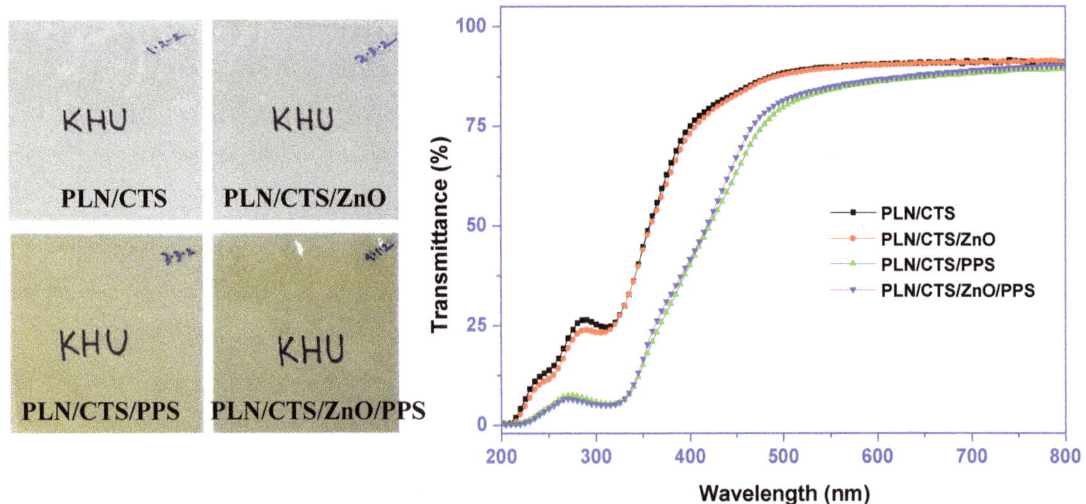

Figure 3. The appearance and transmittance spectra of the pullulan/chitosan-based composite films.

Table 1. Surface color and transmittance of pullulan/chitosan-based bioactive composite films.

Films	L	a	b	ΔE	WI	YI	T_{280} (%)	T_{660} (%)
PLN/CTS	90.7 ± 0.3^{c}	-2.3 ± 0.3^{c}	13.5 ± 1.4^{a}	9.2 ± 1.5^{a}	86.0 ± 1.4^{c}	21.2 ± 2.3^{a}	24.7 ± 1.9^{c}	91.2 ± 0.3^{c}
PLN/CTS/ZnO	90.7 ± 0.1^{c}	-2.1 ± 0.1^{c}	12.2 ± 0.5^{a}	7.9 ± 0.5^{b}	87.2 ± 0.5^{c}	19.2 ± 0.8^{b}	23.1 ± 0.6^{c}	90.6 ± 0.3^{b}
PLN/CTS/PPS	88.1 ± 0.3^{b}	-5.1 ± 0.2^{a}	25.8 ± 1.1^{c}	22.0 ± 1.1^{d}	73.5 ± 1.1^{a}	41.8 ± 1.8^{d}	7.0 ± 0.4^{b}	88.1 ± 0.6^{a}
PLN/CTS/ZnO/PPS	87.0 ± 0.3^{a}	-4.0 ± 0.2^{b}	23.2 ± 1.1^{b}	19.7 ± 1.2^{c}	76.1 ± 1.1^{b}	38.1 ± 2.0^{c}	6.4 ± 0.4^{a}	88.6 ± 0.5^{a}

The data are denoted as a mean ± standard deviation. In the same column, any two means, followed by the same letter, are not meaningfully ($p > 0.05$) dissimilar from Duncan's multiple range test.

The surface color data and related color parameters are shown in Table 1. As anticipated from the UV-vis spectra, the lightness (*L*-value) of all the films was over 87%, indicating that the lightness was not considerably altered due to the reinforcement with fillers. The *a*-value was also not much transformed, although the *b*-value was amplified in the case of propolis, demonstrating increased yellowness. Accordingly, the total color difference (ΔE) of the films also increased significantly ($p < 0.05$) when reinforced with propolis, even if adding ZnONPs alone had shown a reduction in ΔE. The whiteness index (WI) showed a slight increase in the presence of ZnONPs but showed a decreasing trend in the presence of propolis, which is presumably due to the yellowness of propolis. As expected, the yellowness index (YI) of the film was significantly increased by the addition of propolis.

3.2.2. Morphology

The morphological structure of the film was observed using the FE-SEM (surface and cross-section) and AFM, and the data are shown in Figure 4. The SEM surface images showed all films were smooth surfaced, without any outward defects. By alloying, pullulan and chitosan produce a compatible film [22]. The reinforcement with ZnONPs or propolis did not meaningly transform the morphology of the film, indicating their appropriate miscibility in the binary polymer matrices. The cross-section morphology low and high magnification also showed a proper distribution of fillers in the polymers without significantly altering the morphology of the control films. Interestingly, incorporating ZnONPs did not induce any agglomerated particles, suggesting the proper blending of nanofillers in the polymer matrices could contribute to the good adhesion of the filler and the pullulan/chitosan matrix. For more insight into the nanofillers' distribution, elemental mapping was carried out (Figure S2, Supporting Information), and the results showed a uniform distribution of the nanoparticles in the polymer matrices. Moreover, the morphology of the films was checked by AFM micrographs (Figure 4), and the results showed a good dispersion of the fillers analogous to the SEM data. The surface roughness of the film was also determined from the AFM micrographs, and the obtained Ra and Rq values (data not shown) indicate that the increase in roughness was observed only for ZnONPs while adding propolis did not significantly alter the roughness of the pullulan/chitosan films.

3.2.3. Mechanical Properties

The mechanical properties of the pullulan/chitosan-based films were determined in terms of the tensile strength (TS), elongation at break (EB), and elastic modulus (EM), and the results are shown in Table 2. The thickness of the control blend film was 43.1 μm, which slightly decreased to 41.6 μm with the mixing of ZnONPs. This might be due to the formation of H-bonding between ZnONPs and chitosan, which may have pulled the polymer chains closer, resulting in enhanced compactness and decreased thickness [45]. However, when propolis was added to the neat blend films, the thickness increased to 47 μm. The reason for this may be the high molecular weight of polyphenols in propolis, which occupy the interchain spaces in the polymer matrix, pushing them farther apart [46,47]. Also, this phenomenon designates the absence of any chemical interaction among propolis and the polymer chains. When both ZnONPs and propolis were added, the effect of both the additives balanced each other, resulting in a thickness of 45.5 μm. Nevertheless, the variation in the thickness among different films was not statistically significant.

The TS values confirm the H-bonding interaction of the ZnONPs with the biopolymer chains. The TS of the PLN/CTS/ZnO and PLN/CTS/ZnO/PPS films were 61.5 MPa and 62.6 MPa, respectively, indicating statistically significant ~24% and ~26% increments compared to neat PLN/CTS blends. However, the PLN/CTS/PPS film showed an insignificant increase (by 4%) in the TS value. Likewise, the EM values of the the ZnONPs-incorporated films also increased to 2.6 GPa compared to 2.2 GPa for films without ZnONPs. Additionally, the EB values were decreased more for the ZnONPs-incorporated films. Nevertheless, the difference in the EM and EB values was not statistically significant, indicating minor variations in the flexibility and stiffness of the films [48].

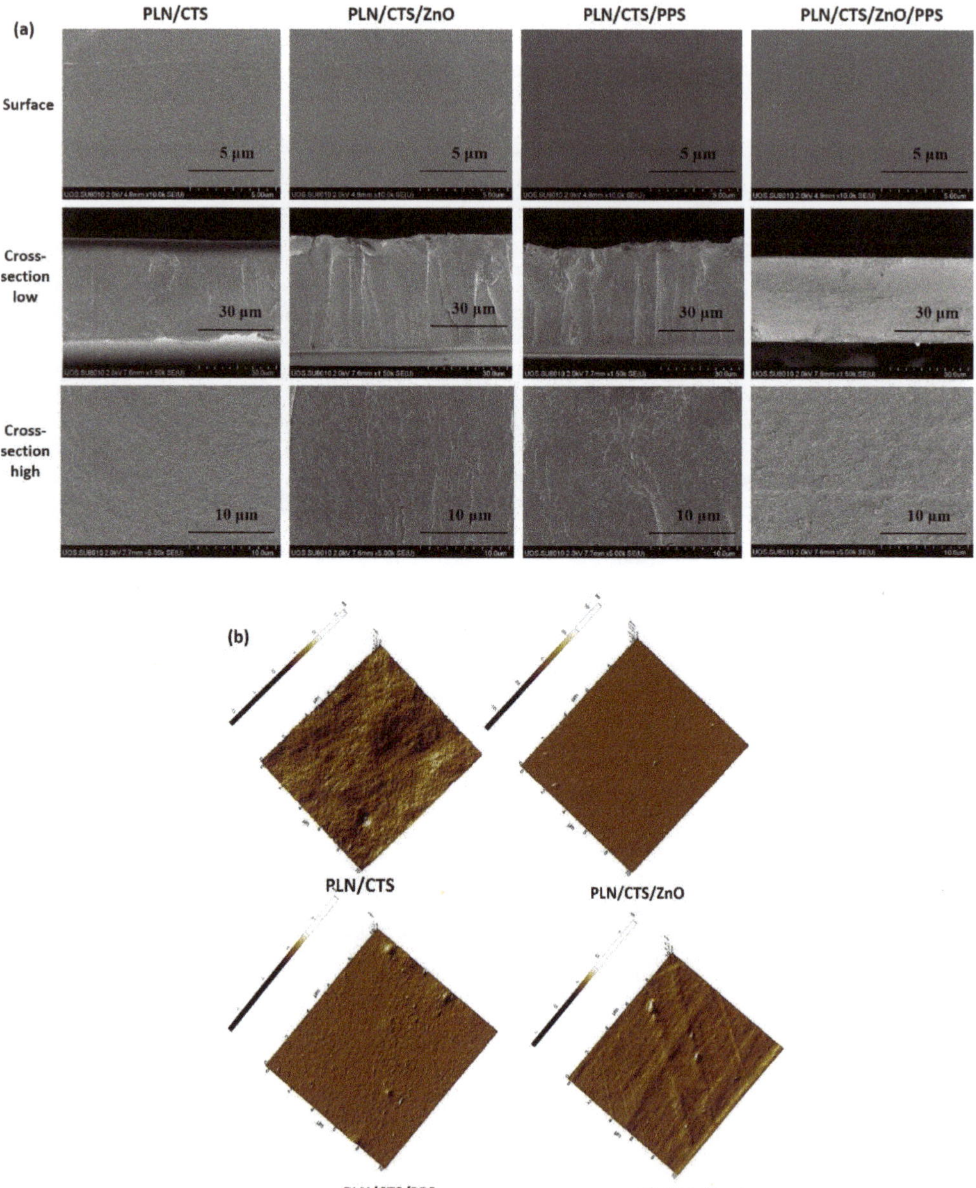

Figure 4. FESEM surface and cross-section micrographs (**a**) and AFM images (**b**) of pullulan/chitosan-based composite films.

Table 2. Mechanical properties, water vapor permeability, and water contact angle of pullulan/chitosan-based bioactive films.

Films	Thickness (μm)	TS (MPa)	EB (%)	EM (GPa)	WVP ($\times10^{-9}$ g.m/m^2.Pa.s)	WCA (deg.)
PLN/CTS	43.1 ± 5.0^{a}	49.7 ± 5.3^{a}	5.9 ± 1.0^{b}	2.2 ± 0.2^{a}	0.77 ± 0.01^{b}	56.5 ± 3.8^{a}
PLN/CTS/ZnO	41.6 ± 3.9^{a}	61.5 ± 4.3^{c}	4.8 ± 0.9^{a}	2.6 ± 0.3^{b}	0.79 ± 0.03^{b}	59.6 ± 1.6^{ab}
PLN/CTS/PPS	47.0 ± 5.2^{a}	51.7 ± 4.8^{b}	5.7 ± 0.5^{b}	2.2 ± 0.2^{a}	0.72 ± 0.04^{a}	64.6 ± 4.1^{c}
PLN/CTS/ZnO/PPS	45.5 ± 4.5^{a}	62.6 ± 4.2^{c}	5.2 ± 0.4^{ab}	2.6 ± 0.2^{b}	0.69 ± 0.01^{a}	61.3 ± 2.0^{b}

The data are denoted as a mean ± standard deviation. In the same column, any two means, followed by the same letter, are not meaningfully ($p > 0.05$) dissimilar from Duncan's multiple range tests.

3.2.4. Water Vapor Permeability (WVP) and Water Contact Angle (WCA)

The changes in the WVP and WCA values upon incorporating ZnONPs and propolis are shown in Table 2. No statistically meaningful change in the WVP value was observed due to incorporating ZnONPs in the blend films. Although it is reported that nanoparticles, when added to the polymer matrix, generally enhance the water vapor barrier properties by occupying the vacant interchain spaces and creating a tortuous path for the water molecules to pass through [45], some reports also indicate the opposite effect due to the hydrophilicity of the ZnO nanoparticles capped with biomolecules [49]. The unperturbed WVP values may be the result of the balance between both these attributes of ZnONPs. Conversely, adding propolis enhanced the water vapor barrier properties, which is attributed to the decreased hygroscopicity of the films resulting from the increased content of hydrophobic components in the propolis [50]. The WCA values displayed a similar trend where the films with propolis displayed high hydrophobicity compared to the neat blend films. However, the ZnONPs-incorporated films displayed increased interaction with water due to their hydrophilic nature. Despite this, all the films displayed substantial increments in the WCA values. However, the films could not be classified as hydrophobic as the WCA values were not above 65°, which is the recommended WCA value for hydrophobic films [51]. Nevertheless, the PLN/CTS/PPS films exhibited a WCA value of 64.6°, making them the most hydrophobic of their counterparts.

3.2.5. FTIR and Thermal Analysis

The results of the structural characterization of the films and their thermal analysis are provided in the Supplementary Information, Figures S3 and S4, respectively.

3.3. Antimicrobial Activity

The antimicrobial activity of the pullulan/chitosan-based films was studied against Gram-positive *L. monocytogenes* and Gram-negative *E. coli* bacteria, and the results are presented in Figure 5. The neat PLN/CTS film displayed significant antimicrobial activity toward the tested bacteria due to chitosan's antimicrobial action [52]. For both the bacterial strains, the growth after 12 h reached > 9 Log CFU/mL for the control group. However, for the groups exposed to the PLN/CTS films, the growth after 12 h was reduced to ~3 Log CFU/mL against both the bacteria, indicating the strong antimicrobial activity of chitosan. With the addition of propolis, no remarkable difference in the bactericidal action was observed against either bacteria, signifying no antimicrobial activity of the polyphenolic extract. Similar results were obtained previously [34]. However, as expected, the antimicrobial action of the films was boosted with the blending of ZnONPs. Against *E. coli*, a 100% bactericidal effect was observed for the PLN/CTS/ZnO film after 9 h. Interestingly, the PLN/CTS/ZnO/PPS film exerted the same effect at a 6 h interval, indicating the possible synergistic bactericidal effect of ZnONPs and propolis. However, this effect was not observed against *L. monocytogenes*, and 100% bacterial inhibition was observed after 12 h for the PLN/CTS/ZnO and PLN/CTS/ZnO/PPS films. The results also indicated a higher antimicrobial activity of ZnONPs toward Gram-negative bacteria than Gram-positive ones, which is consistent with previous reports [26,28].

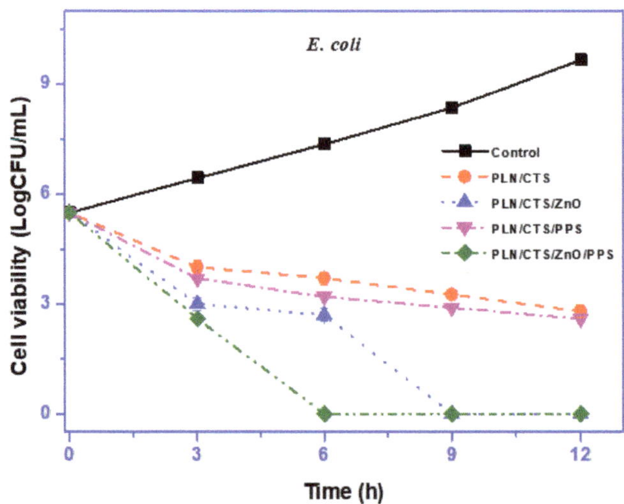

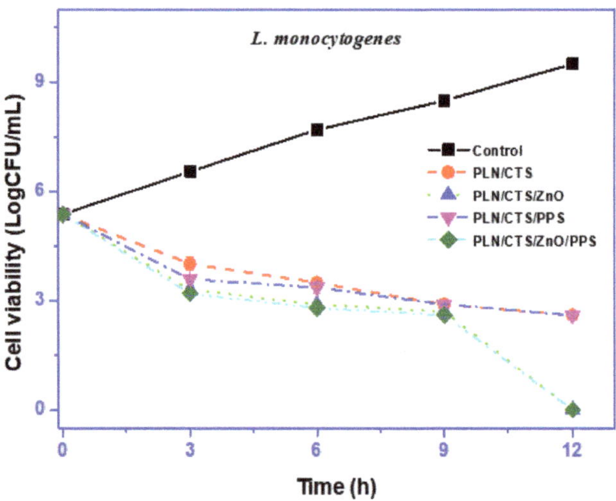

Figure 5. Antimicrobial activity of pullulan/chitosan-based composite films.

3.4. Antioxidant Activity

Figure 6 shows the outcomes of the antioxidant activity of the pullulan/chitosan-based films studied against DPPH and ABTS free radicals. The PLN/CTS film showed ~5% and ~33% antioxidant activity toward DPPH and ABTS free radicals, respectively, due to the presence of chitosan. The antioxidant activity of chitosan has been reported previously and is attributed to the protonated amine group, which stabilizes the free radicals [53]. Additionally, the higher activity against ABTS compared to DPPH occurredbecause hydrophilic chitosan interacts better with ABTS, due to its aqueous solution, than with the methanolic DPPH solution [51]. With the addition of ZnONPs, no variation in the antioxidant activity was detected. However, when propolis was mixed with the pullulan/chitosan films, the antioxidant activity sharply increased to ~30% and ~70% against DPPH and ABTS, respectively. This signifies that propolis imparts a strong antioxidant property to the films due to its polyphenolic composition [34,54]. Previous studies also report the enhanced antioxidant action of bio-based polymer films upon the addition of propolis [34]. Interestingly, for the PLN/CTS/ZnO/PPS film, the antioxidant activity was slightly decreased compared to the PLN/CTS/PPS films. This may be owing to the capability of ZnONPs to bring out reactive oxygen species, including oxidative free radicals, which counter the effect of propolis. This observation was previously reported as well [28].

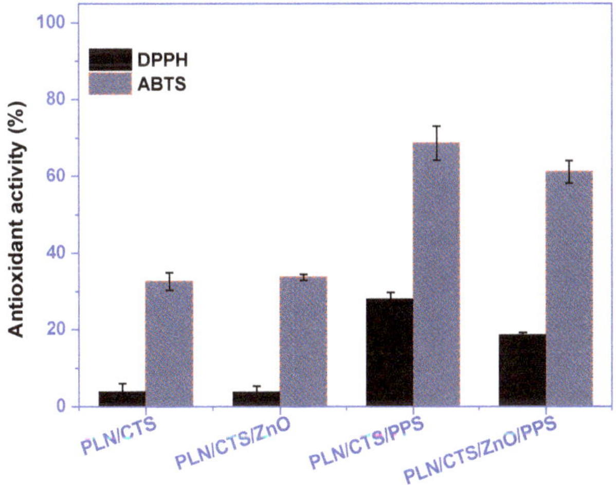

Figure 6. Antioxidant activity of pullulan/chitosan-based composite films.

3.5. Meat Packaging Test

The commonly used preservation method for meat products is refrigeration after skin-tight packaging using PE/PP films. However, certain bacteria can grow and contaminate the packed meat even under these conditions [55]. Hence, estimating the total aerobic bacterial count (TABC) on the packed meat offers critical evidence about its hygiene and safety. The total aerobic bacterial count of the pork loin meat stored at 10 °C was estimated for 8 days, and the results are presented in Figure 7a. The initial TABC on the packed pork loin meat sample was <2 Log CFU/g. A rapid increase in the TABC was observed for the control samples while storage reached ~6 Log CFU/g after 6 days and finally ~9 Log CFU/g within 8 days. As per the International Commission on Microbiological Specifications for Foods (ICMSF) regulations, an upper acceptable microbiological limit of 7 Log CFU/g has been proposed for meat products [55–57]. According to this proposed limit, the packed meat in the control group was deemed consumable for up to 6 days. However, for the meat samples wrapped with the PLN/CTS/ZnO/PPS film before packaging, the TABC value remained at 6.7 Log CFU/g even after 8 days of storage, below the proposed upper acceptable microbiological limit. Additionally, this value was significantly ($p > 0.05$) lower than that obtained for the control group. Several researchers performed studies on various meat types and determined them to be consumable if the TABC value remained below 7 Log CFU/g [28,55,58,59]. Hence, it can be inferred that the PLN/CTS/ZnO/PPS film can act as an extra wrapping material that may expand the shelf life of packed pork meat, acting in the form of hurdle technology.

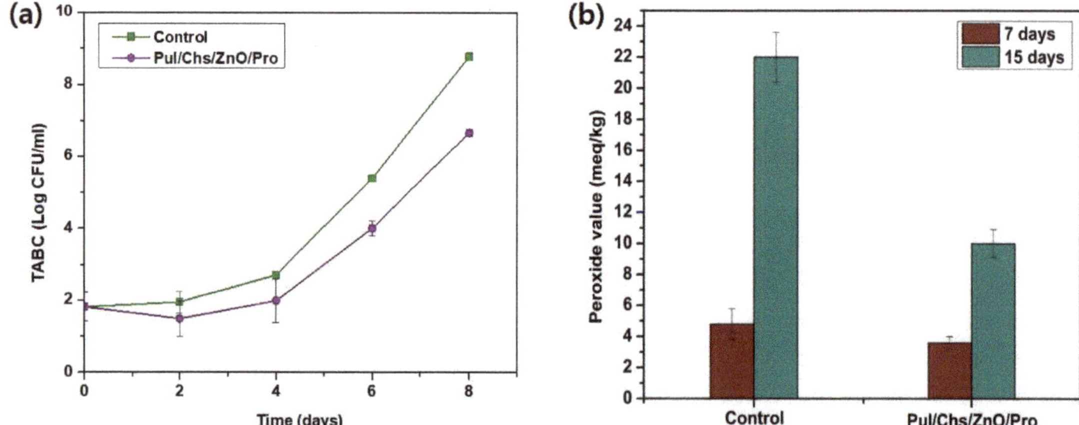

Figure 7. Application of pullulan/chitosan-based composite films for packaging of pork belly meat. (**a**) Total aerobic bacterial count, (**b**) peroxide values stored at 10 °C.

Besides microbial contamination, the lipid oxidation in the packed meat product is another important factor determining its quality and shelf life. The oxidation of meat lipids is inevitable during storage and transportation and renders the product rancid over time, making it unpalatable [55]. Hydroperoxides are the principal lipid peroxidation compounds that cause the development of off-flavors in food [28]. Hence, their assessment provides important evidence about the packed food quality. The peroxide values (PV) of the packed pork loin meat were determined over time, and the results are reported in Figure 7b. The initial value of lipid peroxidation in the packed meat was 0 meq/kg, which amplified over time, depending on the package type. After 7 days of storage, the control and test samples exhibited peroxide values of ~5 meq/kg and ~4 meq/kg, respectively, which were not considerably different statistically ($p < 0.05$). However, after 15 days of storing, while the PV of the meat packed in the control group increased abruptly to 22 meq/kg, the meat wrapped in the PLN/CTS/ZnO/PPS film exhibited a far lower peroxide value of ~10 meq/kg, showing around 55% decreased lipid oxidation, indicating a statistically significant ($p > 0.05$) reduction. This result was consistent with the antioxidant test results for the films (Figure 6), which indicate their oxidative free radical scavenging ability.

4. Conclusions

The main aim of this work was to investigate the effects of green-synthesized ZnONPs and propolis on the properties (physical and functional) of the pullulan/chitosan-based film for meat packaging applications. The ZnONPs were developed using enoki mushroom extract, and the obtained nanoparticles were stable and in a size range of 10–40 nm. Adding ZnONPs and propolis significantly enhanced the UV-light blocking and mechanical properties without significantly changing thermal stability. The alloying of fillers also slightly enhanced the water vapor barrier properties and hydrophobicity compared to the neat films. Additionally, the pullulan/chitosan-based binary composite film presented strong antibacterial activity toward *E. coli* and *L. monocytogenes* and showed excellent antioxidant activity. The PLN/CTS/ZnO/PPS film was used to wrap pork loin meat, and the TABC and PV data clearly showed that the growth of bacteria and lipid peroxidation was hindered in the packaged meat.

Supplementary Materials: The following are available online at https://www.mdpi.com/article/10.3390/foods10112789/s1: Experiment, results, and discussion, Figure S1; DLS analysis (size and Zeta potential) of ZnONPs, Figure S2; Elemental mapping of PLN/CTS/ZnO/PPS film, Figure S3; FTIR spectra of pullulan/chitosan-based bioactive composite films, Figure S4; TGA and DTG thermograms of pullulan/chitosan-based bioactive composite films, References.

Author Contributions: Data curation, S.R. and R.P.; investigation, S.R. and R.P.; software, S.R.; visualization, S.R. and R.P.; writing—original draft, S.R. and R.P.; writing—review & editing, J.-W.R. All authors have read and agreed to the published version of the manuscript.

Funding: This work was supported by the National Research Foundation of Korea (NRF) grant funded by the Korean government (MSIT) (No. 2019R1A2C2084221) and the Brain Pool Program funded by the Ministry of Science, ICT, and Future Planning through the NRF (2019H1D3A1A01070715).

Institutional Review Board Statement: Not Applicable.

Informed Consent Statement: Not Applicable.

Data Availability Statement: The data presented in this study are available on request from the corresponding author.

Conflicts of Interest: The authors declare no conflict of interest.

References

1. Karwowska, M.; Łaba, S.; Szczepański, K. Food Loss and Waste in Meat Sector—Why the Consumption Stage Generates the Most Losses? *Sustainability* **2021**, *13*, 6227. [CrossRef]
2. Fang, Z.; Zhao, Y.; Warner, R.D.; Johnson, S. Active and intelligent packaging in meat industry. *Trends Food Sci. Technol.* **2017**, *61*, 60–71. [CrossRef]
3. Barone, A.S.; Matheus, J.R.V.; de Souza, T.S.P.; Moreira, R.F.A.; Fai, A.E.C. Green-based active packaging: Opportunities beyond COVID-19, food applications, and perspectives in circular economy—A brief review. *Compr. Rev. Food Sci. Food Saf.* **2021**, *20*, 4881–4905. [CrossRef] [PubMed]
4. Qian, M.; Liu, D.; Zhang, X.; Yin, Z.; Ismail, B.B.; Ye, X.; Guo, M. A review of active packaging in bakery products: Applications and future trends. *Trends Food Sci. Technol.* **2021**, *114*, 459–471. [CrossRef]
5. Domínguez, R.; Barba, F.J.; Gómez, B.; Putnik, P.; Kovačević, D.B.; Pateiro, M.; Santos, E.M.; Lorenzo, J.M. Active packaging films with natural antioxidants to be used in meat industry: A review. *Food Res. Int.* **2018**, *113*, 93–101. [CrossRef] [PubMed]
6. Huang, J.; Hu, Z.; Hu, L.; Li, G.; Yao, Q.; Hu, Y. Pectin-based active packaging: A critical review on preparation, physical properties and novel application in food preservation. *Trends Food Sci. Technol.* **2021**, *118*, 167–178. [CrossRef]
7. Wicochea-Rodríguez, J.D.; Chalier, P.; Ruiz, T.; Gastaldi, E. Active Food Packaging Based on Biopolymers and Aroma Compounds: How to Design and Control the Release. *Front. Chem.* **2019**, *7*, 398. [CrossRef]
8. Jamróz, E.; Kopel, P. Polysaccharide and Protein Films with Antimicrobial/Antioxidant Activity in the Food Industry: A Re-view. *Polymers* **2020**, *12*, 1289. [CrossRef]
9. Sivakanthan, S.; Rajendran, S.; Gamage, A.; Madhujith, T.; Mani, S. Antioxidant and antimicrobial applications of biopolymers: A review. *Food Res. Int.* **2020**, *136*, 109327. [CrossRef]
10. Adeyeye, O.A.; Sadiku, E.R.; Reddy, A.B.; Ndamase, A.S.; Makgatho, G.; Sellamuthu, P.S.; Perumal, A.B.; Nambiar, R.B.; Fasiku, V.O.; Ibrahim, I.D.; et al. The Use of Biopolymers in Food Packaging. In *Green Biopolymers and Their Nanocomposites*; Springer: Singapore, 2019; pp. 137–158.
11. Roy, S.; Rhim, J.-W. Anthocyanin food colorant and its application in pH-responsive color change indicator films. *Crit. Rev. Food Sci. Nutr.* **2021**, *61*, 2297–2325. [CrossRef]
12. Roy, S.; Rhim, J.-W. New insight into melanin for food packaging and biotechnology applications. *Crit. Rev. Food Sci. Nutr.* **2021**, 1–27. [CrossRef] [PubMed]
13. Gnanasekaran, D. *Green Biopolymers and Its Nanocomposites in Various Applications: State of the Art*; Springer Science and Business Media LLC: Singapore, 2019; pp. 1–27.
14. Sady, S.; Błaszczyk, A.; Kozak, W.; Boryło, P.; Szindler, M. Quality assessment of innovative chitosan-based biopolymers for edible food packaging applications. *Food Packag. Shelf Life* **2021**, *30*, 100756. [CrossRef]
15. Roy, S.; Rhim, J.-W. Fabrication of bioactive binary composite film based on gelatin/chitosan incorporated with cinnamon essential oil and rutin. *Colloids Surf. B Biointerfaces* **2021**, *204*, 111830. [CrossRef]
16. Rukmanikrishnan, B.; Ramalingam, S.; Rajasekharan, S.K.; Lee, J.; Lee, J. Binary and ternary sustainable composites of gellan gum, hydroxyethyl cellulose and lignin for food packaging applications: Biocompatibility, antioxidant activity, UV and water barrier properties. *Int. J. Biol. Macromol.* **2020**, *153*, 55–62. [CrossRef]
17. Priyadarshi, R.; Kim, S.-M.; Rhim, J.-W. Pectin/pullulan blend films for food packaging: Effect of blending ratio. *Food Chem.* **2021**, *347*, 129022. [CrossRef] [PubMed]

18. Roy, S.; Kim, H.-J.; Rhim, J.-W. Effect of blended colorants of anthocyanin and shikonin on carboxymethyl cellulose/agar-based smart packaging film. *Int. J. Biol. Macromol.* **2021**, *183*, 305–315. [CrossRef] [PubMed]
19. Roy, S.; Rhim, J.-W. Fabrication of pectin/agar blended functional film: Effect of reinforcement of melanin nanoparticles and grapefruit seed extract. *Food Hydrocoll.* **2021**, *118*, 106823. [CrossRef]
20. Wu, J.; Zhong, F.; Li, Y.; Shoemaker, C.; Xia, W. Preparation and characterization of pullulan–chitosan and pullulan–carboxymethyl chitosan blended films. *Food Hydrocoll.* **2013**, *30*, 82–91. [CrossRef]
21. Li, Y.; Yokoyama, W.; Wu, J.; Ma, J.; Zhong, F. Properties of edible films based on pullulan–chitosan blended film-forming solutions at different pH. *RSC Adv.* **2015**, *5*, 105844–105850. [CrossRef]
22. Roy, S.; Rhim, J.-W. Effect of chitosan modified halloysite on the physical and functional properties of pullulan/chitosan biofilm integrated with rutin. *Appl. Clay Sci.* **2021**, *211*, 106205. [CrossRef]
23. Duan, M.; Yu, S.; Sun, J.; Jiang, H.; Zhao, J.; Tong, C.; Hu, Y.; Pang, J.; Wu, C. Development and characterization of electrospun nanofibers based on pullulan/chitin nanofibers containing curcumin and anthocyanins for active-intelligent food packaging. *Int. J. Biol. Macromol.* **2021**, *187*, 332–340. [CrossRef]
24. Li, S.; Yi, J.; Yu, X.; Wang, Z.; Wang, L. Preparation and characterization of pullulan derivative/chitosan composite film for potential antimicrobial applications. *Int. J. Biol. Macromol.* **2020**, *148*, 258–264. [CrossRef]
25. Kim, I.; Viswanathan, K.; Kasi, G.; Thanakkasaranee, S.; Sadeghi, K.; Seo, J. ZnO Nanostructures in Active Antibacterial Food Packaging: Preparation Methods, Antimicrobial Mechanisms, Safety Issues, Future Prospects, and Challenges. *Food Rev. Int.* **2020**, 1–29. [CrossRef]
26. Roy, S.; Kim, H.; Panicker, P.; Rhim, J.-W.; Kim, J. Cellulose Nanofiber-Based Nanocomposite Films Reinforced with Zinc Oxide Nanorods and Grapefruit Seed Extract. *Nanomaterials* **2021**, *11*, 877. [CrossRef]
27. Roy, S.; Rhim, J.-W. Carboxymethyl cellulose-based antioxidant and antimicrobial active packaging film incorporated with curcumin and zinc oxide. *Int. J. Biol. Macromol.* **2020**, *148*, 666–676. [CrossRef]
28. Priyadarshi, R.; Kim, S.-M.; Rhim, J.-W. Carboxymethyl cellulose-based multifunctional film combined with zinc oxide nanoparticles and grape seed extract for the preservation of high-fat meat products. *Sustain. Mater. Technol.* **2021**, *29*, e00325. [CrossRef]
29. Xie, Y.; Pan, Y.; Cai, P. Cellulose-based antimicrobial films incroporated with ZnO nanopillars on surface as biodegradable and antimicrobial packaging. *Food Chem.* **2022**, *368*, 130784. [CrossRef]
30. Roy, S.; Rhim, J.-W. Carrageenan-based antimicrobial bionanocomposite films incorporated with ZnO nanoparticles stabilized by melanin. *Food Hydrocoll.* **2019**, *90*, 500–507. [CrossRef]
31. Arfat, Y.A.; Benjakul, S.; Vongkamjan, K.; Sumpavapol, P.; Yarnpakdee, S. Shelf-life extension of refrigerated sea bass slices wrapped with fish protein isolate/fish skin gelatin-ZnO nanocomposite film incorporated with basil leaf essential oil. *J. Food Sci. Technol.* **2015**, *52*, 6182–6193. [CrossRef]
32. Woźniak, M.; Mrówczyńska, L.; Waśkiewicz, A.; Rogoziński, T.; Ratajczak, I. Phenolic Profile and Antioxidant Activity of Propolis Extracts from Poland. *Nat. Prod. Commun.* **2019**, *14*, 1–7. [CrossRef]
33. Suriyatem, R.; Auras, R.A.; Rachtanapun, C.; Rachtanapun, P. Biodegradable Rice Starch/Carboxymethyl Chitosan Films with Added Propolis Extract for Potential Use as Active Food Packaging. *Polymers* **2018**, *10*, 954. [CrossRef]
34. Roy, S.; Rhim, J.-W. Preparation of Gelatin/Carrageenan-Based Color-Indicator Film Integrated with Shikonin and Propolis for Smart Food Packaging Applications. *ACS Appl. Bio Mater.* **2020**, *4*, 770–779. [CrossRef]
35. Galeotti, F.; Maccari, F.; Fachini, A.; Volpi, N. Chemical Composition and Antioxidant Activity of Propolis Prepared in Different Forms and in Different Solvents Useful for Finished Products. *Foods* **2018**, *7*, 41. [CrossRef]
36. Rabeea, M.A.; Owaid, M.N.; Aziz, A.A.; Jameel, M.S.; Dheyab, M.A. Mycosynthesis of gold nanoparticles using the extract of Flammulina velutipes, Physalacriaceae, and their efficacy for decolorization of methylene blue. *J. Environ. Chem. Eng.* **2020**, *8*, 103841. [CrossRef]
37. Roy, S.; Ezati, P.; Rhim, J.-W. Gelatin/Carrageenan-Based Functional Films with Carbon Dots from Enoki Mushroom for Active Food Packaging Applications. *ACS Appl. Polym. Mater.* **2021**. [CrossRef]
38. Dai, L.; Zhang, J.; Cheng, F. Effects of starches from different botanical sources and modification methods on physicochemical properties of starch-based edible films. *Int. J. Biol. Macromol.* **2019**, *132*, 897–905. [CrossRef]
39. Ren, L.; Yan, X.; Zhou, J.; Tong, J.; Su, X. Influence of chitosan concentration on mechanical and barrier properties of corn starch/chitosan films. *Int. J. Biol. Macromol.* **2017**, *105*, 1636–1643. [CrossRef]
40. Philip, D. Biosynthesis of Au, Ag and Au–Ag nanoparticles using edible mushroom extract. *Spectrochim. Acta Part A Mol. Biomol. Spectrosc.* **2009**, *73*, 374–381. [CrossRef]
41. Faisal, S.; Khan, M.A.; Jan, H.; Shah, S.A.; Abdullah, A.; Shah, S.; Rizwan, M.; Ullah, W.; Akbar, M.T.; Redaina, R. Edible mushroom (*Flammulina velutipes*) as biosource for silver nanoparticles: From synthesis to diverse biomedical and environmental applications. *Nanotechnology* **2021**, *32*, 065101. [CrossRef] [PubMed]
42. Roy, S.; Das, T.K.; Maiti, G.P.; Basu, U. Microbial biosynthesis of nontoxic gold nanoparticles. *Mater. Sci. Eng. B* **2016**, *203*, 41–51. [CrossRef]
43. Ghosh, S.; Roy, S.; Naskar, J.; Kole, R.K. Process optimization for biosynthesis of mono and bimetallic alloy nanoparticle catalysts for degradation of dyes in individual and ternary mixture. *Sci. Rep.* **2020**, *10*, 1–14. [CrossRef]
44. Al-Gaashani, R.; Radiman, S.; Daud, A.; Tabet, N.; Al-Douri, Y. XPS and optical studies of different morphologies of ZnO nanostructures prepared by microwave methods. *Ceram. Int.* **2013**, *39*, 2283–2292. [CrossRef]

45. Priyadarshi, R.; Negi, Y.S. Effect of Varying Filler Concentration on Zinc Oxide Nanoparticle Embedded Chitosan Films as Potential Food Packaging Material. *J. Polym. Environ.* **2016**, *25*, 1087–1098. [CrossRef]
46. Lim, G.-O.; Jang, S.-A.; Bin Song, K. Physical and antimicrobial properties of Gelidium corneum/nano-clay composite film containing grapefruit seed extract or thymol. *J. Food Eng.* **2010**, *98*, 415–420. [CrossRef]
47. Mir, S.A.; Dar, B.N.; Wani, A.A.; Shah, M.A. Effect of plant extracts on the techno-functional properties of biodegradable packaging films. *Trends Food Sci. Technol.* **2018**, *80*, 141–154. [CrossRef]
48. Riahi, Z.; Priyadarshi, R.; Rhim, J.-W.; Bagheri, R. Gelatin-based functional films integrated with grapefruit seed extract and TiO2 for active food packaging applications. *Food Hydrocoll.* **2021**, *112*, 106314. [CrossRef]
49. Shankar, S.; Teng, X.; Li, G.; Rhim, J.-W. Preparation, characterization, and antimicrobial activity of gelatin/ZnO nanocomposite films. *Food Hydrocoll.* **2015**, *45*, 264–271. [CrossRef]
50. Bodini, R.; Sobral, P.; Favaro-Trindade, C.S.; Carvalho, R. Properties of gelatin-based films with added ethanol–propolis extract. *LWT* **2013**, *51*, 104–110. [CrossRef]
51. Ezati, P.; Priyadarshi, R.; Bang, Y.-J.; Rhim, J.-W. CMC and CNF-based intelligent pH-responsive color indicator films integrated with shikonin to monitor fish freshness. *Food Control.* **2021**, *126*, 108046. [CrossRef]
52. Priyadarshi, R.; Kumar, B.; Rhim, J.-W. Green and facile synthesis of carboxymethylcellulose/ZnO nanocomposite hydrogels crosslinked with Zn2+ ions. *Int. J. Biol. Macromol.* **2020**, *162*, 229–235. [CrossRef]
53. Wan, A.; Xu, Q.; Sun, Y.; Li, H. Antioxidant Activity of High Molecular Weight Chitosan and N,O-Quaternized Chitosans. *J. Agric. Food Chem.* **2013**, *61*, 6921–6928. [CrossRef] [PubMed]
54. Priyadarshi, R.; Riahi, Z.; Rhim, J.-W. Antioxidant pectin/pullulan edible coating incorporated with Vitis vinifera grape seed extract for extending the shelf life of peanuts. *Postharvest Biol. Technol.* **2022**, *183*, 111740. [CrossRef]
55. Alizadeh-Sani, M.; Mohammadian, E.; McClements, D.J. Eco-friendly active packaging consisting of nanostructured biopolymer matrix reinforced with TiO2 and essential oil: Application for preservation of refrigerated meat. *Food Chem.* **2020**, *322*, 126782. [CrossRef] [PubMed]
56. Ellis, D.I.; Goodacre, R. Rapid and quantitative detection of the microbial spoilage of muscle foods: Current status and future trends. *Trends Food Sci. Technol.* **2001**, *12*, 414–424. [CrossRef]
57. Panagou, E.Z.; Papadopoulou, O.; Carstensen, J.M.; Nychas, G.-J.E. Potential of multispectral imaging technology for rapid and non-destructive determination of the microbiological quality of beef filets during aerobic storage. *Int. J. Food Microbiol.* **2014**, *174*, 1–11. [CrossRef]
58. Camo, J.; Beltrán, J.A.; Roncalés, P. Extension of the display life of lamb with an antioxidant active packaging. *Meat Sci.* **2008**, *80*, 1086–1091. [CrossRef]
59. Saedi, S.; Shokri, M.; Priyadarshi, R.; Rhim, J.-W. Carrageenan-Based Antimicrobial Films Integrated with Sulfur-Coated Iron Oxide Nanoparticles (Fe3O4@SNP). *ACS Appl. Polym. Mater.* **2021**, *3*, 4913–4923. [CrossRef]

MDPI

Article

Effects of Gaseous Ozone on Microbiological Quality of Andean Blackberries (*Rubus glaucus* Benth)

Sandra Horvitz [1,*], Mirari Arancibia [2], Cristina Arroqui [1], Erika Chonata [2] and Paloma Vírseda [1]

1 Research Institute for Innovation & Sustainable Development in Food Chain, Campus Arrosadía, Public University of Navarre, 31006 Pamplona, Spain; cristina.arroqui@unavarra.es (C.A.); virseda@unavarra.es (P.V.)

2 Food Science and Engineering Faculty, Technical University of Ambato, Av. Los Chasquis y Rio Payamino, Ambato 180206, Ecuador; marancibias@uta.edu.ec (M.A.); erii_2404@hotmail.es (E.C.)

* Correspondence: sandra.horvitz@unavarra.es

Abstract: Andean blackberries are highly perishable due to their susceptibility to water loss, softening, mechanical injuries, and postharvest diseases. In this study, the antimicrobial efficacy of gaseous ozone against spoilage (mesophiles, psychrotrophs, and yeasts and molds) and pathogenic (*E. coli*, *S. enterica*, and *B. cinerea*) microorganisms was evaluated during 10 days of storage at 6 ± 1 °C. Respiration rate and mass loss were also determined. Ozone was applied prior to storage at 0.4, 0.5, 0.6, and 0.7 ppm, for 3 min. The best results were observed with the higher ozone dose, with initial maximum reductions of ~0.5, 1.09, and 0.46 log units for *E. coli*, *S. enterica*, and *B. cinerea*, respectively. For the native microflora, maximum reductions of 1.85, 1.89, and 2.24 log units were achieved on day 1 for the mesophiles, psychrotrophs, and yeasts and molds, respectively, and this effect was maintained throughout storage. In addition, the lower respiration rate and mass loss of the blackberries ozonated at 0.7 ppm indicate that this treatment did not induce physiological damage to the fruit. Gaseous O_3 could be effective in maintaining the postharvest quality of blackberries throughout refrigerated storage but higher doses could be advisable to enhance its antimicrobial activity.

Keywords: blackberry; gray mold; pathogens; storage quality; ozone treatment

Citation: Horvitz, S.; Arancibia, M.; Arroqui, C.; Chonata, E.; Vírseda, P. Effects of Gaseous Ozone on Microbiological Quality of Andean Blackberries (*Rubus glaucus* Benth). *Foods* **2021**, *10*, 2039. https://doi.org/10.3390/foods10092039

Academic Editors: Matteo Alessandro Del Nobile and Amalia Conte

Received: 27 July 2021
Accepted: 25 August 2021
Published: 30 August 2021

1. Introduction

Andean blackberries (*Rubus glaucus* Benth) are mostly cultivated in temperate and cold climates in South America and are usually consumed processed as pulp, jams, juices, and desserts [1]. Nonetheless, the interest in blackberries as fresh fruit has increased in the last years driven by the consumer's interest in berries as sources of bioactive compounds and health benefits [2].

As non-climacteric fruit, blackberries must be harvested at full maturity when they have the best organoleptic and nutritional quality. At this stage, they are also more susceptible to mechanical injuries and microbial attacks, which impair their commercial quality and shorten the postharvest shelf-life [3,4]. One of the main postharvest diseases affecting blackberries is gray mold, caused by *Botrytis cinerea* Pers.: Fr. The contamination with this mold may occur in the field or during the harvest, postharvest, and storage period, even under refrigeration conditions, leading to important economic losses of this crop [5]. In addition, the contamination of berries with pathogens can result in foodborne illness outbreaks and thus, microorganisms such as *Escherichia coli* or *Salmonella* are of great concern for fruit growers and processors.

The most extended methods used to control postharvest decay and guarantee fresh produce safety are the application of synthetic fungicides, washing with chlorine-based sanitizers, and storage at low temperature. Yet, the risk of appearance of pesticide-resistant strains of pathogens and of environmental pollution caused by the residues, together with consumers' demands to minimize chemical use, led to increased restrictions by marketing chains and regulatory agencies on agrochemical use, especially for postharvest applications [6,7].

Likewise, washing before retail is not recommended for delicate fruit such as berries, as their skin may be damaged easily [8]. Furthermore, it is now known that under commercial conditions, postharvest washing can present limited efficacy in fresh produce decontamination, or it can even lead to cross-contamination events between batches [9]. Finally, rapid cooling after harvest and constant cold storage are key factors in delaying microbial growth and extending the postharvest shelf-life of berries [10]. However, cold storage may be insufficient to prevent growth of mold, particularly on berries coming from fields with a high pathogen inoculum [3]. Therefore, the development of alternative sanitizing methods to prolong the storage life of blackberries after harvest is needed.

Conventional thermal treatments are effective in killing foodborne microorganisms, but they can negatively affect the quality of fresh produce. Thus, different individual or combined, physical (hot water, irradiation, UV-light, ozone, cold plasma), chemical (salts, organic acids, natural compounds), and biological treatments have been studied for decay control in fresh fruit and vegetables.

One of these alternatives could be the use of gaseous O_3, which has a high oxidant capacity and can be used for the inactivation of a wide range of microorganisms and for the degradation of chemical contaminants and off-odors in storage rooms [11]. There are several advantages related to the use of O_3 as a sanitizer: it is produced on site and does not require storage, and its precursors are abundant and economically advantageous. In addition, it decomposes almost immediately to oxygen, presenting no safety concerns regarding the consumption of chemical residues [12]. In 2001, the US Food and Drug Administration (FDA) declared ozone to be a generally recognized as safe (GRAS) substance for the commercial use as a disinfectant and sanitizer in food handling [13]. Due to these circumstances, its application in food processing is considered an environmental-friendly technology and it is allowed by organic certification [14].

Due to the lack of a protective skin and berries' surface roughness and sensitivity, gaseous treatments are preferred for these fruit [15]. Moreover, it is a common practice to pick berries straight into the punnets and prepare them in the field for retail. Thus, only gaseous ozone is applicable to them before the punnets are sealed and dispatched [16]. The antimicrobial activity of gaseous ozone against spoilage bacteria and fungi, and pathogens such as *E. coli* and *Salmonella* spp., has been studied in different berries including, among others, strawberries, table grapes, blueberries, blackberries, and raspberries.

Gaseous ozone contributed to reduce sour rot [17], the germination of *Botrytis cinerea* conidia, and the incidence and severity of gray mold during cold storage of table grapes [18]. Similarly, ozone proved efficacy in lowering decay in strawberries [19,20], raspberries [21,22], mulberries [23], blueberries [20,24], and blackberries [25]. On the contrary, after an initial microbial reduction, the exposure to gaseous ozone resulted ineffective in preventing decay after cold storage of strawberries [26], mulberries [27], and blueberries [28,29].

Even though the highest ozone doses often resulted in the highest antimicrobial activity, low concentrations of ozone are preferred in order to minimize exposure of workers to potentially hazardous concentrations of the gas and to reduce the risk of physiological damage to the treated produce [30]. In effect, as different fruit vary in their sensitivity to ozone, an optimal treatment should be established for each particular product. Previous research indicated that storage under continuous ozone prevented fungal decay and extended the shelf-life of blackberries [25]. However, to the best of our knowledge, to date, there are no previous reports dealing with the control of native microflora and inoculated pathogens on Andean blackberries by means of pre-storage treatments with gaseous ozone at low concentrations.

Therefore, the objective of this study was to investigate the effect of low gaseous ozone doses (0.4–0.7 ppm for 3 min) on the growth of native microflora (total aerobic mesophilic bacteria, psychrophiles, and yeast–mold) and inoculated pathogens (*E. coli*, *S. enterica*, and

B. cinerea) on Andean blackberries during refrigerated storage. The respiration rate and mass loss of the fruit were also determined as indicators of possible physiological damage.

2. Materials and Methods

2.1. Plant Material

The plant material used for this study was Andean blackberries (*Rubus glaucus* Benth), hand-harvested in Tungurahua Province, Ecuador at maturity stage 4 (dark red), according to external color of the fruit and following the color chart of the Ecuadorian Quality Standard for fresh blackberries [31]. Immediately after harvest, 30 kg of fruit were transported to the Technical University of Ambato for analyses. Fruit that were uniform in size and color, sound, and free from blemishes and injuries were selected for the study.

2.2. Microbial Strains

The strains of *Escherichia coli* (ATCC 25922) and *Salmonella enterica* (ATCC 9842) were obtained from the American Type Culture Collection (Gaithersburg, MD, USA) and *Botrytis cinerea* was kindly provided by BioSeb Organics Ltd. (Ambato, Ecuador).

2.3. Preparation of Inoculum

A loopful of each stock bacterial culture was individually transferred into 30 mL of brain heart infusion (BHI) broth (Difco, Detroit, MI, USA) and incubated at 37 $\pm$ 2 °C for 24 h prior to experimentation. Cells were used when a concentration of 10^9 CFU mL^{-1} was reached. Microbial concentration was determined according to the McFarland scale at 600 nm (OD600) [32]. An isolate of *B. cinerea* was cultured in a Petri dish on potato dextrose agar (PDA, Difco, Detroit, MI, USA). Streptomycin (1.0 mg mL^{-1}) was added to the media to inhibit bacterial growth and the plate was incubated at 25 $\pm$ 1 °C for 7 days. The fungi spores were scraped with a sterile loop and diluted with sterile distilled water. Conidia were counted using a hemocytometer and the suspension was adjusted to 10^7 CFU mL^{-1}.

2.4. Inoculation of Andean Blackberries

The blackberries were distributed in transparent polyethylene terephthalate (PET) plastic containers with perforated lids. The fruit was placed in a single layer, with each box containing 100 $\pm$ 5 g of fruit. The blackberries of each box were inoculated to reach 10^4 conidia g^{-1} of *Botrytis cinerea* and 10^4 CFU g^{-1} of *E. coli* and *S. enterica* by placing the inoculum on the surface of the fruit with a calibrated micropipette. Three containers/pathogen were prepared for each treatment and evaluation date.

In order to avoid the growth of native fungi and guarantee the growth only of the inoculated microorganism, before the inoculation with *B. cinerea*, the fruit were disinfected with an ethanol solution (70%, v/v) for 10 s, washed with sterile distilled water, and dried at room temperature. Inoculated blackberries were air dried for 1 h at room temperature (20 $\pm$ 2 °C) in a biosafety cabinet to allow the attachment of the microorganisms to the fruit surfaces. Thereafter, the fruit was stored under refrigeration (6 $\pm$ 1 °C) for 24 h until ozone treatment.

2.5. Ozone Treatment

Ozone was generated in situ utilizing a surface discharge ozone generator (COM-SD-30, Anseros GmbH, Tübingen, Germany) and synthetic air as the feeding gas (Figure 1). A fan installed inside the treatment chamber (Precision, Pompano Beach, FL, USA) facilitated an even distribution of the gas. Ozone production was of 30 mg h^{-1} and ozone concentration was continuously monitored and controlled by circulating air from the chamber through an ultraviolet absorption ozone analyzer (Anseros MP; Anseros GmbH, Tübingen, Germany), calibrated in the range of 0–2000 ppm connected to a computer. The software integrated the concentration and time data and when the appropriate dose (concentration $\times$ time) was reached, the ozone generator was stopped.

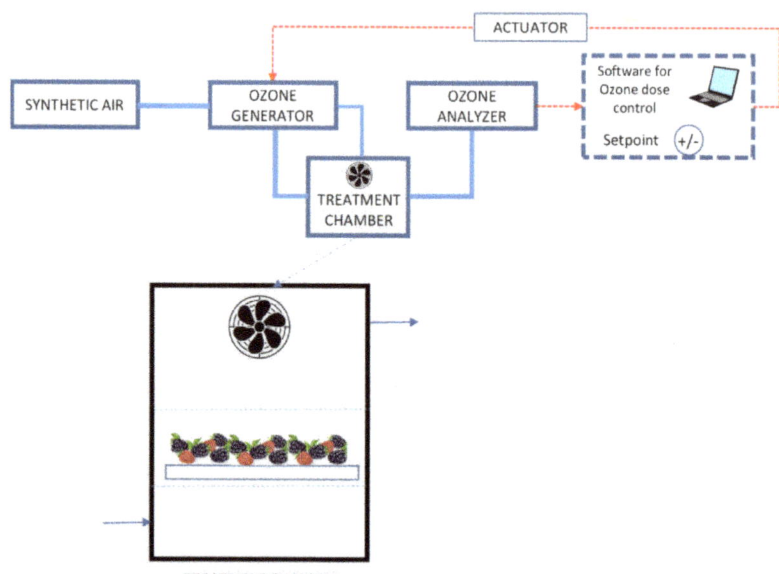

Figure 1. Diagram of gaseous ozone treatment system.

Two independent ozonation processes were performed, one for the inoculated fruit and the second for the non-inoculated blackberries. In both cases, the fruit was divided in five groups: untreated berries represented the control sample, whereas the remaining four groups were subjected to gaseous ozone at 0.4, 0.5, 0.6, and 0.7 ppm, for 3 min.

2.5.1. Inoculated Blackberries

The inoculated fruit was ozonized one day after the inoculation. For this purpose, the containers with the fruit were placed in the treatment chamber, with previous retirement of the lids. For each pathogen, three samples/evaluation date were treated with each of the ozone doses studied.

2.5.2. Non-Inoculated Blackberries

The effect of ozonation on the native microflora of the blackberries was evaluated on non-inoculated samples. Around 2 kg of blackberries were placed on stainless steel mesh trays and taken inside the chamber for the treatment with each O_3 dose.

2.6. Packing and Storage

After the O_3 treatments, the fruit for the native microflora (aerobic mesophiles, psychrotrophs, and molds and yeasts) studies were packaged in the same transparent polyethylene terephthalate (PET) plastic containers (100 $\pm$ 5 g) as those used for the inoculated fruit. All the samples were stored at 6 $\pm$ 1 °C.

2.7. Respiration Rate

The respiration rate of the blackberries was measured using the closed system method [33]. Samples of 100 $\pm$ 10 g of blackberries were placed into glass jars with a hermetic closure and stored open at 6 $\pm$ 1 °C. On each evaluation date, the jars were closed and the internal O_2 and CO_2 were determined after 8 h with the use of an O_2/CO_2 gas analyzer (MAPY 4.0 LE SP, WITT-Gasetechnik, Germany). The results were expressed as mg of CO_2 produced per kilogram per hour (mg kg^{-1} h^{-1}).

2.8. Mass Loss

For mass loss, three trays were randomly selected and individually weighed at the beginning of the experiment, and every two days during the storage period. Results were expressed as percentage of mass loss relative to the initial mass.

2.9. Microbiological Analyses

Microbiological analyses were performed on days 1, 4, 7, and 10 of storage, considering day 1 as the day of ozonation. In addition, the containers were visually controlled daily in order to detect symptoms of microbial development. At each evaluation date, 3 samples/treatment were analyzed.

2.9.1. Inoculated Microorganisms

For the analyses, 5 g of fruit inoculated with each of the pathogens under study was aseptically transferred to an individual filter stomacher bag and homogenized in 45 mL sterile buffered peptone water (Difco, USA) for 120 s at 200 rpm, using a Stomacher 400 circulator (Seward, AK, USA). Serial decimal dilutions of each homogenized sample were made in peptone water. From each dilution, 0.1 mL aliquots were aseptically surface-plated on the following media: Sabouraud dextrose agar plus chloramphenicol, Eosin Methylene Blue (EMB) Agar Levine, and Salmonella Shigella (SS) Agar, for *B. cinerea*, *E. coli*, and *S. enterica*, respectively. All the culture media were from Acumedia (Lansing, MI, USA). Culture conditions were as follows: 37 ± 2 °C for 48 h for *E. coli* and *S. enterica* and 25 ± 1 °C for 5 days in the case of *B. cinerea*.

2.9.2. Native Microflora: Total Aerobic Mesophiles, Psychrotrophs, and Yeasts and Molds

Serial dilutions for these microbial groups were prepared as described above. From each dilution, 1 mL aliquots were aseptically pour-plated for mesophiles and psychrothrophs and 0.1 mL was surface-plated for molds and yeasts analyses. The following media and culture conditions were used: (1) plate count agar (PCA, Difco, Detroit, MI, USA) incubated at 35 ± 2 °C for 48 h and at 7 °C for 7 days, for total mesophilic and psychrotrophic microorganisms, respectively, and (2) Sabouraud dextrose agar plus chloramphenicol media (Acumedia, Lansing, MI, USA) incubated at 25 °C ± 2 for 5 days for yeasts and molds. All the samples were analyzed in duplicate, and microbial counts were expressed as $\log_{10}$ (cfu g^{-1}) of fruit.

2.10. Statistical Analyses

The analyses were conducted in triplicate, considering each container as the experimental unit. Data were subjected to a one-way analysis of variance ($\alpha = 0.05$) using the IBM SPSS Statistics Version 27 software (IBM Corporation, Armonk, NY, USA). When significant differences were observed, mean treatments were compared using Tukey's test.

3. Results and Discussion

3.1. Respiration Rate

Respiration rate (RR) is an indicator of the metabolic activity of fruit and vegetables and thus, an indicator of postharvest shelf-life. The RR of the control and the O_3-treated blackberries is shown in Figure 2.

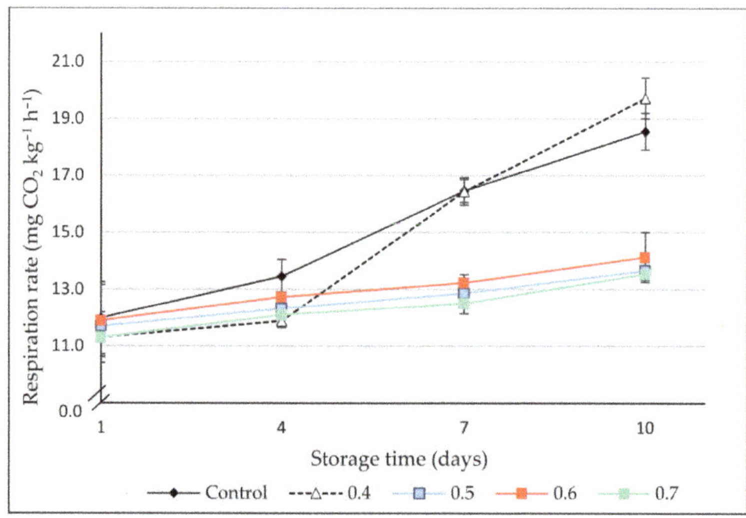

Figure 2. Respiration rate (mg CO_2 kg^{-1} h^{-1}) of control (no ozone treatment) and ozonated (0.4, 0.5, 0.6, and 0.7 ppm gaseous ozone/3 min) blackberries during 10 days of cold storage (6 ± 1 °C). Values are the mean of 3 independent samples and error bars represent the confidence interval (95%) for the mean.

Initially, the respiration rate of the blackberries ranged between 11.30 ± 0.26 (0.4 ppm O_3-treated blackberries) and 11.98 ± 0.51 (control) mg CO_2 kg^{-1} h^{-1}, with no significant differences among treatments ($p > 0.05$). Similar results were observed in cantaloupes [34], mulberries [23], and strawberries [35] treated with low doses of ozone. On the contrary, Forney et al. [36] reported an increase in the CO_2 production of broccoli treated with high doses of this gas (7 ppm), which was attributed to physiological damage caused to the florets. During the storage period, the RR of all the fruit increased continuously. However, the highest CO_2 production occurred in the control and those blackberries treated with 0.4 ppm O_3. In effect, after 10 days, the RR of these fruit was significantly higher than in the blackberries ozonized with the highest doses. Chen et al. [34] and Han et al. [23] found similar inhibitory effects of ozone on respiration rate during storage of melons and mulberries, respectively. The elevated CO_2 production observed in the present study, which represented an increase of around 55% and 75%, in the control and the 0.4 ppm O_3-treated fruit, respectively, could be related to the highest microbial growth observed in these treatments by the end of the storage period.

3.2. Mass Loss

During the postharvest period, mass loss is caused mainly by the respiration and transpiration of the fruit [37]. In the case of blackberries, their high respiration rates together with the lack of a protective peel make these fruit very susceptible to moisture loss. In effect, regardless of the treatment, mass loss increased constantly during storage, with maximum values of around 8% after 10 days of refrigerated storage (Figure 3). This value is above the maximum mass loss acceptable for commercialization of blackberries, reported as 6% [38]. Nevertheless, in all the evaluation dates, the exposure of blackberries to the highest O_3 dose resulted in lower mass loss, indicating that no physiological damage occurred due to ozonation. The lower mass loss in the fruit treated with the highest O_3 concentrations could be associated with the lower respiration rates and microbial counts observed in these blackberries. Similar results were reported for strawberries [39] and winter jujubes [40] washed with aqueous ozone and in table grapes [41], red peppers [42], and blueberries [28] exposed to gaseous ozone. According to Contigiani et al. [6], a thicker and reinforced cuticle in the O_3-treated fruit, which contributes to keeping cell integrity and offers a protective effect against moisture loss, can explain the positive effect of ozone in hindering mass loss.

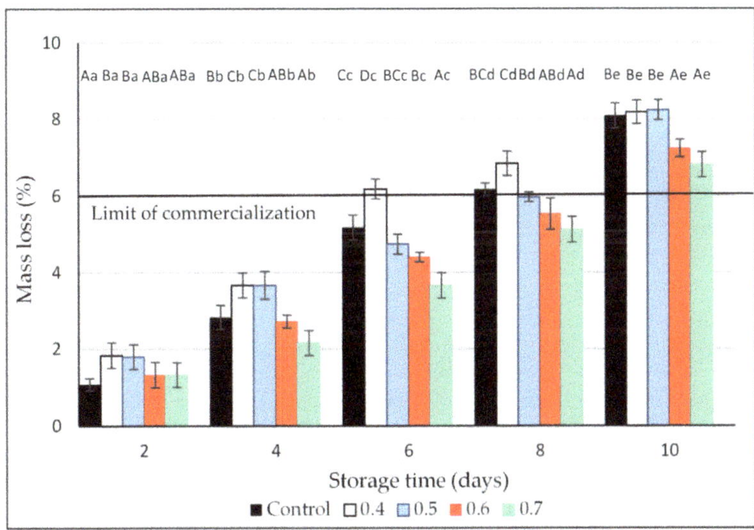

Figure 3. Effect of 3 min exposure to ozone (0, 0.4, 0.5, 0.6, and 0.7 ppm) on mass loss of Andean blackberries during 10 days of cold storage. Values are the mean of 3 independent samples and error bars represent the confidence interval (95%) for the mean. For each evaluation date, different capital letters indicate significant differences among O_3 doses ($p < 0.05$). For each O_3 dose, different lower-case letters indicate significant differences among evaluation dates ($p < 0.05$).

3.3. Microbiological Analyses

3.3.1. *E. coli* and *S. enterica*

The counts for *E. coli* and *S. enterica* in the blackberries as affected by the exposure to gaseous O_3 and storage time at 6 °C are presented in Table 1.

Table 1. *E. coli* and *S. enterica* counts (log cfu g^{-1}) of control (no O_3 treatment) and ozonated (0.4, 0.5, 0.6, and 0.7 ppm gaseous ozone/3 min) blackberries during 10 days of cold storage (6 ± 1 °C).

Microorganism	Day	O_3 Concentration (ppm)				
		0.0	0.4	0.5	0.6	0.7
E. coli	1	4.71 ± 0.02 Dd	4.45 ± 0.02 Cd	4.36 ± 0.05 BCd	4.27 ± 0.06 ABc	4.23 ± 0.13 Ac
	4	4.39 ± 0.05 Cc	4.30 ± 0.02 Bc	4.20 ± 0.08 BAc	4.19 ± 0.04 BAc	4.29 ± 0.04 Bc
	7	4.00 ± 0.02 Bb	3.99 ± 0.01 Bb	3.95 ± 0.03 BBb	3.92 ± 0.06 BBb	3.65 ± 0.09 Ab
	10	3.49 ± 0.06 Ba	3.39 ± 0.08 Ba	3.51 ± 0.08 BBa	3.05 ± 0.09 BAa	2.96 ± 0.11 Aa
S. enterica	1	4.47 ± 0.03 Ed	4.29 ± 0.09 Dd	3.96 ± 0.08 BCd	3.75 ± 0.10 BBd	3.38 ± 0.02 Ad
	4	4.00 ± 0.08 Cc	3.90 ± 0.03 Bc	3.87 ± 0.03 BBc	2.97 ± 0.10 BAc	2.98 ± 0.01 Ac
	7	3.25 ± 0.05 Eb	2.99 ± 0.19 Db	2.15 ± 0.05 BCb	1.04 ± 0.07 BBb	0.80 ± 0.06 Ab
	10	1.58 ± 0.14 Da	0.95 ± 0.16 Ca	0.50 ± 0.00 BBa	0.00 ± 0.00 BAa	0.00 ± 0.00 Aa

Values are the mean ± standard deviation ($n = 3$). For each microorganism and evaluation date, different capital letters indicate significant differences among O_3 doses ($p < 0.05$). For each microorganism and O_3 dose, different lower-case letters indicate significant differences among evaluation dates ($p < 0.05$).

Immediately after exposure to ozone, the counts of *E. coli* and *S. enterica* were significantly reduced in all the ozone-treated blackberries, with the greatest reductions observed with the highest O_3 dose: 0.48 and 1.09 log units for *E. coli* and *Salmonella*, respectively. Ozone's antimicrobial activity is based on its oxidant potential, which provokes injuries to the cell walls and a progressive oxidation of the microorganisms' cellular components [43]. However, it has limited penetration and thus, can be ineffective against latent infections, microbial growth occurring in wounds, and bacteria attached to uneven surfaces of fresh produce, all of which restrict the contact of ozone with the target microorganisms [44]. In this sense, the low microbial reductions observed in this study could be explained by the roughness and irregularities of the blackberries' surface where the inoculated bacteria can remain protected from ozone action and the relatively low O_3 doses used. Similar findings were reported for ground pepper [45], raspberries and strawberries [46], and mushrooms [47] treated with gaseous ozone and demonstrate that surface area is critical regarding the efficiency of O_3 treatments. Under these circumstances, higher O_3 concentrations and/or longer exposures to the gas were necessary to achieve the pathogens' inactivation.

During the storage period, the counts for both pathogens progressively decreased in all the treatments. The final counts of *E. coli* ranged from 2.96 $\pm$ 0.11 (0.7 ppm O_3) to 3.49 $\pm$ 0.06 log units in the control samples, whilst *Salmonella* was not detected on day 10 in the blackberries exposed to 0.6 and 0.7 ppm O_3. Daş et al. [48] reported similar findings in O_3-treated cherry tomatoes inoculated with *Salmonella* after 6 days of cold storage. Both blackberries and tomatoes are acidic fruit and *Salmonella* is very susceptible to acid environments. On the contrary, some strains of *E. coli* remained viable even at pH 2.5 [49]. In this study, and in addition to ozone activity, the low pH (3–3.9) of the blackberries could contribute to limit the growth of this pathogen [50].

3.3.2. *Botrytis cinerea*

The counts of *Botrytis* on non-ozonated blackberry samples or O_3-treated fruit (0.4, 0.5, 0.6, and 0.7 ppm gaseous O_3/3 min) are shown in Figure 4. On day 1, *Botrytis* counts for the control reached 3.84 $\pm$ 0.05 log units and the maximum reduction (0.46 log units) was observed in the blackberries treated with the highest O_3 dose. In effect, only this treatment significantly reduced *Botrytis* counts initially and following the storage period (Figure 4). On the contrary, the treatment with 0.4 ppm O_3 was not effective against this pathogen, with higher counts than the control. During the cold storage period, the growth of *Botrytis* progressively increased, and mycelium could be visually detected in the samples from all the treatments.

The results found in the literature regarding O_3 efficacy to control *B. cinerea* are somehow contradictory. Vlassi et al. [51] found that, regardless of inoculation technique (injection or immersion), treating table grapes with gaseous ozone (15 ppm/60 min) on a daily basis was an effective means of controlling *Botrytis cinerea* during 40 days of cold storage. On the contrary, Sharpe et al. [29] reported that low doses of ozone (450 ppb applied for 48 h before storage) reduced decay incidence in apples and grapes but were ineffective in blueberries, probably due to the inability of ozone to penetrate the fruit tissue to kill latent infections occurring early in the growing season. In the same way, gray mold was controlled by ozone in table grapes but not in citrus or stone fruit [52]. These authors attributed the differences to the fact that in the former, the inoculum was on the surface of the fruit, whilst in the latter, the pathogen was inoculated into wounds, hindering the access of the ozone to the inoculum. Finally, in naturally infected fruit, the treatment with 0.15 (grapes) and 0.7 (strawberries) ppm gaseous ozone, applied continuously in the storage room completely inhibited *Botrytis* development [30]. These apparent discrepancies among studies can be explained by differences in the type of commodity, cultivars, the inoculum level, and the presence of wounds, as well as experimental conditions, such as gas concentration and time of exposure; factors that can influence the antimicrobial efficacy of ozone treatments [15].

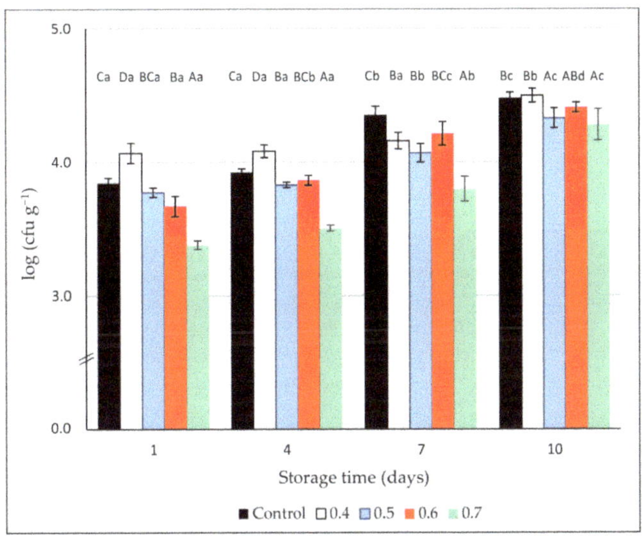

Figure 4. Effect of 3 min exposure to ozone (0, 0.4, 0.5, 0.6, and 0.7 ppm) on *Botrytis cinerea* counts (log (cfu g^{-1})) on inoculated blackberries during 10 days of cold storage. Values are the mean of 3 independent samples and error bars represent the confidence interval (95%) for the mean. For each evaluation date, different capital letters indicate significant differences among O_3 doses ($p < 0.05$). For each O_3 dose, different lower-case letters indicate significant differences among evaluation dates ($p < 0.05$).

3.3.3. Native Microflora: Total Aerobic Mesophiles, Psychrotrophs, and Molds and Yeasts

The microbial counts for the native microflora of the control and the O_3-treated blackberries are shown in Table 2. On the initial day, the control samples presented 2.93 ± 0.03 (aerobic mesophiles), 4.44 ± 0.15 (psychrotrophs), and 5.25 ± 0.09 (molds and yeasts) log unit counts.

Table 2. Aerobic mesophiles, psychrotrophs, and molds and yeasts counts (log cfu g^{-1}) of control (no O_3 treatment) and ozonated (0.4, 0.5, 0.6, and 0.7 ppm gaseous ozone/3 min) blackberries during 10 days of cold storage (6 ± 1 °C).

Microbial Group	Day	O_3 Concentration (ppm)				
		0.0	0.4	0.5	0.6	0.7
Aerobic mesophiles	1	2.93 ± 0.03 Ea	2.66 ± 0.10 Da	2.25 ± 0.06 Ca	1.30 ± 0.19 Ba	1.08 ± 0.12 Aa
	4	3.23 ± 0.10 Cb	2.75 ± 0.16 Ba	2.54 ± 0.21 Bb	2.18 ± 0.03 Ab	2.08 ± 0.08 Ab
	7	4.00 ± 0.06 Dc	3.17 ± 0.02 Cb	2.58 ± 0.11 Bb	2.26 ± 0.12 Ab	2.29 ± 0.06 Ac
	10	4.00 ± 0.10 Ec	3.88 ± 0.02 Dc	3.66 ± 0.04 Cc	3.15 ± 0.07 Bc	2.73 ± 0.02 Ad
Psychrotrophs	1	4.44 ± 0.15 Ca	3.95 ± 0.12 Ba	2.70 ± 0.01 Aa	2.58 ± 0.02 Aa	2.55 ± 0.04 Aa
	4	4.54 ± 0.06 Da	4.07 ± 0.04 Cb	3.72 ± 0.02 Bb	4.15 ± 0.12 Cb	3.46 ± 0.25 Ab
	7	5.10 ± 0.03 Db	4.84 ± 0.00 Cc	4.29 ± 0.01 Bc	4.32 ± 0.23 Bb	4.01 ± 0.01 Ac
	10	6.18 ± 0.05 Ec	5.32 ± 0.07 Dd	4.89 ± 0.05 Cd	4.78 ± 0.03 Bc	4.11 ± 0.03 Ac
Molds and yeasts	1	5.25 ± 0.09 Ea	4.88 ± 0.05 Da	4.21 ± 0.03 Ca	3.22 ± 0.08 Ba	3.01 ± 0.17 Aa
	4	6.18 ± 0.12 Eb	5.30 ± 0.02 Db	5.04 ± 0.02 Cb	4.61 ± 0.03 Bb	4.36 ± 0.03 Ab
	7	6.81 ± 0.10 Ec	6.16 ± 0.01 Dc	5.60 ± 0.04 Cc	5.43 ± 0.02 Bc	5.31 ± 0.01 Ac
	10	7.07 ± 0.04 Dd	6.22 ± 0.00 Cd	6.20 ± 0.03 Cd	6.13 ± 0.02 Bd	5.79 ± 0.02 Ad

Values are the mean ± standard deviation ($n = 3$). For each microorganism and evaluation date, different capital letters indicate significant differences among O_3 doses ($p < 0.05$). For each microorganism and O_3 dose, different lower-case letters indicate significant differences among evaluation dates ($p < 0.05$).

In all the O_3-treated fruit, the counts for the three microbial groups studied were significantly lower ($p < 0.05$) when compared with the control. Furthermore, the extent of the reductions increased with increasing O_3 dose, ranging from 0.27 to 1.85, 0.49 to 1.89, and 0.37 to 2.24 log units for the mesophiles, psychrotrophs, and molds and yeasts, respectively. During the cold storage period, the microbial populations gradually increased regardless of the treatment. Yet, in all the ozone-treated blackberries, the reductions achieved were maintained throughout the storage period. Among the O_3 treatments, the best results were observed when the blackberries were exposed to 0.7 ppm O_3 with counts on day 10 for total aerobic mesophiles and psychrotrophs even lower than the initial counts of the control samples.

Gaseous ozone can be applied either as a pre-storage treatment or continuously or intermittently during the storage period. Similar to our results, the application of gaseous ozone at doses of 130 g m^{-3}/30 min and 0.5 to 2 mg L^{-1}/60 min reduced the total mesophile counts in juniper berries [53] and greenhouse tomatoes [54], with the inhibitory effects being maintained during storage. In the same way, Alves et al. [55] found that 18 and 14 mg L^{-1} gaseous ozone applied to strawberries for 30 min were effective in controlling aerobic mesophiles and molds and yeasts, respectively, during 4 days of storage. In studies involving the application of ozone during storage, the exposure of sweet cherries to 2 ppm gaseous ozone for 30 min every 6 days delayed decay development and lowered decay incidence on the fruit for up to 18 days [56]. As well, treating raspberries with 8–10 ppm O_3 in cycles of 30 min once every 12 h reduced the aerobic mesophilic bacteria and fungi counts on this fruit during 3 days of storage at room temperature [22]. On the contrary, both intermittent (0.1 ppm applied every 30 min) and continuous (0.35 ppm, 3 days) O_3 applications were only partially effective in preventing fungal growth on table grapes and strawberries, respectively [26,57].

These variable and somehow contradictory results could be explained by differences in the type of product, differences in the ozone application methods and in the doses used (time and concentration), the microbial type and microbial load, as well as variations in environmental conditions (temperature, relative humidity), all of which can affect the efficacy of ozone as a sanitizer [58].

4. Conclusions

The application of low doses of gaseous ozone prior to storage was studied as an environmental-friendly alternative to guarantee Andean blackberries' quality and safety. The best results were obtained after the exposure of the fruit to 0.7 ppm gaseous O_3 for 3 min. This treatment slowed down the respiration and mass loss rates, indicating that no physiological damage occurred in the treated fruit. Moreover, it was proven effective in reducing both the native microflora and the inoculated pathogens on the fruit throughout the storage period. Therefore, gaseous ozone could be considered a promising processing technology for prolonging the postharvest life of fresh Andean blackberries during refrigerated storage. However, as the log reduction in microbial populations observed in this study may not be enough to ensure the safety of the product, further studies would be necessary to determine the optimal treatment conditions for this fruit. As well, the effects of the treatment on the bioactive compounds and the physicochemical and sensory quality of the blackberries would be assessed.

Author Contributions: Conceptualization, S.H. and M.A.; Data curation, C.A. and E.C.; Formal analysis, C.A. and E.C.; Funding acquisition, S.H. and P.V.; Investigation, S.H., M.A., C.A., E.C. and P.V.; Methodology, S.H., M.A. and P.V.; Project administration, S.H. and M.A.; Resources, M.A., C.A. and P.V.; Supervision, S.H., M.A., C.A. and P.V.; Writing—original draft, E.C. and P.V.; Writing—review and editing, S.H., M.A., C.A. and P.V. All authors have read and agreed to the published version of the manuscript.

Funding: This research was funded by the Research and Development Department (DIDE) of the Technical University of Ambato through the project "Development of new technologies for berries postharvest conditioning", grant 1302-CU-P-2015.

Data Availability Statement: Not applicable.

Conflicts of Interest: The authors declare no conflict of interest. The funders had no role in the design of the study; in the collection, analyses, or interpretation of data; in the writing of the manuscript, or in the decision to publish the results.

References

1. Martínez, A.; Beltrán, O.; Velastegui, G.; Ayala, G.; Jácome, R.; Yánez, M.; Luciano, E. *Manual del Cultivo de la Mora de Castilla*; INIAP: Quito, Ecuador, 2007; pp. 9–16. (In Spanish)
2. Clark, J.R.; Finn, C.E. Blackberry cultivation in the world. *Rev. Bras. Frutic.* **2014**, *36*, 46–57. [CrossRef]
3. Gabioud Rebeaud, S.; Varone, V.; Vuong, L.; Cotter, P.Y.; Ançay, A.; Christen, D. Postharvest ozone treatment on raspberries. *Acta Hortic.* **2020**, *1277*, 449–454. [CrossRef]
4. Horvitz, S.; Chanaguano, D. Microbial and sensory quality of an Andean blackberry (*Rubus glaucus* Benth) cultivar. *Acta Hortic.* **2020**, *1275*, 121–124. [CrossRef]
5. Junqueira-Gonçalves, M.P.; Alarcón, E.; Niranjan, K. The efficacy of potassium sorbate-coated packaging to control postharvest gray mold in raspberries, blackberries and blueberries. *Postharvest Biol. Technol.* **2016**, *11*, 205–208. [CrossRef]
6. Contigiani, E.V.; Jaramillo-Sánchez, G.; Castro, M.A.; Gómez, P.L.; Alzamora, S.M. Postharvest Quality of Strawberry Fruit (*Fragaria* × *Ananassa* Duch cv. Albion) as Affected by Ozone Washing: Fungal Spoilage, Mechanical Properties, and Structure. *Food Bioproc. Tech.* **2018**, *11*, 1639–1650. [CrossRef]
7. Romanazzi, G.; Smilanick, J.L.; Feliziani, E.; Droby, S. Integrated management of postharvest gray mold on fruit crops. *Postharvest Biol. Technol.* **2016**, *113*, 69–76. [CrossRef]
8. Vardar, C.; Ilhan, K.; Karabulut, O.A. The application of various disinfectants by fogging for decreasing postharvest diseases of strawberry. *Postharvest Biol. Technol.* **2012**, *66*, 30–34. [CrossRef]
9. Murray, K.; Wu, F.; Shi, J.; Jun Xue, S.; Warriner, K. Challenges in the microbiological food safety of fresh produce: Limitations of post-harvest washing and the need for alternative interventions. *Food Qual. Saf.* **2017**, *1*, 289–301. [CrossRef]
10. Horvitz, S.; Chanaguano, D.; Arozarena, I. Andean blackberries (*Rubus glaucus* Benth) quality as affected by harvest maturity and storage conditions. *Sci. Hortic.* **2017**, *226*, 293–301. [CrossRef]
11. Torlak, E. Use of gaseous ozone for reduction of ochratoxin A and fungal populations on sultanas. *Aust. J. Grape Wine Res.* **2019**, *25*, 25–29. [CrossRef]
12. Pandiselvam, R.; Sunoj, S.; Manikantan, M.R.; Kothakota, A.; Hebbar, K.B. Application and Kinetics of Ozone in Food Preservation. *Ozone Sci Eng.* **2017**, *39*, 115–126. [CrossRef]
13. United States Food and Drug Administration. *Rules and Regulations. Part 173-Secondary Direct Food Additives Permitted in Food for Human Consumption (21 CFR Part 173 Authority: 21 USC. 321, 342, 348)*; Federal Register; United States Food and Drug Administration: Silver Spring, MD, USA, 2001; Volume 66, p. 123.
14. Zainuri, J.; Sauqi, A.; Sjah, T.; Desiana, R.Y. Combination of ozone and packaging treatments maintained the quality and improved the shelf life of tomato fruit. *IOP Conf. Ser. Earth Environ. Sci.* **2018**, *102*, 012027. [CrossRef]
15. Tzortzakis, N.; Chrysargyris, A. Postharvest ozone application for the preservation of fruits and vegetables. *Food Rev Int.* **2017**, *33*, 270–315. [CrossRef]
16. Glowacz, M.; Rees, D. The practicality of using ozone with fruit and vegetables. *J. Sci. Food Agric.* **2016**, *96*, 4637–4643. [CrossRef]
17. Pinto, L.; Caputo, L.; Quintieri, L.; de Candia, S.; Baruzzi, F. Efficacy of gaseous ozone to counteract postharvest table grape sour rot. *Food Microbiol.* **2017**, *66*, 190–198. [CrossRef] [PubMed]
18. Ozkan, R.; Smilanick, J.L.; Karabulut, O.A. Toxicity of ozone gas to conidia of *Penicillium digitatum*, *Penicillium italicum*, and *Botrytis cinerea* and control of gray mold on table grapes. *Postharvest Biol. Technol.* **2011**, *60*, 47–51. [CrossRef]
19. Lopes-Morais, M.; Alvinhão, J.E.O.; Vilela-Franco, D.; Silva, E.d.B.; Dessimoni-Pinto, N.A.V. Application of ozone aiming to keep the quality of strawberries using a low cost reactor. *Rev Bras Frutic.* **2015**, *37*, 559–567. [CrossRef]
20. Pinto, L.; Palma, A.; Cefola, M.; Pace, B.; D'Aquino, S.; Carboni, C.; Baruzzi, F. Effect of modified atmosphere packaging (MAP) and gaseous ozone pre-packaging treatment on the physico-chemical, microbiological and sensory quality of small berry fruit. *Food Packag. Shelf Life* **2020**, *26*, 100573. [CrossRef]
21. Onopiuk, A.; Półtorak, A.; Moczkowska, M.; Szpicer, A.; Wierzbicka, A. The impact of ozone on health-promoting, microbiological, and colour properties of *Rubus ideaus* raspberries. *CyTA-J. Food* **2017**, *15*, 563–573. [CrossRef]
22. Piechowiak, T.; Antos, P.; Kosowski, P.; Skrobacz, K.; Józefczyk, R.; Balawejder, M. Impact of ozonation process on the microbiological and antioxidant status of raspberries (*Rubus ideaeus* L.) during storage at room temperature. *Agric Food Sci.* **2019**, *28*, 35–44. [CrossRef]
23. Han, Q.; Gao, H.; Chen, H.; Fang, X.; Wu, W. Precooling and ozone treatments affects postharvest quality of black mulberry (*Morus nigra*) fruits. *Food Chem.* **2017**, *221*, 1947–1953. [CrossRef] [PubMed]
24. Piechowiak, T.; Antos, P.; Józefczyk, R.; Kosowski, P.; Skrobacz, K.; Balawejder, M. Impact of Ozonation Process on the Microbiological Contamination and Antioxidant Capacity of Highbush Blueberry (*Vaccinum corymbosum* L.) Fruit during Cold Storage. *Ozone Sci. Eng.* **2019**, *41*, 376–385. [CrossRef]
25. Barth, M.M.; Zhou, C.; Mercier, J.; Payne, F.A. Ozone storage effects on anthocyanin content and fungal growth in blackberries. *J. Food Sci.* **1995**, *60*, 1286–1288. [CrossRef]

26. Pérez, A.G.; Sanz, C.; Ríos, J.J.; Olías, R.; Olías, J.M. Effects of ozone treatment on postharvest strawberry quality. *J. Agric. Food Chem.* **1999**, *47*, 1652–1656. [CrossRef]
27. Tabakoglu, N.; Karaca, H. Effects of ozone-enriched storage atmosphere on postharvest quality of black mulberry fruits (*Morus nigra* L.). *LWT* **2018**, *92*, 276–281. [CrossRef]
28. Concha-Meyer, A.; Eifert, J.D.; Williams, R.C.; Marcy, J.E.; Welbaum, G.E. Shelf Life Determination of Fresh Blueberries (*Vaccinium corymbosum*) Stored under Controlled Atmosphere and Ozone. *Int. J. Food Sci.* **2015**, *2015*, 164143. [CrossRef]
29. Sharpe, D.; Fan, L.; McRae, K.; Walker, B.; MaCKay, R.; Doucette, C. Effects of ozone treatment on *Botrytis cinerea* and *Sclerotinia sclerotiorum* in relation to horticultural product quality. *J. Food Sci.* **2009**, *74*, M250–M257. [CrossRef]
30. Chilosi, G.; Tagliavento, V.; Simonelli, R. Application of ozone gas at low doses in the cold storage of fruit and vegetables. *Acta Hortic.* **2015**, *1071*, 681–686. [CrossRef]
31. Instituto Ecuatoriano de Normalización. *INEN 2427:2010: Frutas Frescas. Mora. Requisitos*; Instituto Ecuatoriano de Normalización: Quito, Ecuador, 2010. (In Spanish)
32. Brodowska, A.J.; Nowak, A.; Kondratiuk-Janyska, A.; Piątkowski, M.; Śmigielski, K. Modelling the Ozone-Based Treatments for Inactivation of Microorganisms. *Int. J. Environ. Res. Public Health* **2017**, *14*, 1196. [CrossRef]
33. Pérez-Gallardo, A.; García-Almendárez, B.; Barbosa-Cánovas, G.; Pimentel-González, D.; Reyes-González, L.R.; Regalado, C. Effect of starch-beeswax coatings on quality parameters of blackberries (*Rubus* spp.). *J. Food Sci. Technol.* **2015**, *52*, 5601–5610. [CrossRef]
34. Chen, C.; Zhang, H.; Zhang, X.; Dong, C.; Xue, W.; Xu, W. The effect of different doses of ozone treatments on the postharvest quality and biodiversity of cantaloupes. *Postharvest Biol. Technol.* **2020**, *163*, 111124. [CrossRef]
35. Allende, A.; Marín, A.; Buendía, B.; Tomás-Barberán, F.; Gil, M.I. Impact of combined postharvest treatments (UV-C light, gaseous O_3, superatmospheric O_2 and high CO_2) on health promoting compounds and shelf-life of strawberries. *Postharvest Biol. Technol.* **2007**, *46*, 201–211. [CrossRef]
36. Forney, C.F.; Song, J.; Fan, L.; Hildebrand, P.D.; Jordan, M.A. Ozone and 1-MCP alter the postharvest quality of broccoli. *J. Amer. Soc. Hort. Sci.* **2003**, *128*, 403–408. [CrossRef]
37. Kumar, P.; Sethi, S.; Sharma, R.R.; Srivastav, M.; Varghese, E. Effect of chitosan coating on postharvest life and quality of plum during storage at low temperature. *Sci. Hortic.* **2017**, *226*, 104–109. [CrossRef]
38. Joo, M.; Lewandowski, N.; Auras, R.; Harte, J.; Almenar, E. Comparative shelf life study of blackberry fruit in bio-based and petroleum-based containers under retail storage conditions. *Food Chem.* **2011**, *126*, 1734–1740. [CrossRef]
39. Nayak, S.L.; Sethi, S.; Sharma, R.R.; Sharma, R.M.; Singh, S.; Singh, D. Aqueous ozone controls decay and maintains quality attributes of strawberry (*Fragaria* × *ananassa* Duch.). *J. Food Sci. Technol.* **2020**, *57*, 319–326. [CrossRef]
40. Li, H.; Xiong, Z.; Gui, D.; Li, X. Effect of aqueous ozone on quality and shelf life of Chinese winter jujube. *J. Food Process. Preserv.* **2019**, *43*, e14244. [CrossRef]
41. Wieczynska, J.; Lovino, R.; Lamaj, F.; De Cillis, M.F.; Baser, N.; Ismaili, K.; Verrastro, V.; Tarricone, L.; Simeone, V. Effect of O_3 and high CO_2 application during cold storage on quality of organic table grape (*Vitis vinifera* L. 'Italia'). *Acta Hortic.* **2015**, *1071*, 575–581. [CrossRef]
42. Horvitz, S.; Cantalejo, M.J. The Effects of Gaseous Ozone and Chlorine on Quality and Shelf-life of Minimally Processed Red Pepper. *Acta Hortic.* **2010**, *877*, 583–589. [CrossRef]
43. Niveditha, A.; Pandiselvam, R.; Prasath, V.A.; Singh, S.K.; Gul, K.; Kothakota, A. Application of cold plasma and ozone technology for decontamination of *Escherichia coli* in foods—A review. *Food Control* **2021**, *130*, 108338. [CrossRef]
44. Ali, A.; Yeoh, W.K.; Forney, C.; Siddiqui, M.W. Advances in postharvest technologies to extend the storage life of minimally processed fruits and vegetables. *Crit. Rev. Food Sci. Nutr.* **2018**, *58*, 2632–2649. [CrossRef]
45. Emer, Z.; Akbas, M.Y.; Ozdemir, M. Bactericidal activity of ozone against *E. coli* in whole and ground black peppers. *J. Food Prot.* **2008**, *71*, 914–917. [CrossRef]
46. Bialka, K.L.; Demirci, A. Utilization of gaseous ozone for the decontamination of *Escherichia coli* O157:H7 and *Salmonella* on raspberries and strawberries. *J. Food Prot.* **2007**, *70*, 1093–1098. [CrossRef]
47. Akata, I.; Torlak, E.; Erci, F. Efficacy of gaseous ozone for reducing microflora and foodborne pathogens on button mushroom. *Postharvest Biol. Technol.* **2015**, *109*, 40–44. [CrossRef]
48. Daş, E.; Candan-Gürakan, G.; Bayındırlı, A. Effect of controlled atmosphere storage, modified atmosphere packaging and gaseous ozone treatment on the survival of *Salmonella enteritidis* on cherry tomatoes. *Food Microbiol.* **2006**, *23*, 430–438. [CrossRef]
49. Gibson, K.E.; Almeida, G.; Jones, S.L.; Wright, K.; Lee, J.A. Inactivation of bacteria on fresh produce by batch wash ozone sanitation. *Food Control* **2019**, *106*, 106747. [CrossRef]
50. Jacques, A.C.; Zambiazi, R.C.; Ávila-Gandra, E.; Krumreich, F.; Rickies da Luz, S.; Ribeiro Galvão Machado, M. Sanitization by Chlorine compounds and Ozone: Effect on Bioactive Compounds in Blackberry (*Rubus fruticosus*) cv. Tupy. *Rev. Ceres* **2015**, *62*, 507–515. [CrossRef]
51. Vlassi, E.; Vlachos, P.; Kornaros, M. Effect of ozonation on table grapes preservation in cold storage. *J. Food Sci. Technol.* **2018**, *55*, 2031–2038. [CrossRef]
52. Smilanick, J.L.; Margosan, D.M.; Mlikota, F. Impact of ozonated water on the quality and shelf-life of fresh citrus fruit, stone fruit and table grapes. *Ozone Sci. Eng.* **2002**, *24*, 343–356. [CrossRef]

53. Brodowska, A.J.; Śmigielski, K.; Nowak, A.; Czyżowska, A.; Otlewska, A. The Impact of Ozone Treatment in Dynamic Bed Parameters on Changes in Biologically Active Substances of Juniper Berries. *PLoS ONE* **2015**, *10*, e0144855. [CrossRef]
54. Karakosta, E.S.; Karabagias, I.K.; Riganakos, K.A. Shelf Life Extension of Greenhouse Tomatoes Using Ozonation in Combination with Packaging under Refrigeration. *Ozone Sci. Eng.* **2019**, *41*, 389–397. [CrossRef]
55. Alves, H.; Alencar, E.R.; Ferreira, W.F.S.; Silva, C.R.; Ribeiro, J.L. Microbiological and physical-chemical aspects of strawberry exposed to ozone gas at different concentrations during storage. *Braz. J. Food Technol.* **2019**, *22*, e2018002. [CrossRef]
56. Liu, Z.; Li, W.; Zhai, X.; Li, X. Combination of precooling with ozone fumigation or low fluctuation of temperature for the quality modifications of postharvest sweet cherries. *J. Food Process. Preserv.* **2021**, *45*, e15504. [CrossRef]
57. Artés-Hernández, F.; Aguayo, E.; Artés, F. Alternative atmosphere treatments for keeping quality of 'Autumn seedless' table grapes during long-term cold storage. *Postharvest Biol. Technol.* **2004**, *31*, 59–67. [CrossRef]
58. Horvitz, S.; Cantalejo, M.J. Application of ozone for the postharvest treatment of fruits and vegetables. *Crit. Rev. Food Sci. Nutr.* **2014**, *54*, 312–339. [CrossRef] [PubMed]

Article

Domestic Use Simulation and Secondary Shelf Life Assessment of Industrial *Pesto alla genovese*

Carola Nicosia [1], Patrizia Fava [1,2], Andrea Pulvirenti [1,2], Andrea Antonelli [1,2] and Fabio Licciardello [1,2,*]

1 Department of Life Sciences, University of Modena and Reggio Emilia, 42122 Reggio Emilia, Italy; carola.nicosia@unimore.it (C.N.); patrizia.fava@unimore.it (P.F.); andrea.pulvirenti@unimore.it (A.P.); andrea.antonelli@unimore.it (A.A.)

2 Interdepartmental Research Centre for the Improvement of Agro-Food Biological Resources (BIOGEST-SITEIA), University of Modena and Reggio Emilia, 42122 Reggio Emilia, Italy

* Correspondence: fabio.licciardello@unimore.it

Abstract: The secondary shelf life (SSL) is defined as the time after package opening during which the food product retains a required level of quality. The SSL, indicated in labels as "best if used within *x* days after opening", could lead to domestic food waste if not correctly evaluated. In this context, the SSL of two brands of industrial shelf-stable pesto products (with an indicated SSL of 5 days) was studied through a domestic use simulation performed in five households under two scenarios simulating real opening and storage conditions. The quality of pesto after opening was assessed through microbiological and sensory analyses, determination of instrumental colour parameters, pH and volatiles profiling. For both pesto sauces tested, a SSL $\geq$ 20 days was proven. Irrespective of the intensity of use (scenarios 1 and 2), the pesto was microbiologically stable: the maximum count for total aerobic mesophilic bacteria (TMB) observed during 20 days of storage was $9.64 \pm 1.7 \times 10^2$ CFU/g, starting from a commercially stable product. Colour parameters L* and ΔE did not change significantly during storage ($p > 0.05$), while the a* and BI values significantly changed ($p < 0.05$) during the first 5 days, and then stabilized during the rest of the household storage. Nevertheless, the slight colour modifications were not perceived by the sensory panel. Moreover, sensory assessors were not able to discern pesto samples stored for up to 20 days after first opening, from a just-opened reference sample, proving that the sensory appreciation of pesto was not influenced by the time after opening. The results of this study suggest the possibility to significantly extend or even omit the SSL indications for industrial pesto sauces. The objective assessment of SSL could have impressive practical outcomes both for the industry and the end user. The elongation of the SSL on the food label might increase food sustainability, thanks to the potential reduction of food wastes, thus giving added value to the commercial products. In addition, the end user could benefit the increase of the useful period for the food consumption after first opening, with significant domestic food waste reduction, reduced household stock turnover and consequent cost savings.

Keywords: household food waste; stability evaluation; sensory acceptability; period after opening (PAO)

Citation: Nicosia, C.; Fava, P.; Pulvirenti, A.; Antonelli, A.; Licciardello, F. Domestic Use Simulation and Secondary Shelf Life Assessment of Industrial *Pesto alla genovese*. *Foods* **2021**, *10*, 1948. https://doi.org/10.3390/foods10081948

Academic Editors: Matteo Alessandro Del Nobile and Amalia Conte

Received: 29 July 2021
Accepted: 18 August 2021
Published: 21 August 2021

1. Introduction

In the context of sustainability improvement in the food sector, various measures have been proposed, especially related to the mitigation of environmental impacts of processes and materials, the optimization of distribution and logistics, and the minimization of food losses and wastes (FLW) along the food chain [1]. It is widely accepted in the scientific community that FLW are responsible for a high fraction of global environmental impacts [2,3]. FLW represent an economic, social, and environmental issue, and for this reason, the EU has targeted the halving of food wastes by 2030, according to Sustainable Development Goals (SDGs) [4]. Specifically, SDG 12.3 aims to "by 2030, halve per capita global food waste at the retail and consumer levels and reduce food losses along production and supply chains, including post-harvest losses". The most recent estimates report that 17% of overall food production is wasted [5]. Most of this waste is produced in the downstream of the chain, especially at the household level: available data range from 33–38% [6], through 45% [7] to 61% of total FLW [5]. Recent evidence [5] has shown that consumer food waste has been significantly underestimated and that figures related to food waste at consumer and food service level (also referred to as "avoidable food wastes") are more than twice as much as previously estimated [6].

It should be highlighted that domestic food waste: (i) cannot find alternative uses; (ii) represents the highest fraction of total FLW and (iii) is responsible for the highest environmental impact, since wasted products represent the highest level of resource consumption and emissions. Indeed, the resource consumption necessary for food production is in vain when food is lost or wasted and misses its goal of human consumption [8,9]. Given the need for FLW prevention and based on the awareness of the importance of domestic food waste [10], it is urgent to adopt effective mitigation measures.

Household food waste reduction could be achieved by the downsizing of packages. Smaller packages, indeed, reduce the probability of not being able to consume the product within an appropriate time. However, this approach brings about a higher consumption of packaging materials per food unit. In many cases, it is the overly short secondary shelf life (SSL), i.e., the timespan from the first opening to unacceptability, that turns food products into wastes. This timespan, also referred to as "period after opening" (PAO) is mandatory for some cosmetics [11]; for foods, it is usually communicated to consumers through the package label in terms of instructions for use after the first opening with sentences such as "after opening, store refrigerated and consume within *x* days", where *x* ranges from 24 h to a few days, depending on the product category and, within the same product category, depending on the producer. Such indications, however, does not have any scientific support and may even mislead consumers, thus contributing to foods which are still perfectly suitable for consumption being discarded.

Pesto alla genovese is a traditional Italian sauce made from basil, olive oil, grated hard cheese, pine seeds, salt, and garlic commonly used as a dressing for pasta. It is widely available as a shelf-stable product in glass jars, with a shelf life of 2–3 years; however, its stability after opening has never been assessed. Therefore, the aim of this work was to estimate the secondary shelf life of Italian *pesto alla genovese* by simulating two levels of domestic use and storage in five different house environments and through monitoring microbiological, sensory and chemical quality descriptors. The study took into consideration two popular commercial brands and compared the experimental results with the respective label-indicated secondary shelf lives.

2. Materials and Methods

2.1. Products

Two shelf-stable commercial pesto sauces from different manufacturers, identified here as P1 and P2, were used in this study, both packaged in a 190-g glass jar with a metal cap and having a similar composition. The composition listed on the label of P1 was: sunflower oil, fresh basil 30%, cashews, Parmigiano Reggiano PDO 5%, corn fibre, whey powder, salt, milk protein, extra virgin olive oil, sugar, basil extract, natural flavours, lactic acid, garlic. P2 had the following composition: Genoese basil PDO 35%, sunflower oil, cashews, extra virgin olive oil 10%, Grana Padano PDO 6%, Pecorino Romano PDO 4%, salt, pine nuts, lactic acid, garlic and ascorbic acid.

Pictures of P1 and P2 are available in Figure S2, in the Supplementary Material. The average nutritional values of the two commercial pesto sauces are shown in Table 1. Both products had a residual primary (commercial) shelf life of at least one year when the experiments started. The SSL reported on the labels of both P1 and P2 was 5 days from opening, under refrigerated storage.

Table 1. Mean nutritional values of the used pesto brands (g per 100 g).

Constituents	Pesto 1 (P1)	Pesto 2 (P2)
Fat	46	48
Carbohydrate	9.8	3.2
Fibre	5.0	1.5
Protein	4.7	6.5
Salt	3.25	2.4

2.2. Simulation of Domestic Use and Sampling Plan

The SSL assessment was performed using a deterministic approach, as suggested by Nicoli and Calligaris [12], which consists in storing the opened food product under the expected environmental conditions (i.e., at home in the case of shelf-stable pesto) and the worst-case scenario. The pesto samples were divided into five lots, each referred to a different domestic environment (coded from A to E). Pesto jars were stored at ambient temperature until use, according to protocols described hereafter. Among the possible way to simulate the worst-case found in literature, repeating the container opening and closure during the household storage was the selected approach for this study [12]. Two levels of use (referred to as "scenarios", based on the number of openings and duration of each opening) were tested in the five home environments mentioned above, to simulate a real utilization by the consumer (Table 2), while the analytical determinations were carried out at the laboratories of the Department of Life Sciences of the University of Modena and Reggio Emilia. This is an innovative feature that distinguishes this study from most of the research in the field of SSL [13–15].

Table 2. Scenarios simulating consumer's utilization of pesto sauce.

	Duration of Opening (min)	Removed Amount	Time (min) at T_{amb} after Closing	Repetitions
Scenario 1	2	2 tablespoons	0	1
Scenario 2	3	2 tablespoons	30	3

Scenario 1 (S1) consisted of a single opening for each jar, corresponding to the beginning of SSL, hence referred to as time zero (t_0). Two tablespoons of pesto were removed from each jar, which remained uncapped for 2 min, then the jars were closed and stored in the domestic refrigerator, where they were kept for 20 days.

Scenario 2 (S2) consisted of three openings of each jar at 2-days intervals. At each opening, two tablespoons of pesto were removed from each jar, which remained uncapped for 3 min, then the jars were closed and kept at ambient temperature for 30 min before placing in the domestic refrigerator, where they were stored for 20 days.

Dataloggers (mini-TH, XS Instruments, Carpi, Italy) were used to monitor conditions of the five domestic refrigerators where samples were stored, by recording temperature and humidity at 1-h intervals up to 168 h (one week).

The opening procedure described above for each scenario was performed simultaneously on four pesto jars in each home environment. Following the opening protocols, pesto jars were stored in domestic refrigerators and delivered to the laboratory on day 5, 11, 16, and 20 after the first opening. On the scheduled day, microbiological, sensory, and chemical-physical analyses were performed on the pesto samples from the five domestic environments.

2.3. Microbiological Analysis

Microbiological analyses were performed to assess the degree of contamination of pesto samples during refrigerated storage after first opening, through total aerobic mesophilic and *Clostridium* counts.

Ten grams of each sample were diluted with 90 g of sterile physiological solution (0.9% NaCl) in sterile stomacher bags and homogenized in a laboratory Stomacher 400 blender

(Seward Limited, Worthing, UK) at high speed for 60 s. Serial dilutions were prepared in sterile physiological solution. The pour plate method was performed for the aerobic mesophilic count, using Plate Count Agar (PCA, Tryptic Glucose Yeast Agar, Biolife, Milan, Italy) and the plates were incubated at 30 °C for 48–72 h. For *Clostridium* counts, the prepared samples underwent thermal shock at 95 °C for 10 min, before pouring in plate with *Clostridium* Agar (Biolife). The plates were incubated anaerobically at 30 °C for 48 h. Anaerobic conditions were obtained using Oxoid AnaeroGen bags (Thermo Fisher Scientific Inc., Waltham, MA, USA) inside anaerobic jars. All experiments were performed in duplicate. The results are expressed as colony-forming units (CFU) per gram of pesto.

2.4. Sensory Evaluation

The sensory test was performed on day 5, 11, 16, and 20 after the first opening. A panel of 12 judges (six males and six females, aged 23 to 63) was selected among the staff of the Department. The panelists were familiar with sensory evaluation and had previous experience in testing food through the triangular test. The sensory analysis aimed to understand the acceptance of samples during household storage. Triangular tests were performed on pesto samples from each of the five domestic conditions, presented together with a reference sample, which belonged to the same production batch (same formulation, processing, and storage conditions) but freshly opened. The reference jar, even if not opened, was also stored in the refrigerator before the sensory evaluation. Each assessor was asked to evaluate five triplets of randomly coded samples. Each set contained two identical samples and one different sample, which was randomly selected between the stored pesto (from home environment A, B, C, D or E) and the control one. The objective of the sensory evaluation was to assess: (i) whether judges were able to discriminate the stored samples from the control (just opened) one, and (ii) which was the degree of acceptance of samples throughout the time after the first opening.

Randomization was applied in the presentation order for all the panelists. All the samples were served at the same temperature to avoid the "stimulus error" due to the preparation of the samples. The participants were provided with unsalted crackers and water to be used during the tasting session. The judges were asked to select the different sample within each triplet, and to express its overall acceptability on a scale from 0 (extremely dislike) to 10 (extremely like). Figure S1, found in the Supplementary Material, shows the pesto samples prepared for the sensory analysis.

A significance level of 5% was chosen ($\alpha = 0.05$), which means accepting a 5% risk of finding a difference when there is none. Using an α value of 0.05 and a 12-member panel, the minimum number of correct answers to reject the assumption of no difference is 8 [16].

2.5. Measurement of Colour Parameters

The determination of colour was carried out at day 5, 11, 16, and 20 after first opening. Colour was measured using a chroma-meter CR-400/410 (Minolta Camera, Co., Ltd., Osaka, Japan) with a window diameter of 8 mm, equipped with a standard illuminant D65, and 10° observer angle. The values of the CIEL*a*b* parameters L* (lightness), a* (redness/greenness), and b* (yellowness/blueness) were recorded.

The chroma-meter was calibrated using a white ceramic tile (Y: 93.7, x: 0.3134, y: 0.3195) before the determination. For the colour assessment, a standardized amount of pesto (8 g) was weighed, and the colour was measured on three different points of each sample and reported as mean value ± standard deviation.

The total colour difference (ΔE) and the browning index (BI) of samples were calculated as follows [17–19]:

$$\Delta E = \sqrt{(L^*{}_0 - L^*{}_t)^2 + (a^*{}_0 - a^*{}_t)^2 + (b^*{}_0 - b^*{}_t)^2} \quad (1)$$

$$BI = \frac{100 \times \left[\left[\frac{a^*_t + (1.75 \times L^*_t)}{(5.645 \times L^*_t) + a^*_0 - (3.012 \times b^*_t)}\right] - 0.31\right]}{0.17} \quad (2)$$

2.6. Determination of a_w and pH

Water activity (a_w) and pH were determined on the products at the time of first opening (t_0) and pH was also monitored at day 5, 11, 16, and 20 after first opening. The a_w of pesto samples was measured using the dew point AquaLab 4TE water activity meter (Meter Group, Inc., Pullman, WA, USA). A standardized amount of sample (4 g) was weighed and brought to ambient temperature before the a_w determination.

To measure the pH, the samples were homogenized with the physiological solution and the pH was measured using a CyberScan pH 310 (Thermo Fisher Scientific Inc., Waltham, MA, USA) after calibration with buffer solutions at pH 7.00 and 4.00 (Sigma-Aldrich, Merck KGaA, Darmstadt, Germany).

The a_w and pH were additionally measured on ten commercially available shelf-stable pesto sauces, to have an overview of these parameters within the product category. All a_w and pH determinations were conducted in triplicate.

2.7. HS-SPME-GC/MS Volatiles Profiling

Volatile organic compounds (VOCs) of pesto sauce were analyzed by headspace solid-phase micro extraction (HS-SPME) followed by gas-chromatography/mass spectrometry (GC-MS) analysis just after opening and following refrigerated storage for 20 d from the first opening, in order to assess possible changes in the volatile fraction.

Two grams of just-opened P1 and P2 pesto and of each of the five samples (one from each household) stored for 20 d after opening, were weighted into 25-mL screw-cap glass vials provided with Mininert© valves (Merck KGaA, Darmstadt, Germany). Vials were conditioned at 50 °C for 15 min in a thermoblock (Falc Instruments, Treviglio, Italy), then a divinylbenzene/carboxen/polydimethylsiloxane (DVB/CAR/PDMS) SPME fibre was exposed in the headspace for 30 min at the same temperature for the extraction of volatile compounds. Chromatographic separation of analytes was carried out by an Agilent GC-MSD (7890A/5975C, Agilent Technologies Inc., Santa Clara, CA, USA) provided with a Stabilwax-DA (0.25 mm i.d. × 30 m × 0.25 μm) capillary column (Restek, Milan, Italy). After extraction, fibers were desorbed for 3 min into the GC injector port set in splitless mode at 240 °C. The GC carrier gas used was helium at 1 mL/min and the detector temperature was set at 240 °C. GC oven temperature program was: start at 50 °C for 3 min, 5 °C/min until 160 °C, hold at 160 °C for 2 min, 20 °C/min until 240 °C, hold at 240 °C for 2 min. Peak identification was carried out by comparison with system libraries (Wiley, NIST). The analyses were performed in duplicate.

2.8. Statistical Analysis

Statistical analysis was performed by one-way analysis of variance (ANOVA) followed by Tukey's multiple range test ($p < 0.05$) using the SPSS statistical software (SPSS 20 for Windows, SPSS Inc., IBM, New York, NY, USA). The results were expressed as mean ± standard deviation (SD).

3. Results and Discussion

3.1. Simulation of Domestic Use

Jars of *pesto alla genovese* were opened, used, and stored in five households (labelled from A to E), as described in the Materials and Methods section. For each home environment, the refrigerator temperatures were monitored for one week using dataloggers. The temperature profiles during one week and the mean temperatures of the five households are reported in Figure 1.

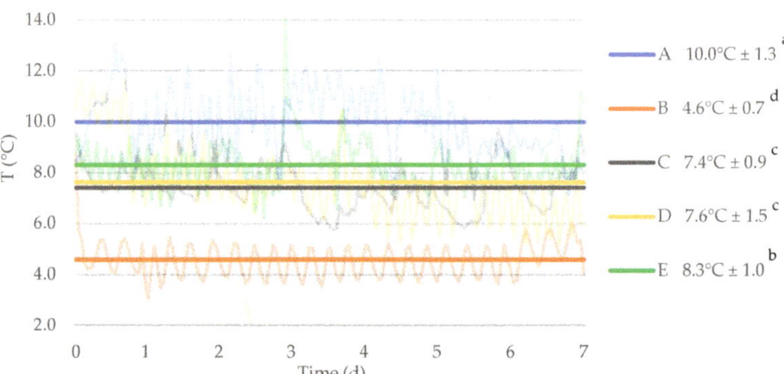

Figure 1. Temperatures (°C) recorded in the five domestic refrigerators (A to E) during one week. Values on the right are given as mean ± SD ($n = 168$). Faded lines represent the temperature profiles (data recorded at 1-h intervals), while the bold horizontal lines indicate the mean values measured for each household. Different superscript letters in the legend indicate significant differences ($p < 0.05$) among mean values.

The values measured in the refrigerators of the five households ranged between 3.1 °C (in household B) and 14.1 °C (in household E). As regards mean values calculated across the monitoring period, the minimum recorded temperature was 4.6 °C ± 0.7 (household B) while the highest mean value was 10.0 °C ± 1.3 (household A). Interestingly, significant differences ($p < 0.05$) were observed among the mean temperatures of the five domestic refrigerators. Some of the domestic refrigerators, such as the ones of households D and A, showed a higher fluctuation of temperatures (hence higher standard deviations) than others, whose temperatures oscillated in a tighter range, as for domestic refrigerators B and C. Overall, the households used for this study covered a wide range of temperatures and could be well representative of different possible conditions of domestic storage of food products after opening.

3.2. Microbiological Analyses

For the sake of conciseness, results of microbiological analyses have been condensed in Table 3, which shows the highest counts for total aerobic mesophilic bacteria (TMB) and Clostridia, expressed as CFU/g, observed in the five home environments on each sampling day. The microbial counts in the table are shown based on the scenario, type of pesto (conditions S1P1, S2P1, S1P2, S2P2), and category of microorganisms (total aerobic mesophilic bacteria and Clostridia).

Table 3. Maximum observed counts for total aerobic mesophilic bacteria (TMB) and Clostridia (Cl) amongst the five domestic environments, for each day of sampling, scenario, and pesto. S: Scenario (1, 2); P: Pesto brand (1, 2).

Time after Opening (d)	TMB (CFU/g × 10^1)				Cl (CFU/g)			
	S1 P1	S2 P1	S1 P2	S2 P2	S1 P1	S2 P1	S1 P2	S2 P2
5	96.4 ± 16.6 ^{a}	7.7 ± 7.0 ^{b}	10.4 ± 1.9 ^{b}	31.5 ± 1.3 ^{a}	n.d.	5.0 ± 7.1	n.d.	n.d.
11	2.7 ± 1.3 ^{b}	96.4 ± 3.8 ^{a}	1.8 ± 0.0 ^{b}	79.3 ± 14.0 ^{a}	n.d.	20.0 ± 0.0	n.d.	n.d.
16	3.2 ± 1.9 ^{ab}	13.5 ± 5.1 ^{a}	4.1 ± 0.6 ^{ab}	1.8 ± 0.0 ^{b}	n.d.	55.0 ± 21.2 ^{a}	n.d.	5.0 ± 7.1 ^{b}
20	2.3 ± 1.9 ^{b}	3.6 ± 0.0 ^{b}	0.9 ± 0.0 ^{b}	8.1 ± 0.0 ^{a}	5.0 ± 7.1	n.d.	n.d.	n.d.

Values are given as mean ± SD ($n = 2$). Different letters in the same line indicate significant differences ($p < 0.05$). If letters are not provided, no significant difference was observed. n.d.: not detected.

Since the shelf-stability of commercial pesto sauce results from a sterilizing heat process able to destroy all vegetative forms of microorganisms, the product remains commercially sterile until the first opening. For this reason, the degree of contamination of the samples depends on the contamination of the domestic environments in which the opening took place, while the storage temperature could influence the growth rate of microorganisms.

As it can be inferred, the maximum TMB count found within S1P1 was 964 CFU/g for sample C at 5 days after opening. For the same pesto and scenario (S1P1), on the following days, TMB counts did not exceed 32 CFU/g. This behaviour, in which the growth of microorganisms was not increasing throughout the storage time, was noticed more than once in this study. Probably, it could be due to the utilization of a different jar of pesto for each day of sampling, although they were opened at the same time, and treated and stored in the same conditions. In the second scenario of P1, the maximum TMB count (964 CFU/g) was recorded on the 11th day of household storage on sample E. It is worth recalling that each sample was analysed only once, and that at each sampling time a different jar was taken. Even if the approach does not allow to follow growth kinetics, the general trend of counts observed in both scenarios allows to state that *pesto alla genovese*, during the period after opening, has high microbial stability and that the contamination by environmental microbes occurring upon opening and use, is not followed by exponential growth at any storage temperature.

In the first scenario of P2, very low TMB counts were observed, with the highest count of 104 CFU/g for sample E at 5 d after the first opening. As for P1, also P2 showed lower TMB counts during storage, thus confirming the high stability of this product category. In the second scenario of P2, higher TMB counts were observed compared to S1, 793 CFU/g being the maximum values recorded for sample E after 11 d of household storage. Despite the higher degree of use in S2, consisting of 3 openings with product withdrawal, TMB counts remained at very low levels for any of the household conditions. This suggests that neither the household environment contamination nor the storage temperature, even in the worst scenario, can significantly affect the microbiological quality of *pesto alla genovese*.

The highest TMB count observed amongst all samples was 964 CFU/g, which is more than three orders of magnitude lower than the maximum acceptable microbial growth in food of 10^6 [20] and even lower than the threshold of 10^7 CFU/g, which indicates spoiled foods [21]. For the sake of completeness, also the maximum recorded counts were subjected to comparison by ANOVA, which highlighted significant differences among treatments, without any clear indication on the effect of the scenario on the microbial load. Significant changes were not found on the TMB counts amongst the scenarios and the type of pesto used ($p = 0.8$). On the other hand, despite significant changes during storage time were found ($p < 0.05$), it should be noted that in most of the cases the microbial load observed 20 d after opening was not significantly different from the one detected at day 5.

The *Clostridium* growth in S1P1 was observed only at day 20 of storage at a very small extent (5 CFU/g) in three samples out of five (A, B, C). In S2 it was slightly higher than the one observed in S1, with a maximum of 55 CFU/g after 16 d in sample D.

In both Pesto sauces P1 and P2, higher *Clostridium* counts were observed in the second scenario, possibly because of the higher number of openings and the longer exposure to the environment (jar kept uncapped), which could have increased the probability of contamination. Interestingly, in P1 it was noticed a higher *Clostridium* growth than the one observed on P2, in which only one sample (A at t_{16}) showed Clostridia. The different pH and a_w values of the pesto sauces, however, do not influence the degree of contamination by *Clostridium*, but the germination of spores that are already present. The different degree of *Clostridium* contamination between the two pesto sauces could be due to the household environment. To corroborate this hypothesis, *Clostridium* counts for S2P1 were only observed in household D, suggesting that this environment is the source of contamination during the openings of the jars.

The selective parameters (pH and a_w) of both pesto sauces are able to prevent the germination of *Clostridium* spores, and the *Clostridium* counts observed are a consequence of the germination induced by the thermal shock. The limiting pH for germination of *Cl. botulinum* spores is 4.8–5.0 [22] hence, the very low pH of P2 (4.18) avoids the germination of any *Clostridium* spore. On the other hand, the pH of P1 (5.64) alone could allow spore germination, but this event is unlikely thanks to the very low a_w value (0.9046), taking into account that in optimum conditions (30–40 °C and pH 7.0) the a_w limiting *Clostridium* spores' germination is 0.94–0.97 and by decreasing the pH of the medium, the limiting a_w increases [22].

3.3. Sensory Evaluation

Sensory evaluation was included in this study since sensory perception throughout the period after opening is complementary to hygienic quality assessment: indeed, the end of SSL could depend on microbiological and/or sensory thresholds, as well as primary shelf life can be determined as the lowest value between microbial acceptability limit and sensory acceptability limit [23].

The judges were asked to indicate the different sample within each triplet, anyway, the discrimination of the aged samples from the reference did not necessarily indicate non-acceptability; for this reason, panelists were asked to rate the acceptability of the sample recognized as different.

For P1, it was found that sample E was perceived as different from the reference (just-opened) sample after 11 d from opening, with eight correct answers out of 12. Nevertheless, the evaluations of the judges were still positive: 100% of the eight panelists graded the overall acceptability higher or equal to 6, with mean overall acceptability of 7.6. Moreover, 37.5% of the judges described the sample as: "creamy and with a delicate flavour", "softer flavour", "very fragrant". The level of acceptability within every single triangular test was compared with the acceptability value of the just-opened sample taken as a reference. At the other sampling times, S1P1 samples from the five households were not discriminated from the reference pesto. Concerning the second scenario (S2P1), only one sample (E, 20 d after first opening) was significantly perceived as different. As for S1P1, the sample recognized as different scored in all cases $\geq$6 for overall acceptability, with a mean value of 7.8. Two judges gave a grade of 9 and described the sample as and "very tasty" "having more delicate flavour". Based on the panelists assessment, the only two samples of P1 which were recognized as different were not perceived as worse than the control.

In the first scenario of P2, 5 d after opening, three samples (A, D, and E) were discriminated from the reference pesto. However, assessors who were able to recognize the difference indicated the aged samples as "less acid", "less bitter", "less sour", and "more fragrant" than the reference pesto. Similarly, 11 d after opening, sample B was recognized as different from the reference, but its overall acceptability (6.7) was even higher than that of the reference sample (5.6). At 16 and 20 d after opening, none of the samples resulted significantly different from the reference product. In S2P2 the only differences correctly recognized by the panelists (8–9 corrected answers out of 12) were after 11 (sample B) and 20 d (samples C and D) from opening. Anyway, the first sample received overall acceptability of 5.2, only 0.5 points less than the reference (5.7). Samples C and D were graded 6.3 and 7.2, respectively, compared to 6.2 of the reference.

As for P1, the analysis of the judges' evaluations for P2 showed that, in most cases, aged samples could not be discriminated from the reference (just-opened) sample and that differences correctly assigned did not indicate a decrease of sensory quality of pesto during domestic storage in none of the tested scenarios.

In general, the sensory analysis proved that assessors were generally not able to discriminate just-opened pesto from the product which had been opened up to 20 days before, irrespective of the degree of use. For both pesto brands, some of the judges found it difficult to assess differences amongst the samples, even in the worst scenario (S2) and after 20 d from first opening. Figure S2 (Supplementary Material) shows pictures of the two commercial pesto sauces stored for 20 days after first opening.

No relation between the degree of microbial contamination and sensory perception of pesto samples was observed: indeed, on one hand, some samples with null growth or very low counts were correctly discriminated (8 corrected answers out of 12) while, on the other hand, the samples with the highest microbial contamination were not discriminated from the reference pesto by the sensory analysis.

3.4. Colour Measurement

Colour modifications during storage are mainly due to degradative processes such as phenols oxidation, chlorophylls degradation, and nonenzymatic browning [20]. The instrumental colour parameters L* and a* and the derived parameters ΔE, and BI were assessed to evaluate colour modifications of pesto during 20 d of household storage after first opening. The b* parameter was not considered since several authors suggested using a* for green fruits and vegetables, because of its correlation to both green intensity and consumer acceptance [24,25].

Significant changes in lightness (L*) (Table 4) were found during domestic storage after opening only in a few samples of P1 and, in particular, for sample D in S1 and samples D and E in S2 ($p < 0.05$); however, this change was not perceived by the judges during the sensory evaluation. No significant difference was found between the two scenarios ($p = 0.08$). Similarly, for P2 lightness did not differ significantly between the two scenarios ($p = 0.44$) and significant differences of L* during storage occurred only for samples A and D in S1 ($p < 0.05$). It should be noted, however, that the L* value after 20 d from opening was not significantly different from the reference (just-opened) sample.

Table 4. Instrumental colour parameters (L* and a*) at time of first opening (t_0) and after 5, 11, 16 and 20 d after first opening (t_5, t_{11}, t_{16} and t_{20}) for both pesto sauces (P1 and P2) managed in 5 different home environments (A–E) under two scenarios (S1 and S2).

L*			t_5	t_{11}	t_{16}	t_{20}
P1	S1	A	43.44 ± 3.99	46.51 ± 0.40	44.28 ± 5.09	45.21 ± 4.69
t_0		B	45.43 ± 1.47	46.14 ± 1.49	46.57 ± 1.16	45.91 ± 1.90
47.98 ± 0.62 ^{a}		C	45.07 ± 1.69	48.04 ± 1.87	46.71 ± 0.98	45.79 ± 2.21
		D	43.79 ± 1.78 ^{b}	45.85 ± 0.40 ^{ab}	46.51 ± 0.58 ^{ab}	44.17 ± 1.97 ^{b}
		E	43.65 ± 3.91	45.51 ± 1.72	47.55 ± 0.62	45.03 ± 2.51
	S2	A	46.68 ± 3.47	44.04 ± 1.08	45.47 ± 2.50	45.41 ± 0.37
		B	47.31 ± 0.28	45.02 ± 2.78	46.54 ± 0.37	47.65 ± 2.61
		C	46.40 ± 1.57	46.69 ± 0.69	48.53 ± 0.44	48.09 ± 0.59
		D	46.70 ± 1.14 ^{ab}	44.32 ± 0.67 ^{b}	45.56 ± 1.39 ^{ab}	45.96 ± 1.48 ^{ab}
		E	45.25 ± 0.45 ^{bc}	40.87 ± 0.86 ^{d}	47.74 ± 0.59 ^{ab}	43.80 ± 1.84 ^{c}
P2	S1	A	47.29 ± 1.26 ^{a}	45.48 ± 0.89 ^{ab}	45.68 ± 0.69 ^{ab}	44.04 ± 1.95 ^{b}
t_0		B	44.54 ± 1.23	45.73 ± 1.11	45.70 ± 0.90	44.02 ± 0.42
44.34 ± 0.59 ^{ab}		C	46.04 ± 2.04	46.29 ± 0.58	46.42 ± 1.86	44.13 ± 1.37
		D	46.60 ± 1.01 ^{a}	45.45 ± 1.4 ^{ab}	43.45 ± 1.45 ^{b}	44.42 ± 0.67 ^{ab}
		E	45.23 ± 0.64	45.87 ± 0.74	43.28 ± 1.87	45.27 ± 1.50
	S2	A	44.92 ± 2.15	46.08 ± 1.87	45.43 ± 2.19	45.20 ± 2.14
		B	45.52 ± 2.54	44.95 ± 1.37	45.93 ± 1.07	45.71 ± 1.62
		C	46.13 ± 0.85	44.97 ± 3.31	46.33 ± 2.96	48.08 ± 0.62
		D	46.77 ± 2.78	45.70 ± 2.47	45.82 ± 0.87	46.40 ± 1.69
		E	45.24 ± 3.63	44.74 ± 2.83	44.73 ± 1.27	46.06 ± 0.76
a*			t_5	t_{11}	t_{16}	t_{20}
P1	S1	A	−4.16 ± 0.41	−3.79 ± 0.13	−4.09 ± 0.38	−3.46 ± 0.45
t_0		B	−3.99 ± 0.20	−3.90 ± 0.22	−3.99 ± 0.36	−3.54 ± 0.11
−3.82 ± 0.12 ^{ab}		C	−4.17 ± 0.18 ^{b}	−3.98 ± 0.11 ^{b}	−3.89 ± 0.40 ^{ab}	−3.32 ± 0.20 ^{a}
		D	−3.63 ± 0.16	−3.55 ± 0.37	−3.66 ± 0.35	−3.78 ± 0.25
		E	−3.97 ± 0.09	−3.84 ± 0.44	−3.81 ± 0.20	−3.90 ± 0.11
	S2	A	−4.14 ± 0.35	−3.81 ± 0.41	−3.23 ± 0.30	−4.03 ± 0.14
−3.82 ± 0.12 ^{bc}		B	−4.00 ± 0.12 ^{c}	−3.41 ± 0.10 ^{ab}	−3.28 ± 0.32 ^{a}	−3.83 ± 0.12 ^{bc}
		C	−3.94 ± 0.35	−3.82 ± 0.21	−3.84 ± 0.29	−3.60 ± 0.01
−3.82 ± 0.12 ^{b}		D	−3.57 ± 0.14 ^{ab}	−3.50 ± 0.26 ^{ab}	−3.17 ± 0.29 ^{a}	−3.66 ± 0.24 ^{ab}
		E	−3.52 ± 0.31	−4.26 ± 0.23	−3.34 ± 0.29	−3.55 ± 0.63

Table 4. *Cont.*

a*			t_5	t_{11}	t_{16}	t_{20}
P2	S1	A	−3.85 ± 0.27^{b}	−3.73 ± 0.24^{b}	−3.69 ± 0.07^{b}	−3.50 ± 0.42^{ab}
t_0		B	−3.53 ± 0.20^{ab}	−3.62 ± 0.44^{ab}	−3.84 ± 0.20^{b}	−3.36 ± 0.20^{ab}
−3.0 ± 0.13^{a}		C	−3.65 ± 0.40^{ab}	−3.54 ± 0.14^{ab}	−4.01 ± 0.14^{b}	−3.57 ± 0.37^{ab}
		D	−3.86 ± 0.22^{b}	−3.67 ± 0.10^{b}	−3.61 ± 0.32^{b}	−3.58 ± 0.12^{b}
		E	−3.45 ± 0.32	−3.59 ± 0.16	−3.47 ± 0.22	−3.42 ± 0.57
	S2	A	−3.54 ± 0.24^{b}	−3.63 ± 0.13^{b}	−3.27 ± 0.18^{ab}	−3.33 ± 0.10^{ab}
		B	−3.68 ± 0.46	−3.33 ± 0.18	−3.38 ± 0.17	−3.59 ± 0.34
		C	−3.75 ± 0.11^{bc}	−3.50 ± 0.17^{b}	−3.40 ± 0.23^{b}	−3.97 ± 0.15^{c}
		D	−3.63 ± 0.23^{ab}	−3.89 ± 0.30^{b}	−3.65 ± 0.27^{ab}	−3.46 ± 0.33^{ab}
		E	−3.90 ± 0.18	−3.14 ± 0.54	−3.36 ± 0.36	−3.40 ± 0.33

The t_0 values were repeated only for samples having significance letters different than displayed. Values are given as mean ± SD (n = 3). Different letters in the same line indicate significant differences ($p < 0.05$). If letters are not provided, no significant difference was observed.

Since no significant difference was observed between scenarios for neither of the two pesto sauces, it can be inferred that the conditions of use did not influence the lightness value during the time after opening. Overall, the lightness value of the two pesto sauces underwent only slight modifications in a few samples, which however were perceived as acceptable by the sensory panel, hence not influencing the suitability for consumption of the products.

The parameter a* represents the red-green axis of the CIEL*a*b* colour system, in which negative values indicate the greenness colour intensity [19]. This value was used as a quality parameter to evaluate the colour changes of fresh pesto [24] and in the shelf life assessment of pesto [20]. Values of a* of P1 and P2 during storage after opening are reported in Table 4.

Significant changes ($p < 0.05$) were found during domestic storage after opening only in sample C for S1P1 and in samples B and D for S2P1. Anyway, by considering altogether the values of the samples from the five domestic environments and of both scenarios, the a* parameter of P1 was not significantly different during household storage (p = 0.08). A significant difference during time after opening was observed for S1P2 in samples A, B, C, and D and for S2P2 in samples A, C, and D. In general, changes in a* occurring in P2 highlight a slight decrease (i.e, a slight increase of greenness) during the first 5 days, followed by stabilization. A comparison between a* values from the two scenarios highlighted no significant differences at each sampling time (p = 0.46).

Zardetto and Barbanti [20] studied the variation of greenness in fresh pesto, finding that the a* values increased during the primary shelf life. The difference with our findings may be due to the different products since the authors focused on a refrigerated, minimally processed pesto. On the other side, the pesto sauces used in this study had undergone a more intense thermal treatment (allowing shelf-stability) with subsequent degradation of chlorophyll and colour modification: as described by Zeppa and Turon [24], indeed, the thermal treatment causes an increase of the a* value of pesto.

Often, derived colour indexes have proved to be more useful to assess colour changes compared to primary parameters; for this reason, the overall colour variation (ΔE) and the browning index (BI) were calculated and the trend throughout storage has been reported in Figure 2. For both P1 and P2, the ΔE values did not differ significantly between the two scenarios (p = 0.29 and 0.16, respectively), whereby the ΔE values of both scenarios of the same pesto sauce have been condensed and shown in a single broken line, each for one type of pesto. Statistical analysis revealed differences throughout the period after opening only for P1, however, mean values at 5 and 20 d were not significantly different. In general, it can be inferred that ΔE for both pesto sauces remained stable during refrigerated storage after opening.

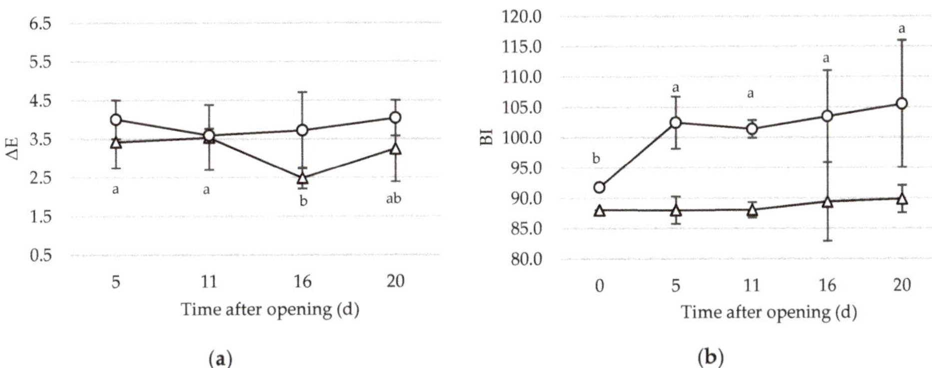

Figure 2. (**a**) Overall color variation (ΔE) and (**b**) Browning Index (BI) for pesto 1 (triangle) and pesto 2 (circle) during household storage after first opening. Each line reports the mean value of samples from the five households and managed under the two scenarios. Different letters within each line indicate significant differences ($p < 0.05$). If letters are not provided, no significant difference was observed.

BI can reveal a possible browning of the product, which in pesto sauce is a consequence of a nonenzymatic process, due to the reaction between proteins and oxidized lipids [20]. In P1, no significant difference of BI values between S1 and S2 was observed ($p = 0.27$), as well as during the period after opening ($p = 0.32$). On the other hand, the BI of P2 was found to change significantly during the first 5 days of storage after opening ($p < 0.05$). Similarly to P1, no significant difference ($p = 0.13$) was observed between the scenarios. The difference in BI between the two types of pesto may be due to their different a_w value, which is considerably lower in P1 (0.9046) as compared with P2 (0.9554): indeed, as suggested by Severini et al. [26], nonenzymatic browning is strictly related to a_w values.

Overall, results suggest that the primary and derived colour parameters of both shelf-stable pesto brands after opening were not influenced by the level of domestic use. Moreover, colour changes are hardly noticeable throughout the period after opening, and for this reason, they were not perceived by the sensory panel.

3.5. Chemical Analysis

The lowering of a_w and the reduction of pH are crucial factors for the stabilization of pesto sauce, to achieve microbiological stability and lowering degradative reactions such as browning, lipid oxidation, enzymatic reactions and protein denaturation [26]. The stabilization of pesto sauce is reached both by lowering the pH and/or the a_w, depending on the producer strategy and on possible taste consequences of the formulation. Low values of a_w are generally reached by the addition of humectants such as sugars, salts, polyols and protein derivates [26], while the pH is lowered by formulation with lactic, citric and ascorbic acid, and glucono-δ-lactone.

Along with the two pesto brands object of this study, the pH and a_w of 8 other commercial brands of shelf-stable *pesto alla genovese* sauces were determined. Results are provided in Table 5, which show a_w ranging from 0.8447 (in P10) to 0.9667 (P9) and pH values ranging from 4.00 (P3) to 5.64 (P1). The a_w values of the pesto sauces used in this study are higher than reported in the literature by Fabiano et al. [27] for industrial pesto (0.82), while they agreed with data reported by Severini et al. [26] (ranging from 0.914 to 0.956). As it can be inferred, the pH and a_w values of the samples object of the study (P1 and P2) fall in the range of commercial products, except for the pH of P1 which was the highest among commercial brands. Hence, based on stability-related parameters, P1 and P2 well represent the pesto category, and the observed results for these two brands may be extended to the commercial product category of pesto sauces.

Table 5. a_w and pH values of ten types of commercial shelf-stable pesto sauces. Values are given as mean ± SD ($n = 3$). Different letters within each column indicate significant differences ($p < 0.05$). P1 and P2 were used in this study for the secondary shelf life assessment.

Pesto Brand (Code)	pH	a_w	Indication of SSL (d)
P1	5.64 ± 0.07 ^{a}	0.9046 ± 0.0004 ^{g}	5
P2	4.18 ± 0.05 ^{f}	0.9554 ± 0.0010 ^{b}	5
P3	4.00 ± 0.01 ^{g}	0.9579 ± 0.0006 ^{b}	7
P4	4.51 ± 0.03 ^{e}	0.9358 ± 0.0008 ^{d}	14
P5	4.47 ± 0.04 ^{e}	0.9239 ± 0.0006 ^{e}	few days
P6	5.45 ± 0.02 ^{b}	0.9574 ± 0.0012 ^{b}	3–4
P7	4.72 ± 0.04 ^{d}	0.9207 ± 0.0015 ^{f}	5
P8	4.43 ± 0.02 ^{e}	0.9508 ± 0.0008 ^{c}	4
P9	4.44 ± 0.02 ^{e}	0.9667 ± 0.0013 ^{a}	3
P10	4.91 ± 0.02 ^{c}	0.8447 ± 0.0008 ^{h}	10

The monitoring of pH (Figure 3) did not show significant differences based on the level of domestic use in both pesto sauces ($p = 0.21$ and 0.06 for P1 and P2, respectively), therefore the values from both scenarios were analyzed together. Very small changes, though statistically significant, were shown during the period after first opening ($p < 0.05$) in both pesto brands. In spite of that, the pH of Pesto samples after 20 d from the first opening was very similar to the initial one, thus confirming the intrinsic stability of the products.

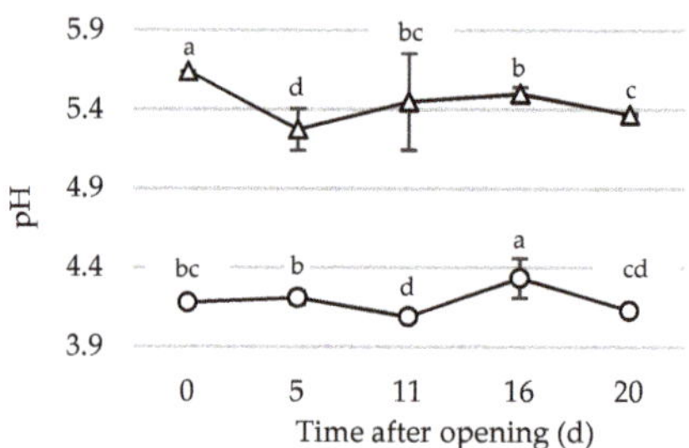

Figure 3. Trend of pH values for Pesto samples (P1: triangle; P2: circle) during household storage after first opening. Each line reports the mean value of samples from the five households and managed under the two scenarios. Different letters within each line indicate significant differences ($p < 0.05$).

3.6. Volatile Organic Compounds Profiling

Pesto alla genovese is appreciated for its flavour, which results from its specific formulation based on basil, hard cheese, garlic, and nuts. The sensory perception of Pesto might change after opening due to loss of key odorants, and oxidation reactions of volatiles and/or fatty components. The sensory assessment of pesto samples during the period after opening did not reveal significant changes perceived by tasters. A further investigation on the volatile fraction was aimed at confirming the stability of pesto during the time after opening and at determining possible changes which are not detected by sensory assessment. Hence, the volatile profiles of P1 and P2 sampled at the time of first opening were compared with those of samples stored refrigerated for 20 d after first opening, managed under scenario S2, which implied a more intensive level of use (worst case).

Figure 4 shows typical chromatograms of pesto (Figure 4a for P1 and Figure 4b for P2, respectively). Overall, more than 70 peaks were identified, mostly related with basil volatile composition. Slight differences occurring in the volatile profile of P1 and P2 result from the differences in formulation (as reported in the Materials and Methods section). Among the major peaks, linalool, estragole (or methyl chavicol, only in P1), eugenol, eucalyptol, *trans*-α-bergamotene, *trans*-β-ocimene, β-myrcene arise from basil [28,29], as the major ingredient in pesto formulation, representing about 72–83% of the overall chromatographic area.

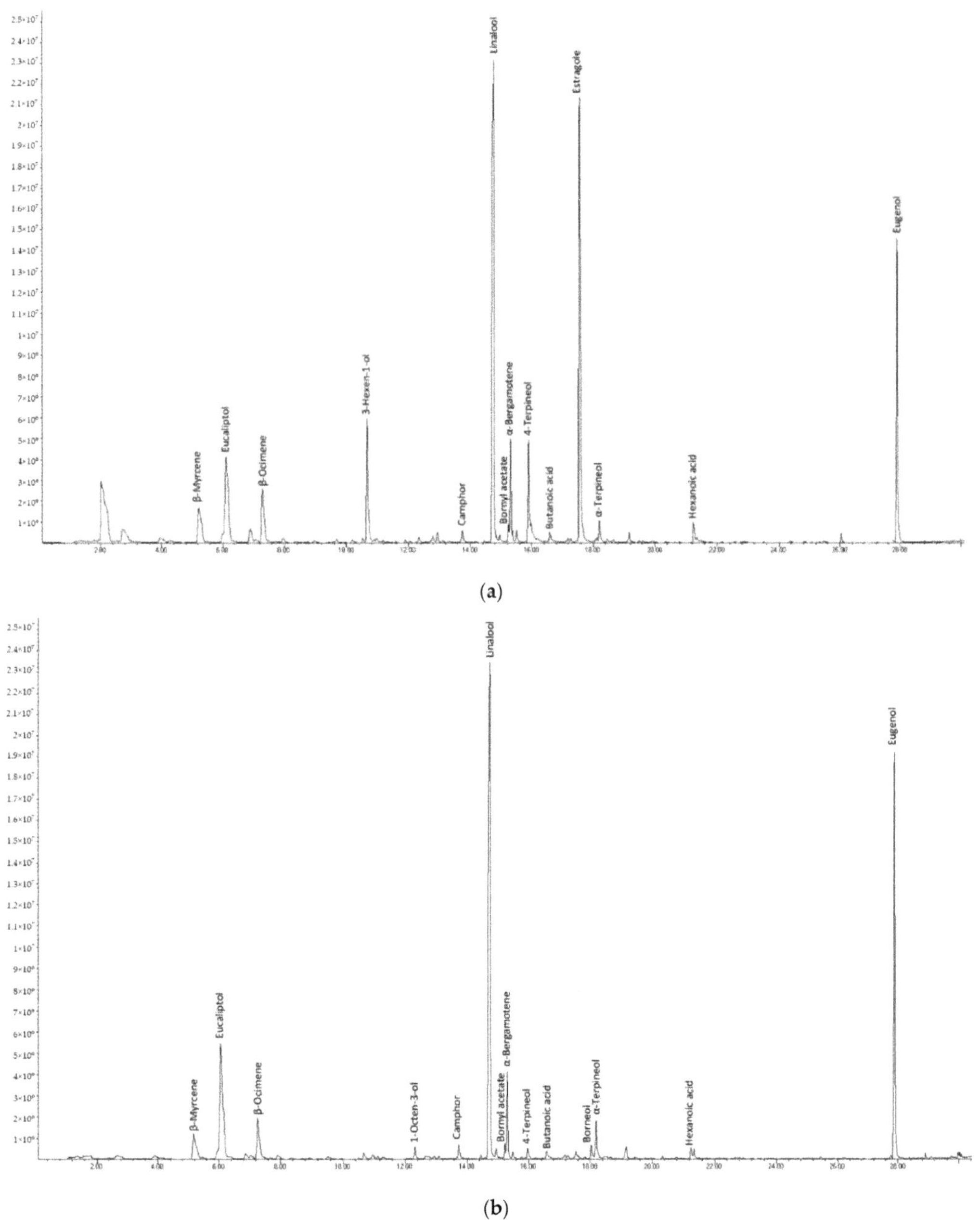

Figure 4. Typical chromatogram of the volatile fraction of pesto (**a**) P1 and (**b**) P2 as obtained by SPME-GC/MS.

The most representative volatile compounds, contributing about 86–90% of total chromatographic area, were selected for comparison between the reference (just-opened) sample and aged (20 d from first opening) sample for P1 (Figure 5a) and P2 (Figure 5b). As it can be inferred, the relative abundance of dominant compounds for aged samples basically reflects the profile of just-opened samples, thus confirming the substantial stability assessed through the sensory evaluation. For P1, the relative abundance of the key volatile compounds determined at the time of opening and after 20 d did not show significant difference ($p < 0.05$). For P2, instead, some significant differences ($p < 0.05$) were highlighted by comparison of the relative abundances at the time of opening and after 20 d, such as for linalool, which decreased with ageing, and for eucalyptol, limonene, and *trans*-β-ocimene which, instead, increased.

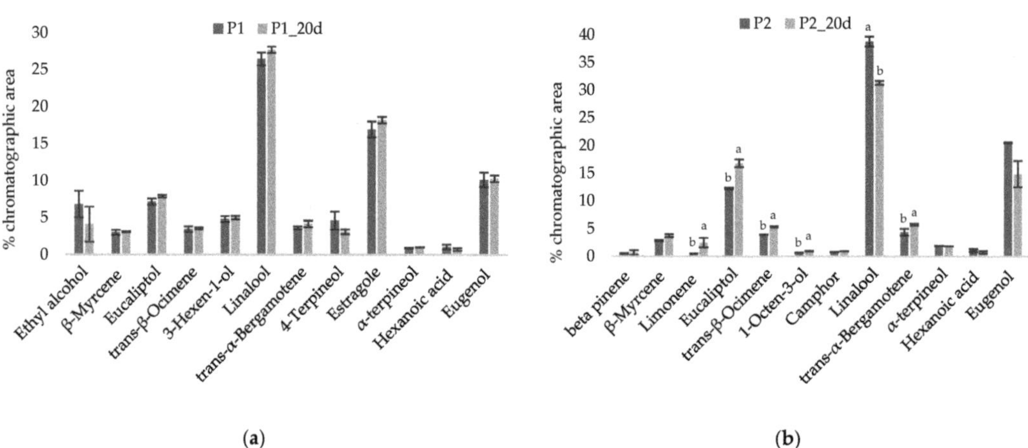

Figure 5. Comparison of the main volatile components (relative abundance) of (**a**) P1 and (**b**) P2 analyzed at the time of first opening and after 20 days from first opening (under S2). Different letters for each compound indicate significant differences ($p < 0.05$). If letters are not provided, no significant difference was observed.

4. Conclusions

This study sought to investigate possible changes occurring in industrial pesto sauce after package opening, by using a mixed approach consisting of a simulation of household conditions and a rigorous scientific investigation of microbiological, sensory, and chemical parameters. This work could offer a first reference methodology for the SSL assessment of a wide range of shelf-stable products.

Results demonstrate that industrial shelf-stable pesto, irrespective of the intensity of domestic use, can be still suitable for consumption after 20 d from the first opening, upon refrigerated household storage. Therefore, our findings suggest the possibility to extend the indication of SSL for the studied pesto sauces from 5 to 20 d. Since the chosen samples well represent the array of industrial pesto sauces commercially available, based on the comparison of pH and a_w, results could apply to the category of shelf-stable pesto sauces, having similar intrinsic parameters.

In a wider context, this work could have relevant practical outcomes both for the industry and the end consumer. The consciousness of the suitability for consumption even after the end of the indicated SSL could have consequences on the producer's decision concerning the SSL to be declared on the food label. Indeed, selling a product with an increased SSL means paying attention to food waste, giving the image of a sustainable business. Furthermore, the extension of the labelled SSL based on its objective assessment would contribute to add value to the packaged food product, without anyway modifying ingredients, formulation, or production process. This innovation might increase competitiveness, leading the consumer to choose the product which lasts longer, rather than other products of the same commercial category but with a lower duration after opening. Overall, a company might improve its market positioning through the reassessment of the SSL of its products.

In addition, the increase of the useful period for food consumption after first opening would bring advantages for the end user, leading to an improvement of the household food management, with consequent cost savings. The enhanced consumer awareness following the modification of SSL in the label might lead to lower food wastes generation.

In conclusion, through the objective assessment of SSL and its effective communication to the end user, this study could have practical potential on domestic food waste reduction and on the overall sustainability of food chains.

Supplementary Materials: The following are available online at https://www.mdpi.com/article/10.3390/foods10081948/s1, Figure S1: Samples of pesto sauce P2 prepared for the sensory analysis. Each of the five triplets consisted of two identical samples and one different sample, which was randomly selected between the stored pesto (11 days after opening, second scenario) and the control one, Figure S2: Pictures of the commercial pesto sauces (a) P1 and (b) P2, at the time of opening (t_0) and after 5, 11, 16 and 20 days of household storage.

Author Contributions: Conceptualization, F.L. and C.N.; methodology, F.L., C.N., A.P.; formal analysis, C.N. and F.L.; investigation, C.N., F.L., P.F., A.P.; resources, F.L., P.F. and A.P.; data curation, C.N. and F.L.; writing—original draft preparation, C.N. and F.L.; writing—review and editing, C.N., F.L., A.P., P.F.; A.A.; supervision, F.L., P.F., A.P.; funding acquisition, F.L., P.F., A.P., A.A. All authors have read and agreed to the published version of the manuscript.

Funding: This research was funded by the Italian Ministry of Agricultural, Food and Forestry Policy (MIPAAF), through the call n. 6146 of 20 November 2019.

Conflicts of Interest: The authors declare no conflict of interest.

References

1. Ingrao, C.; Nikkhah, A.; Dewulf, J.; Siracusa, V.; Ghnimi, S.; Rosentrater, K.A. Introduction to the special issue "Sustainability Issues of Food Processing and Packaging: The Role of Life Cycle Assessment". *Int. J. Life Cycle Assess.* **2021**, *26*, 726–737. [CrossRef]
2. Read, Q.D.; Brown, S.; Cuéllar, A.D.; Finn, S.M.; Gephart, J.A.; Marston, L.T.; Meyer, E.; Weitz, K.A.; Muth, M.K. Assessing the environmental impacts of halving food loss and waste along the food supply chain. *Sci. Total Environ.* **2020**, *72*, 136255. [CrossRef] [PubMed]
3. Tonini, D.; Albizzati, P.F.; Astrup, T.F. Environmental impacts of food waste: Learnings and challenges from a case study on UK. *Waste Manag.* **2018**, *76*, 744–766. [CrossRef] [PubMed]
4. General Assembly of the United Nations. *Transforming Our World: The 2030 Agenda for Sustainable Development*; UN: New York, NY, USA, 2015.
5. Forbes, H.; Quested, T.; O'Connor, C. *Food Waste Index Report 2021*; United Nations Environment Programme: Nairobi, Kenya, 2021.
6. Gustavsson, J.; Cederberg, C.; Sonesson, U.; van Otterdijk, R.; Meybeck, A. *Global Food Losses and Food Waste*; Food and Agriculture Organization of the United Nations: Rome, Italy, 2011.
7. Beretta, C.; Stoessel, F.; Baier, U.; Hellweg, S. Quantifying food losses and the potential for reduction in Switzerland. *Waste Manag.* **2013**, *33*, 764–773. [CrossRef] [PubMed]
8. Wohner, B.; Pauer, E.; Heinrich, V.; Tacker, M. Packaging-related food losses and waste: An overview of drivers and issues. *Sustainability* **2019**, *11*, 264. [CrossRef]
9. Licciardello, F.; Piergiovanni, L. Packaging and food sustainability. In *The Interaction of Food Industry and Environment*; Galanakis, C., Ed.; Elsevier Inc.: Amsterdam, The Netherlands, 2021; pp. 191–222. ISBN 9780128164495.
10. Licciardello, F. Packaging, blessing in disguise. Review on its diverse contribution to food sustainability. *Trends Food Sci. Technol.* **2017**, *65*, 32–39. [CrossRef]
11. UNION, P. Regulation (EC) No 1223/2009 of the european parliament and of the council. *Off. J. Eur. Union L* **2009**, *342*, 59.
12. Nicoli, M.C.; Calligaris, S. Secondary shelf life: An underestimated issue. *Food Eng. Rev.* **2018**, *10*, 57–65. [CrossRef]
13. Benković, M.; Tušek, A. Regression Models for description of roasted ground coffee powder color change during secondary shelf-life as related to storage conditions and packaging material. *Beverages* **2018**, *4*, 16. [CrossRef]
14. Manzocco, L.; Romano, G.; Calligaris, S.; Nicoli, M.C. Modeling the effect of the oxidation status of the ingredient oil on stability and shelf life of low-moisture bakery products: The case study of crackers. *Foods* **2020**, *9*, 749. [CrossRef]
15. Condurso, C.; Cincotta, F.; Merlino, M.; Stanton, C.; Verzera, A. Stability of powdered infant formula during secondary shelf-life and domestic practices. *Int. Dairy J.* **2020**, *109*, 104761. [CrossRef]

16. ISO. *ISO 4120:2004 Sensory Analysis-Methodology-Triangle Test*; International Organization for Standardization: Geneva, Switzerland, 2004.
17. Cefola, M.; D'Antuono, I.; Pace, B.; Calabrese, N.; Carito, A.; Linsalata, V.; Cardinali, A. Biochemical relationships and browning index for assessing the storage suitability of artichoke genotypes. *Food Res. Int.* **2012**, *48*, 397–403. [CrossRef]
18. Dadali, G.; Demirhan, E.; Özbek, B. Color change kinetics of spinach undergoing microwave drying. *Dry. Technol.* **2007**, *25*, 1713–1723. [CrossRef]
19. Mokrzycki, W.; Tatol, M. Colour difference ΔE—A survey. Faculty of mathematics and informatics. *Mach. Graph. Vis.* **2011**, *20*, 383–411.
20. Zardetto, S.; Barbanti, D. Shelf life assessment of fresh green pesto using an accelerated test approach. *Food Packag. Shelf Life* **2020**, *25*, 100524. [CrossRef]
21. Wijtzes, T.; Van'T Riet, K.; Huis In'T Veld, J.H.J.; Zwietering, M.H. A decision support system for the prediction of microbial food safety and food quality. *Int. J. Food Microbiol.* **1998**, *42*, 79–90. [CrossRef]
22. Baird-Parker, A.C.; Freame, B. Combined effect of water activity, pH and temperature on the growth of clostridium botulinum from spore and vegetative cell inocula. *J. Appl. Bacteriol.* **1967**, *30*, 420–429. [CrossRef]
23. Lucera, A.; Conte, A.; Del Nobile, M.A. Shelf life of fresh-cut green beans as affected by packaging systems. *Int. J. Food Sci. Technol.* **2011**, *46*, 2351–2357. [CrossRef]
24. Zeppa, G.; Turon, C. Valutazione mediante "Central Composite design" della degradazione della Clorofilla durante la pastorizzazione del pesto. *Ind. Aliment.* **2014**, *549*, 5–11.
25. Steet, J.A.; Tong, C.H. Degradation kinetics of green color and chlorophylls in peas by colorimetry and HPLC. *J. Food Sci.* **1996**, *61*, 924–928. [CrossRef]
26. Severini, C.; Corbo, M.R.; Derossi, A.; Bevilacqua, A.; Giuliani, R. Use of humectants for the stabilization of pesto sauce. *Int. J. Food Sci. Technol.* **2008**, *43*, 1041–1046. [CrossRef]
27. Fabiano, B.; Perego, P.; Pastorino, R.; Del Borghi, M. The extension of the shelf-life of "pesto" sauce by a combination of modified atmosphere packaging and refrigeration. *Int. J. Food Sci. Technol.* **2000**, *35*, 293–303. [CrossRef]
28. Licciardello, F.; Muratore, G.; Mercea, P.; Tosa, V.; Nerin, C. Diffusional behaviour of essential oil components in active packaging polypropylene films by multiple headspace solid phase microextraction–gas chromatography. *Packag. Technol. Sci.* **2013**, *26*, 173–185. [CrossRef]
29. Klimánková, E.; Holadová, K.; Hajšlová, J.; Čajka, T.; Poustka, J.; Koudela, M. Aroma profiles of five basil (*Ocimum basilicum* L.) cultivars grown under conventional and organic conditions. *Food Chem.* **2008**, *107*, 464–472. [CrossRef]

 *foods*

Article

Shelf Life Extension of Chilled Pork by Optimal Ultrasonicated Ceylon Spinach (*Basella alba*) Extracts: Physicochemical and Microbial Properties

Yuthana Phimolsiripol [1,2,*], Srirana Buadoktoom [1], Pimporn Leelapornpisid [3], Kittisak Jantanasakulwong [1,2], Phisit Seesuriyachan [1,2], Thanongsak Chaiyaso [1,2], Noppol Leksawasdi [1,2], Pornchai Rachtanapun [1,2], Nareekan Chaiwong [1,2], Sarana Rose Sommano [2,4], Charles S. Brennan [5] and Joe M. Regenstein [6]

1 Faculty of Agro-Industry, Chiang Mai University, Chiang Mai 50100, Thailand; s.buadoktoom@gmail.com (S.B.); kittisak.jan@cmu.ac.th (K.J.); phisit.seesuriyachan@gmail.com (P.S.); thanongsak.c@cmu.ac.th (T.C.); noppol@hotmail.com (N.L.); pornchai.r@cmu.ac.th (P.R.); meen.nareekan@gmail.com (N.C.)
2 Cluster of Agro Bio-Circular-Green Industry, Chiang Mai University, Chiang Mai 50100, Thailand; sarana.s@cmu.ac.th
3 Faculty of Pharmacy, Chiang Mai University, Chiang Mai 50200, Thailand; pim_leela@hotmail.com
4 Faculty of Agriculture, Chiang Mai University, Chiang Mai 50200, Thailand
5 School of Science, STEM College, RMIT University, Melbourne 3000, Australia; charles.brennan@rmit.edu.au
6 Department of Food Science, Cornell University, Ithaca, NY 14853-7201, USA; jmr9@cornell.edu
* Correspondence: yuthana.p@cmu.ac.th; Tel.: +665-394-8236; Fax: +665-394-8230

Citation: Phimolsiripol, Y.; Buadoktoom, S.; Leelapornpisid, P.; Jantanasakulwong, K.; Seesuriyachan, P.; Chaiyaso, T.; Leksawasdi, N.; Rachtanapun, P.; Chaiwong, N.; Sommano, S.R.; et al. Shelf Life Extension of Chilled Pork by Optimal Ultrasonicated Ceylon Spinach (*Basella alba*) Extracts: Physicochemical and Microbial Properties. *Foods* **2021**, *10*, 1241. https://doi.org/10.3390/foods10061241

Academic Editors: Matteo Alessandro Del Nobile and Amalia Conte

Received: 4 April 2021
Accepted: 26 May 2021
Published: 29 May 2021

Abstract: The effect of ultrasonication on the antioxidant and antibacterial properties of Ceylon spinach (*Basella alba*) extracts (CE) and the shelf life of chilled pork with CE were studied. The CE were ultrasonicated at different power levels (60–100%) for 10–40 min in an ultrasonic bath with the rise of antioxidant activities ($p \leq 0.05$) proportional to the ultrasonication time. The additional investigation of antibacterial activities showed that the ultrasonicated extracts (100 mg/mL) could inhibit and inactivate *Staphylococcus aureus* and *Escherichia coli* with the optimal condition of 80% power for 40 min. For shelf life testing, fresh pork treated with the ultrasonicated extracts at 100 and 120 mg/mL had lower values of thiobarbituric acid reactive substances (TBARS) than the control (without dipping). For food safety as measured by the total microbial count, the fresh pork dipped with 100–120 mg/mL CE extract could be kept at 0 °C for 7 days, 2 to 3 days longer than control meat at 0 and 4 °C, respectively. A sensory evaluation using a nine-point hedonic scale showed that fresh pork dipped with 100-mg/mL CE extracts was accepted by consumers. It is suggested that CE extracts can be applied in the food industry to enhance the quality and extend the shelf life of meat products.

Keywords: Ceylon spinach; *Basella alba*; ultrasonication; antioxidant; antibacterial activity; pork; *Sus scrofa*

1. Introduction

Pig meat (pork) is one of the most eaten meats in the world, and pork is a human food cooked or processed. The Food and Agriculture Organization (FAO) forecasted global pork production, the direction of pork production is predicted to rise to 131 million tons in 2028 from 121 million tons in 2018 [1]. The meat industry is focused on consumer awareness of meat production for food safety to prevent foodborne diseases, and microbial growth can lead to food spoilage [2]. Antioxidants have been applied in meat and meat products to reduce oxidation [3]. The interaction of natural antibacterial-active extracts and packaging or storage methods appears to be the most economically appropriate technology known as bio-preservation strategies [4,5]. Moreover, lipid and protein oxidation cause the loss of meat quality and a shorter shelf life. Lipid oxidation can produce effects in meat by changing the sensory properties [6]. In addition, de Souza de Azevedo et al. [7] also applied nisin by dipping or spraying for the shelf life extension of pork meat. The utilization of plant extracts as alternatives for meat preservation, including burgers during storage, will be beneficial for both the industry and consumers [8].

Ultrasonication is a green extraction technology that is cost-effective, adaptable, efficient, and effective for extracting natural food ingredients [9]. The extraction times and high temperatures can be mitigated with improved yields ensuring the preservation of the active ingredients [10]. Ultrasound-assisted extracts have been shown to possess greater antioxidant and antimicrobial properties than conventional extraction samples [11]. The acoustic cavitation of an ultrasound facilitates the cell permeability of solvents through damaged cells walls [12]. For example, Thai propolis, which was extracted using an ultrasound for 30 and 60 min, showed increased antibacterial activities against *Micrococcus luteus*, *Listeria monocytogenes*, and *Escherichia coli* [13], with the lowest IC_{50} (50% inhibitory concentration) for the scavenging DPPH radicals when ultrasonicated for 15 min. The ultrasonication method could also produce extracts with higher antibacterial activities against *S. aureus* and *Bacillus subtilis* than the extraction using maceration [14]. Moreover, combined Soxhlet and ultrasonication has also been used for oleaginous seed extraction, which can improve the conventional Soxhlet extraction, resulting in higher yields and shorter extraction times [15].

Ceylon spinach (*Basella alba*) is a popular local vegetable in Thailand rich in vitamins A and C, phenolic compounds, and several other antioxidants. It is low in calories (by volume) and high in protein [16]. *Basella alba* has a long history of use as an additive for food preservation [17], as well as medicinal compounds that are used in astringents, demulcents, laxatives, and soothing agents [18]. Kumar et al. [19] reported in vitro assays in preclinical and clinical studies that have shown that *Basella* has antibacterial, antihyperglycemic, anti-inflammatory, and antiproliferative activity and is cytotoxic. Maran et al. [20] investigated the extraction of *Basella rubra* L. pigments using an ultrasound. They confirmed that an extraction with 94-W ultrasound power at 54 °C for 32 min with a solid:liquid ratio of 1:17 g/mL resulted in the maximum yield of betacyanin (1.43 mg/g) and betaxanthin (5.37 mg/g). Furthermore, Adesina et al. [21] found that the levels of the phospholipids of *Basella alba* and *Basella rubra*, using a Soxhlet extraction, were 1680 and 1920 mg/100 g, respectively. However, the antioxidant and antibacterial activities of Ceylon spinach extracts using Soxhlet combined with ultrasonication and its application in the shelf life of pork have not been investigated. Therefore, the objectives of this study were to investigate the effects of ultrasonic power and time on the antioxidant and antibacterial properties of *Basella alba* stems and to investigate the shelf life extensions of chilled, fresh pork mixed with stem extracts.

2. Materials and Methods

2.1. Plant Materials

Fresh Ceylon spinach (*Basella alba*) stems at 1.5–2 months after planting were collected from the College of Agriculture and Technology field plots (Mueang, Chiang Mai, Thailand) during February 2019. The samples were washed with tap water and dried in a hot-air oven (UNB 400, Memmert, Eagles, WI, USA) at 50 °C for 24 h. The dried plants were ground using an electric grinder (BL3071AD, Tefal, Bangkok, Thailand) and stored at −20 °C before extracting within 3 months.

2.2. Chemicals and Reagents

2,2-Diphenyl-1-picrylhydrazyl (DPPH), 2,2′-azino-bis (3-ethylbenzothiazoline-6-sulfonic acid (ABTS), 2,4,6-tris(2-pyridyl)-s-triazine (TPTZ), gallic acid (GA), and 2-thiobarbituric acid were purchased from Sigma–Aldrich (Singapore, Singapore). Ferrous sulfate, Na_2CO_3, and the Folin–Ciocalteu reagent were bought from Loba Chemie (Mumbai, India). Nutrient broth, Mueller–Hinton broth (MHB), and Mueller–Hinton agar (MHA) were purchased from Himedia (Mumbai, India). Plate count agar (PCA) and peptone water were purchased from

Difco (Cockeysville, MD, USA). Other chemicals were analytical grade and obtained from RCI Labscan (Bangkok, Thailand).

2.3. Preparation of Extracts

Ceylon spinach powder (5 g) were extracted using a Soxhlet apparatus (Quicklet, Northern Ireland, UK) with 200 mL of 95% (*v*/*v*) ethanol at 80 °C for 4 h [16]. The extract was concentrated at 175-mbar reduced pressure in a water bath (B-490, Buchi, Saint Gallen, Switzerland) at 60 °C using a rotary evaporator (CH-9230, Buchi) to evaporate the ethanol, and then, the evaporated sample was further dried using a vacuum oven at 50 °C (VD53, Binder, Tuttlingen, Germany) to obtain dry Ceylon spinach extracts (CE) following the method of Sulaiman et al. [22]. The CE (100-mg of dry, crude extract after the evaporation/mL of distilled water), which corresponded to 1:10 (*w*/*v*), were ultrasonicated following the method adapted from Hashemi et al. [23]. The experiment had two factors: the extraction time (10–40 min) and power (60–100%) and were arranged as a 3 × 3 factorial in a completely randomized design (CRD) with duplicates, as shown in Figure 1. The CE samples in the test tube were ultrasonicated using an ultrasonic bath (40 kHz, 150 W, SB25-12DTD, Drawell, Jacksonville, FL, USA) at 25 °C. The untreated ultrasonication sample was used as the control.

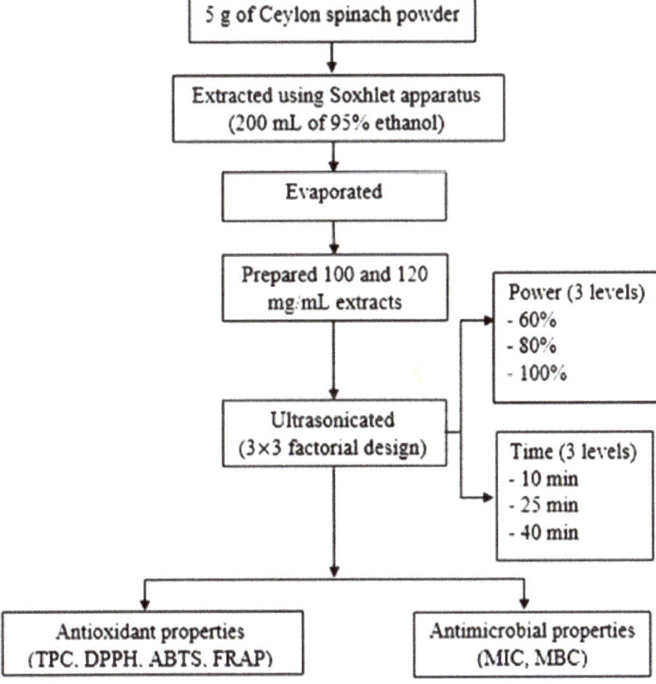

Figure 1. Design of the ultrasonic experiment.

2.4. Antioxidant Properties of Ultrasonicated CE

2.4.1. Total Phenolic Compounds (TPC)

TPC were determined as described by Oluwakemi et al. [24]. Briefly, 500 µL of the sample (100 mg/mL) was mixed with 2.5 mL of 10% Folin–Ciocalteu reagent and 2 mL of 7.5% Na_2CO_3. After incubation in the dark at 30 °C for 30 min, the absorption was measured at 765 nm using a multi-mode microplate reader (SpectraMax® i3x, Molecular

Devices, San Jose, CA, USA). Aqueous solutions of GA were used to prepare the calibration curve. Results (n = 3) were expressed as GA equivalents (E).

2.4.2. DPPH and ABTS Radical Scavenging Activity and Ferric-Reducing Antioxidant Power (FRAP)

Free radical scavenging activity was determined using the DPPH radical assay of Surin et al. [25]. The sample solution (2 mL) was mixed with 2 mL of 0.2-mmol/L DPPH solution. After incubation in the dark at 30 °C for 30 min, the absorbance was measured at 517 nm. The percentage of inhibition of the DPPH radical (n = 3) was calculated. The results were expressed as IC_{50} values, the lowest concentration of the sample required to inhibit 50% of the radicals.

For the ABTS radical scavenging assay of Chaiwong et al. [26], 100 µL of the sample from 0.5–5.0 mg/mL was mixed with 900 µL of 7-mM ABTS reagent. After incubation in the dark at 30 °C for 6 min, the absorbance was measured at 734 nm. The percentage of inhibition of the ABTS radical was calculated. The results (n = 3) were expressed as the IC_{50} values.

The reducing power was determined using the FRAP assay of Surin et al. [27], with some modifications. Briefly, 100 µL of the sample (100 mg/mL) was mixed with 1900 µL of FRAP reagent, which consisted of 2.5 mL of 10-mM TPTZ in 40-mM HCl, 2.5 mL of 20-mM $FeCl_3$ and 25 mL of 0.3-M acetate buffer (pH 3.6). The absorption was measured at 595 nm, and aqueous $FeSO_4$ solutions were used to prepare the calibration curve. The measurements were done in triplicate (n = 3). The untreated ultrasonication sample was used as the control.

2.5. Antibacterial Activities of Ultrasonicated CE Using the Minimum Inhibitory Concentration (MIC) and Minimum Bactericidal Concentration (MBC)

The MIC is the concentration of CE required to inhibit the growth of the tested microorganism. The ultrasonicated samples were prepared at 100 mg/mL of crude extract after evaporation, and serial dilutions were done to obtain solutions at 50, 25, 12.5, 6.25, and 3.125 mg/mL using the modified method of Kumar et al. [28]. The tested microorganisms—namely, *S. aureus* TISTR 2320, *E. coli* TISTR 527, *Salmonella* Typhimurium TISTR 1469, and *Pseudomonas aeruginosa* TISTR 2370—were obtained from the Thailand Institute of Scientific and Technological Research (TISTR, Bangkok, Thailand). The initial microbial samples were, at ~1×10^8 CFU/mL, obtained by adjusting the turbidity to match a 0.5 McFarland standard. Each sample solution (250 µL) was diluted with 250 µL of sterile MHB. The solution was inoculated with 250 µL of microbial suspension and then incubated at 37 °C for 24 h. Changes in the turbidity were measured at 600 nm for comparison with the control.

The MBC is the lowest concentration of extract with the ability to inactivate the tested microorganisms [28]. Six different concentrations of ultrasonicated CE were tested by streaking on MHA plates that were then incubated at 37 °C for 24 h. The lowest concentration of the plant extract required to inactivate the test microorganism was designated as the MBC value.

2.6. Shelf-Life Evaluation of Fresh, Chilled Pork

2.6.1. Preparation of Fresh, Chilled Pork

Fresh center-cut pork sirloins (Longissimus thoracis et lumborum, LTL) were selected and purchased from Charoen Pokphand Foods (Chiang Mai, Thailand) 1 day after slaughter and transported to the laboratory at 0–4 °C within 2 h. A Certificate of Analysis as a standard control for the fresh pork quality was provided by the company to ensure the safety of the product from the aerobic plate count, coliform bacteria, *E. coli*, *S. aureus*, *Clostridium perfringens*, *Salmonella* spp., *Enterococcus* spp., *L. monooxygenase*, *Campylobacter jejuni*, and yeast and molds following the standard US Food and Drug Administration protocols. The pork was sliced perpendicular to the long axis of the muscle with a knife into pieces of ~25 g with 2-cm thickness. Pork samples were dipped into two CE solutions with different concentrations (100 or 120 mg/mL) for 1 min. The preparation of the 100 or 120 mg/mL CE solution was done by dissolving 100 or 120 mg of dry ultrasonicated CE with 1 mL of distilled water. The optimum condition for ultrasonication was 80% power for 40 min, which was selected for the CE preparation. After dipping, the excess surface liquid on the samples were drained away and air-dried on a wire mesh at 20 °C for 10 min in a clean room before packing. The samples were kept in polyethylene (PE) trays (one/tray) and wrapped with food-grade polyvinyl chloride wrapping films (MMP, Bangkok, Thailand) with the experimental design shown in Figure 2. All samples were stored in two refrigerators (KB400, Binder, Bohemia, NY, USA) at 0 ± 1 and at 4 ± 1 °C and kept away from light for a period of 8 days and collected (10 pieces/treatment each time for different analyses) every day prior to comparison with the undipped control. Three pieces were used for the microbial analysis; another 3 pieces for the physicochemical analysis (color, pH, and TBARS); and the remaining 4 pieces were used for the sensory analysis.

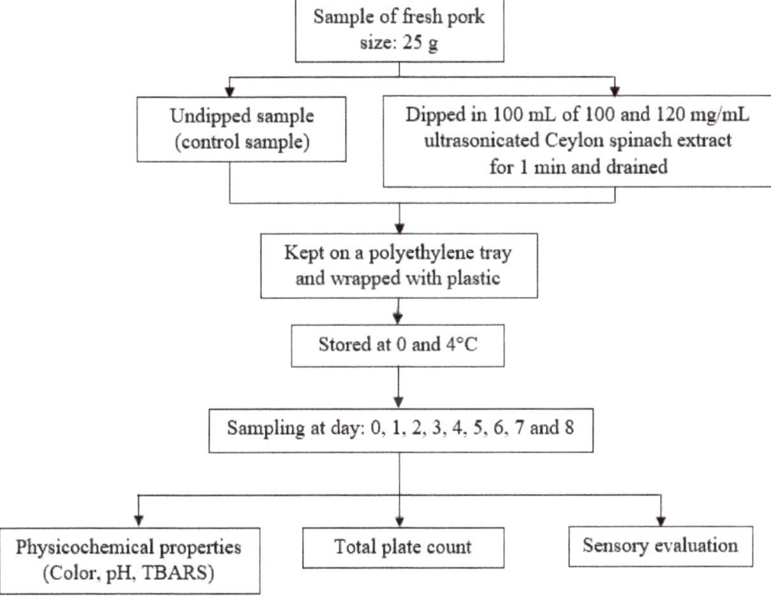

Figure 2. Design of the experiments for the shelf life study.

2.6.2. Color, pH, and TBARS Measurements

The color prior to blooming of the packed and chilled fresh pork after being unwrapped (n = 3), as mentioned by Sen et al. [29], was measured using the CIELAB system using a Chroma meter (CR-410, Konica-Minolta, Tokyo, Japan) and illuminant D65 observer angle of 2°; aperture size of 50 mm; and expressed as L*, a*, and b*. The lightness (L*) value indicates black at 0 and white at 100, the a* value represents the red (+60)–green (−60) color, and the b* value describes the yellow (+60)–blue (−60) color. A 25-g meat sample with 225-mL distilled water was homogenized with a hand blender (MSM64110, Bosch, Bangkok, Thailand) for 30 s at 30 °C before pH measurements (n = 3) using a pH meter (FiveEasy F20, Mettler Toledo, Greifensee, Switzerland). The pH meter was calibrated at 30 °C using pH buffers at 4.0 and 7.0 (RCI Labscan, Ltd., Bangkok, Thailand). The TBARS in chilled pork was determined using the method of Lekjing and Venkatachalam [30]. The meat (1 g) was homogenized using a Vortex mixer (VTX-3000L, LMS, Tokyo, Japan) in a 10-mL mixture of 0.375% (w/v) thiobarbituric acid, 15% (w/v) trichloroacetic acid, and 0.875% (w/v) of 0.25-M HCl. The mixture was heated at 100 °C for 10 min to develop a pink color and cooled down with tap water. The mixture was centrifuged (Rotina 380R, Hettich, Tuttlingen, Germany) at 1520× g at 25 °C for 15 min. The supernatant was measured at 532 nm. The results were determined from a standard curve (0–3000-μM malondialdehyde—MDA) and expressed as the mg MDA/kg of chilled pork ($n = 3$).

2.6.3. Total Plate Count Analysis

Pork samples (25 g) were added to 225 mL of 0.1% peptone water and blended in a Stomacher (IUL-Instruments, Barcelona, Spain) for 2 min ($n = 3$). Subsequent dilutions were prepared using 9 mL of 0.1% peptone water and 1 mL of sample prior to the application of the total plate count technique by pouring on PCA plates and incubated at 37 °C for 24 h [31,32]. The results were expressed as log CFU/g.

2.6.4. Sensory Evaluation

The sensory evaluation was carried out using 50 untrained panelists (male and female, 50:50) following the American Meat Science Association [33] and Vilar et al. [34] recommendations. The ages of the consumers ranged from 19–50 years old. The consumer preferences were evaluated using a 9-point hedonic scale (from 1 = extremely dislikely to 9 = extremely likely), as described by Phimolsiripol et al. [35] and Chokumnoyporn et al. [36]. Consumer sensory panels were done in the sensory normalized testing room at the Chiang Mai University Sensory Research Unit, which conformed to the international standards (ISO) [37]. The analysis was done for 6 samples of fresh, chilled pork, evaluated in two sensory sessions within the same day, which were kept at 0 and 4 °C for up to 8 days of storage. Samples were served on polyethylene trays at 25 °C in a random order. Each sample was coded with a three-digit random number. Water was available for use between samples. The attributes of the unwrapped pork samples were appearance, color, odor, and overall liking.

2.7. Statistical Analysis

The statistical analysis was applied following the method of Biffin et al. [38], with a slight modification. The antioxidant properties, including TPC, DPPH IC_{50}, ABTS IC_{50}, and FRAP, were analyzed and compared to determine the effects of the power and time of the ultrasonication compared to the non-ultrasonicated sample using an analysis of variance (ANOVA) at the 95% confidence level ($p \leq 0.05$). The experimental design was a factorial in the CRD model using the Statistical Package for the Social Sciences (SPSS version 17.0, SPSS, IBM Corp., Armonk, NY, USA). Mean comparisons were done using Duncan's post-hoc test at $p \leq 0.05$. Fixed effects in the full models included the power and time of the ultrasonication (treatments). The relationship or interaction terms between the responses as a function of the power and time of the ultrasonication and optimized optimal conditions were calculated using Design-Expert (Version 6.0.2, Stat-Ease, Inc., Minneapolis, MN, USA). Random terms for all the models included the extraction processing day and replication. Mean values and standard errors of the data were then reported.

For shelf life testing, the physical properties, including the color (L*, a*, and b*); pH; TBARS; and sensory data (appearance, color, odor, and overall liking), were calculated and compared using the ANOVA at the 95% confidence level ($p \leq 0.05$) using SPSS. The fixed variables for the full models included the concentration of ultrasonicated CE, storage temperature, and storage time treatments. Random terms were grouped according to their relations to the sensory panel (test day, session number, testing order, and panelist). The microbial analysis was regressed to a linear equation showing the kinetic reaction rate (k) to delineate the trends of the microbial populations after storage at varied ultrasonicated CE concentration and storage temperature levels, as described by Phimolsiripol et al. [39].

3. Results and Discussion

3.1. Antioxidant Properties of the Ultrasonicated Extracts

3.1.1. Total Phenolic Compounds (TPC)

The TPC of the CE ultrasonicated at 60% power for 40 min significantly increased ($p \leq 0.05$) when compared to the control, as shown in Figure 3a. Increasing the time of the ultrasonication increased the TPC, which is consistent with previous studies [10]. The results indicated disruption of the plant cell walls by ultrasound waves. The phenolic compounds were released from within the solid matrix. The 100% ultrasonication power resulted in a lower TPC when compared to the control and 60% or 80% ultrasonication power. An increase in the power of the ultrasonication decreased the TPC, presumably due to the destruction of some of the extracted phenolic compounds [40], resulting in a reduction in the TPC when applied with too strong an ultrasonic power.

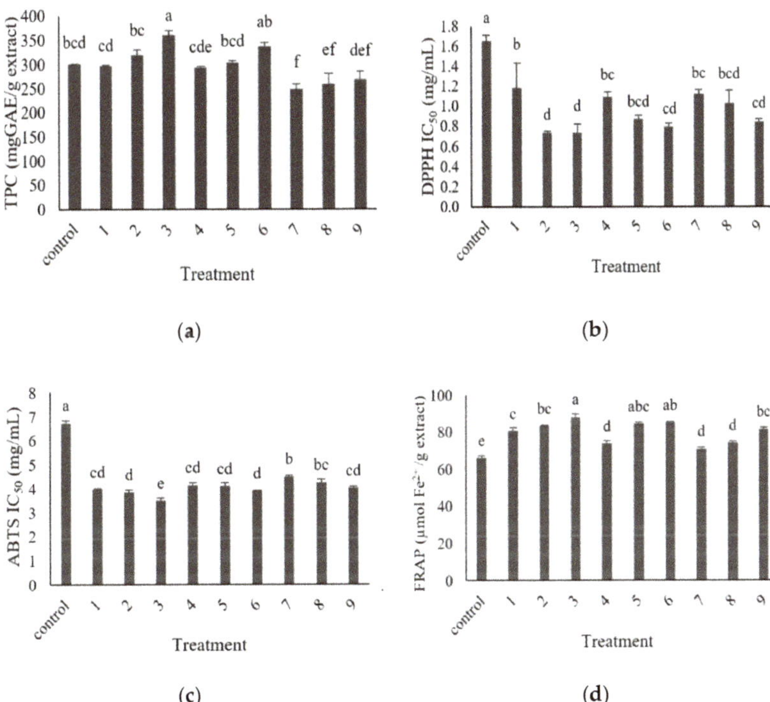

Figure 3. Means and standard errors of the antioxidant properties of the sonicated extracts' (**a**) Total phenolic compounds (TPC), (**b**) IC_{50} values of the 2,2-Diphenyl-1-picrylhydrazyl (DPPH) radical, (**c**) IC_{50} values of the 2,2′-azino-bis (3-ethylbenzothiazoline-6-sulfonic acid (ABTS) radical, and (**d**) FRAP—ferric-reducing antioxidant power (control = non-sonicated extract, 1 = 60% 10 min, 2 = 60% 25 min, 3 = 60% 40 min, 4 = 80% 10 min, 5 = 80% 25 min, 6 = 80% 40 min, 7 = 100% 10 min, 8 = 100% 25 min, and 9 = 100% 40 min). Different letters indicate significant differences between the treatments ($p \leq 0.05$).

3.1.2. DPPH and ABTS Radical Scavenging Activity and Ferric-Reducing Antioxidant Power (FRAP)

There was a significant variation in the DPPH IC_{50} values of the ultrasonicated extracts, as shown in Figure 3b. All ultrasonicated extracts showed higher DPPH radical scavenging activities than the non-ultrasonicated extracts (control). The extract with the highest DPPH radical scavenging activity was obtained by ultrasonication at 60% power for 25 and 40 min. Ultrasonication at 60% power for 10 min resulted in the extract with the lowest ($p \leq 0.05$)

DPPH antioxidant capacity. Increasing the time of the ultrasonication led to increasing the DPPH antioxidant activity for 60% power. These results were consistent with an earlier report by Altemimi et al. [41] that showed the degradation of phenolic compounds when using high powers and high temperatures, thereby producing a cavitation bubble collapse.

The ABTS radical scavenging activity was quantified as the reduction in $ABTS^+$ radicals and expressed as IC_{50} values (Figure 3c). The extracts with the highest and lowest potentials to inhibit the ABTS radicals were obtained by ultrasonication at 60% power for 40 min and 100% power for 10 min, respectively. The data showed that increasing the time of the ultrasonication increased the ABTS radical scavenging activity for the extracts ultrasonicated at 60% and 100% power. Higher ultrasonication times have been associated with higher flavonoid yields at the same ultrasonic power [10]. A longer extraction time permits more contact time for the cavitation bubbles to rupture more plant cells, in turn increasing the TPC extraction [40]. Therefore, the antioxidant capacities will increase. Ultrasound water baths produce enough cavitation to create shear forces to break the cell walls. Furthermore, ultrasonication increases the diffusion of cell contents into the extraction solution [42].

The FRAP of the extracts are shown in Figure 3d. The extract ultrasonicated at 60% power for 40 min showed the highest ($p \leq 0.05$) potential to decrease the ferric ions (Fe^{3+}), while the lowest potential to decrease the Fe^{3+} ions was observed with the extract ultrasonicated at 100% power for 10 min, suggesting that increasing the time of ultrasonication also increased the FRAP. Compared with the control (non-ultrasonicated sample), there was a significant increase (7–33%, $p \leq 0.05$) in the FRAP values after ultrasonication. This trend was consistent with Ilghami et al. [43]. It might be due to an increase in the ultrasonic times, which can increase the diffusivity of the solvent into cells and enhance the desorption and solubility of the target compounds from the cells, thereby improving the antioxidant efficacy [44].

3.2. Antibacterial Activities of Ultrasonicated Extracts Using MIC and MBC

The antibacterial properties of the ultrasonicated extracts were measured by determining the MIC values (Table 1). The lowest MIC value of all nine extracts able to inhibit the growth of *S. aureus* and *E. coli* was 100-mg/mL. When using 60% power, the extracts inhibited *S. typhimurium* and *P. aeruginosa* at a MIC of 100 mg/mL. When the power was increased, the MIC values against *S. typhimurium* and *P. aeruginosa* were 50 mg/mL. It was evident that increasing the power resulted in increasing the antibacterial activities against *S. typhimurium* and *P. aeruginosa*. The MBC defined the lowest concentration of the extract that could inhibit the tested microorganisms. The MBC values showed the results of an in vitro test in which the fixed concentration of the extracts was being tested against an initially fixed concentration of microorganism [45]. All extracts had an MBC of a 100 mg/mL against *S. aureus* and *E. coli*. When ultrasonicated at 60% power for 10, 25, and 40 min, the MBC values against *S. typhimurium* and *P. aeruginosa* were 100 mg/mL. At 80% and 100% power for 10, 25, and 40 min, the MBC values against *S. typhimurium* and *P. aeruginosa* decreased to 50 mg/mL (Table 1).

A comparison between the ultrasonicated and non-ultrasonicated (control) extracts showed that the antibacterial activities against *S. aureus* and *E. coli* were similar. Whereas the antibacterial activities against *S. typhimurium* and *P. aeruginosa* increased when the ultrasonication power was increased to 80% and 100% for 10, 25, and 40 min (Table 1). Ultrasonic waves cause pressure and cavitation with the disruption of cell walls, so that the components of interest can be released [46] and the extracts are more easily released [47]. The Gram-positive and Gram-negative bacteria showed different sensitivities due to their different cell wall structures. The Gram-negative cell envelope is a thin structure that is covered by an outer membrane. On the other hand, Gram-positive bacteria lack the outer cell membrane, and the cell wall is typically much thicker, with multiple peptidoglycan layers [48]. Due to these distinct differences, it may be easier to inhibit Gram-negative bacteria than Gram-positive bacteria. Therefore, the MBC values of the extracts against

S. typhimurium and *P. aeruginosa* were lower than the MBC value against *S. aureus*, which is a Gram-positive bacterium. However, the Gram-negative bacteria *E. coli* could tolerate a higher concentration of extracts than the other Gram-negative bacteria, because *E. coli* has the potential to form a dense biofilm around its cells, thus giving them protection against antibacterial agents [49]. As a result, the MBC values of the extracts against *E. coli* were higher than those for the other Gram-negative bacteria. Annatto dye has also been shown to have a greater antibacterial effect on Gram-positive bacteria (lower MIC and MBC) compared with Gram-negative bacteria [50]. These results were probably due to the presence of lipopolysaccharide in the cell wall of Gram-negative bacteria, which can prevent the influx of active compounds into the cytoplasmic membrane of these bacteria [51].

Table 1. Minimum inhibitory concentrations (MIC) and minimum bactericidal concentrations (MBC) of the ultrasonicated extracts.

Power (%)	Time (min)	MIC (mg/mL)				MBC (mg/mL)			
		S. aureus	*E. coli*	*S. typhimurium*	*P. aeruginosa*	*S. aureus*	*E. coli*	*S. typhimurium*	*P. aeruginosa*
60	10	100	100	100	100	100	100	100	100
60	25	100	100	100	100	100	100	100	100
60	40	100	100	100	100	100	100	100	100
80	10	100	100	50	50	100	100	50	50
80	25	100	100	50	50	100	100	50	50
80	40	100	100	50	50	100	100	50	50
100	10	100	100	50	50	100	100	50	50
100	25	100	100	50	50	100	100	50	50
100	40	100	100	50	50	100	100	50	50
Control (Non-ultrasonicated extract)		100	100	100	100	100	100	100	100

3.3. Optimization of Ultrasonication

The response surface plots (Figure 4) were used to visually observe the relationship between the responses and the various power and times of ultrasonication. The responses studied were TPC, IC_{50} values for the DPPH and ABTS radical scavenging capacities, and FRAP. The response surfaces (Figure 4a–d) were evaluated to predict the optimum power and time of ultrasonication. The optimum condition of ultrasonication was 80% power for 40 min, as shown in Figure 4e. With these conditions, the MIC and MBC values for *S. typhimurium* and *P. aeruginosa* were 50 mg/mL. The content of the TPC was 332-mg GAE/g extract. The IC_{50} values for DPPH and ABTS were 0.77 and 3.84 mg/mL, respectively, and FRAP was 84.7-µmol Fe^{2+}/g extract.

3.4. Shelf Life Evaluation of Fresh Pork

3.4.1. Color and pH Measurements

The L* values of all the chilled, fresh pork decreased significantly ($p \leq 0.05$, Figure 5a) with the increasing storage time. Fresh pork dipped in 120-mg/mL ultrasonicated extract showed lower L* values when compared to fresh pork dipped with 100-mg/mL ultrasonicated extract and the control (nontreated). In addition, the L* value tended to decrease during the increased storage time. Reduction of the L* values reflected pigment materials in the extracts, including chlorophylls [52]. Polyphenol oxidases could oxidize the phenolic compounds to quinones and quinones, which are likely to be condensed to form darker compounds [53]. The control samples (nontreated) showed greater a* values than fresh pork dipped in 100 and 120-mg/mL ultrasonicated extracts at 0–8 day of storage ($p \leq 0.05$, Figure 5b). Increasing the storage time resulted in an increasing redness (a*) of the control sample at day 8 of storage. The control samples showed lower b* values when compared to fresh pork dipped in the ultrasonicated extract. The values of b* were significantly decreased ($p \leq 0.05$) with the increasing storage time, as seen in Figure 5c. Increasing the lipid oxidation could lead to a decrease in the a* and b* values [54]. The changes of the pH values of pork when kept at 0 and 4 °C for 8 days are shown in Figure 5d. The pH values of pork significantly increased ($p \leq 0.05$) with the increased storage time. A similar pH trend was also reported by Lu et al. [31]. The increase of the pH values is associated with bacterial spoilage, related to the action of microbial enzymes, e.g., proteases and lipases, which increase the breakdown of nitrogenous compounds [55]. Changes in the pH may also be related to the spoilage of meat products. The pH usually decreases consequently with the bacterial growth and production of acid from lactic acid bacteria [56].

3.4.2. TBARS Measurement

The TBARS values of all the chilled, fresh pork increased significantly as the storage time increased ($p \leq 0.05$, Figure 6a,b). The changes in TBARS in Figure 6a,b show the natural logarithm plot, which indicated a first order-type reaction. The elevation of TBARS with the storage time was observed. Lu et al. [31] mentioned that increasing the storage time resulted in increasing the TBARS values. Fresh pork dipped in 100 and 120-mg/mL ultrasonicated extract showed lower TBARS values compared to the control sample. The results showed that the antioxidative properties of the ultrasonicated extract had the potential to retard the lipid oxidation in pork. Lekjing and Venkatachalam [30] investigated the effects of a chitosan-based coating at 2% incorporated with 1.5% clove oil (CS + CO) on cooked pork sausage samples. They showed that the TBARS values of the CS + CO-treated samples were lower than those of the 2% chitosan-treated samples at 10 and 15 days of storage. Lorenzo et al. [57] also reported that the TBARS values of refrigerated pork patties treated with BHT, green tea extract, seaweed extract, or grape seed extract were lower than in the control samples. The antioxidant properties of the plant extracts could be used to retard the lipid oxidation [58]. Chilled, fresh pork stored at 0 °C had lower TBARS values when compared to fresh pork stored at 4 °C. The lower temperature was effective in decreasing the lipid oxidation [31]

3.4.3. Total Plate Count Analysis

As mentioned by Kim and Jang [32], the total plate count is an important parameter for fresh pork shelf life testing. The total plate counts of chilled, fresh pork with the ultrasonicated extract dipping and nontreated fresh pork (control) are shown in Table 2. The initial microbial loads of the nontreated and treated samples were similar. The total plate counts of all the chilled, fresh pork samples increased linearly with the increased storage time ($p \leq 0.05$). For storage at 0 °C, the kinetic reaction rate (k) values for the control, 100, and 120 mg/mL were 0.698, 0.636, and 0.497 log CFU/g.day, respectively. While, for pork storage at 4 °C, the k values for the control, 100, and 120 mg/mL were 0.774, 0.700, and 0.681 log CFU/g.day, respectively. It showed that the 120-mg/mL CE had significantly delayed the growth of the total bacteria. For fresh, chilled pork, the total plate count standard [59] was 6.7 log CFU/g. The total plate counts of the control samples and fresh pork dipped in 100 and 120-mg/mL CE stored at 0 °C reached 6.7 log CFU/g by the 5th, 6th, and 7th days of storage, respectively, with visible signs of spoilage. Meanwhile, the control samples and fresh pork dipped in 100 and 120-mg/mL CE stored at 4 °C had total plate count values above the standard limit on the 4th, 5th, and 6th days of storage, respectively. A lower temperature could extend the storage time, as reported by Akoğlu et al. [60]. It was confirmed that the shelf life of chilled pork was prolonged by dipping with either 100 or 120-mg/mL of CE at a lower temperature (0 °C). This was probably due to the phenolic compounds in the ultrasonicated extracts. The phenolic compounds showed that the antimicrobial activity can increasingly against many Gram-positive bacteria. The Gram-positive bacteria showed a better susceptibility to antimicrobial activity, because the outer membrane of the Gram-negative bacteria represented a reduced absorption of the phenolic compounds and barrier of permeability [61]. They can denature the proteins of microbial cell membranes, leading to inactivation or death [62]. The microbial activities were consistent with the reports for different natural additives and their extracts in sausages [63] and bison meat [64]. The authors de Souza de Azevedo et al. [7] found that nisin can be used as an antimicrobial agent for shelf life extension in pork meat. Ranucci et al. [65] also reported sausage made from pork meat with a mix of *Punica granatum* and *Citrus* spp. extracts. They found that the extracts could extend the shelf life of pork by controlling the microbial growth and oxidation during refrigerated storage at 4 ± 1 °C.

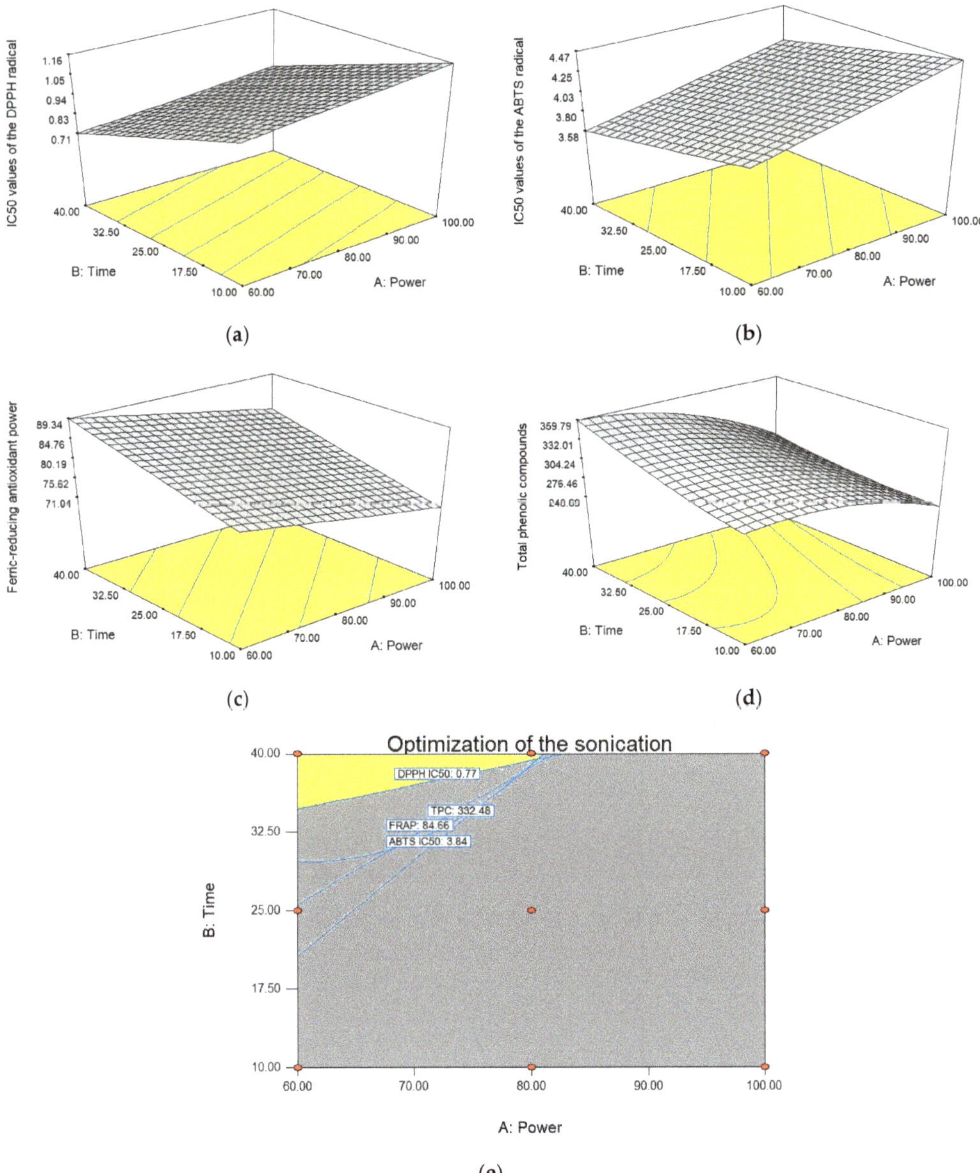

Figure 4. Response surfaces. (**a**) IC_{50} values of the DPPH radical, (**b**) IC_{50} values of the ABTS radical, (**c**) ferric-reducing antioxidant power, (**d**) total phenolic compounds, and (**e**) optimization of the sonication.

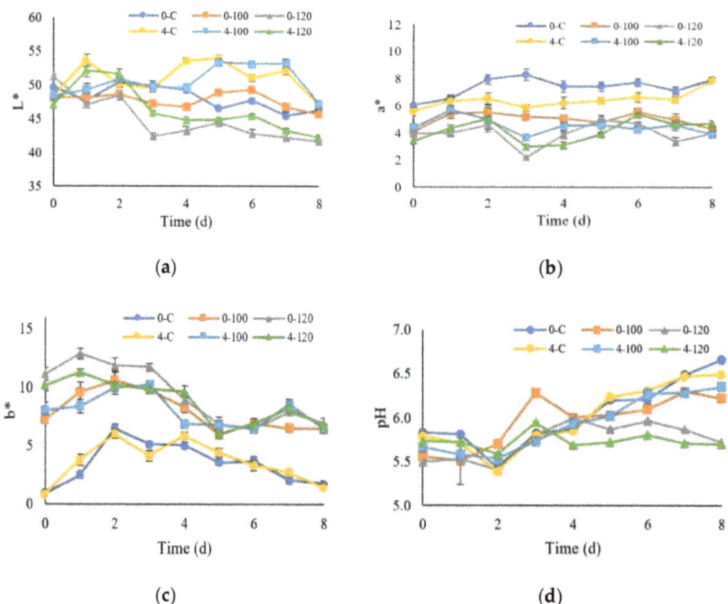

Figure 5. Changes in the means and standard errors of the colors (**a**) Lightness (L*), (**b**) Redness-greenness (a*), and (**c**) Yellowness-blueness (b*) and (**d**) pH when stored at 0 and 4 °C. (0-C = nondipped sample at 0 °C, 0-100 = 100-mg/mL dipped sample at 0 °C, 0-120 = 120-mg/mL dipped sample at 0 °C, 4-C = nondipped sample at 4 °C, 4-100 = 100-mg/mL dipped sample at 4 °C, and 4-120 = 120-mg/mL dipped sample at 4 °C.)

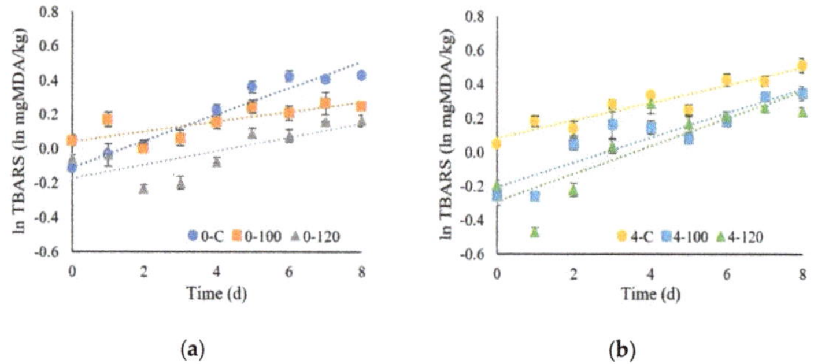

Figure 6. Means and standard errors of the TBARS values of pork sample when stored at (**a**) 0 °C and (**b**) 4 °C. (0-C = nondipped sample at 0 °C, 0-100 = 100-mg/mL dipped sample at 0 °C, 0-120 = 120-mg/mL dipped sample at 0 °C, 4-C = nondipped sample at 4 °C, 4-100 = 100-mg/mL dipped sample at 4 °C, and 4-120 = 120-mg/mL dipped sample at 4 °C.)

3.4.4. Sensory Evaluation

The sensory evaluation showed that the scores for the appearance, color, odor, and overall liking from all the chilled pork samples showed significant decreases ($p \leq 0.05$) with the increasing storage time, as shown in Figure 7. The scores for the appearance and color from the control were higher than fresh pork dipped in the ultrasonicated extracts. The result showed that the fresh pork dipped in 120-mg/mL at 0 and 4 °C had lower consumer acceptance (<5 hedonic score) due to too strong a greenish appearance and color, affected by too much CE extract. The control samples at 0 and 4 °C had a spoiled odor after a storage period of 4–8 days, which resulted in lower scores of the overall liking. The consumers preferred fresh pork dipped in ultrasonicated extracts that received higher odor and overall liking scores than the control samples. This was probably due to the antimicrobial effect of the CE, as confirmed by the increased microbial population found in the experiments reported in Section 3.4.3. In addition, Ramírez-Rojo et al. [66] showed that pork patties treated with an ethanol extract of Mesquite leaves could increase the shelf life with acceptable sensory properties. Due to the sensory acceptance limitations of 120-mg/mL CE, 100-mg/mL CE should be selected to apply for shelf life extensions instead of 120-mg/mL CE with better acceptance by consumers.

Table 2. Total plate count (log CFU/g) of the pork samples dipped with 100 and 120-mg/mL CE in comparison with the undipped samples (control) during storage at 0 and 4 °C.

Storage Day	0 °C			4 °C		
	Control	100 mg/mL	120 mg/mL	Control	100 mg/mL	120 mg/mL
0	3.33 ± 0.11 ^{aG}	3.21 ± 0.26 ^{aG}	2.99 ± 0.07 ^{aE}	3.09 ± 0.39 ^{aF}	3.05 ± 0.18 ^{aG}	3.17 ± 0.13 ^{aG}
1	3.67 ± 0.06 ^{bF}	2.79 ± 0.01 ^{dH}	3.08 ± 0.15 ^{cE}	3.64 ± 0.03 ^{bE}	4.09 ± 0.04 ^{aE}	2.74 ± 0.09 ^{dH}
2	3.21 ± 0.08 ^{cG}	3.57 ± 0.04 ^{bF}	2.88 ± 0.11 ^{dEF}	4.03 ± 0.01 ^{aD}	3.66 ± 0.06 ^{bF}	3.34 ± 0.06 ^{cG}
3	4.41 ± 0.02 ^{bE}	3.86 ± 0.07 ^{dE}	2.71 ± 0.03 ^{fF}	5.35 ± 0.10 ^{aC}	4.27 ± 0.05 ^{cE}	3.58 ± 0.03 ^{eF}
4	4.36 ± 0.03 ^{bE}	4.95 ± 0.02 ^{aC}	4.07 ± 0.23 ^{cD}	5.15 ± 0.06 ^{aC}	4.10 ± 0.10 ^{bcE}	3.83 ± 0.02 ^{cE}
5	5.93 ± 0.22 ^{cD}	4.68 ± 0.08 ^{eD}	4.05 ± 0.04 ^{fD}	7.21 ± 0.03 ^{aB}	6.30 ± 0.03 ^{bD}	5.36 ± 0.07 ^{dD}
6	6.70 ± 0.08 ^{cC}	6.13 ± 0.01 ^{eB}	5.21 ± 0.03 ^{fC}	8.48 ± 0.01 ^{aA}	7.08 ± 0.03 ^{bC}	6.24 ± 0.02 ^{dC}
7	8.04 ± 0.02 ^{bB}	6.21 ± 0.01 ^{eB}	6.44 ± 0.05 ^{dB}	8.48 ± 0.01 ^{aA}	7.88 ± 0.09 ^{bB}	7.42 ± 0.16 ^{cB}
8	8.48 ± 0.01 ^{aA}	7.15 ± 0.10 ^{bA}	7.22 ± 0.06 ^{bA}	8.48 ± 0.01 ^{aA}	8.48 ± 0.01 ^{aA}	8.48 ± 0.01 ^{aA}

Different lowercase letters (a–f) indicate significant differences between treatments ($p \leq 0.05$), and different uppercase letters (A–H) indicate significant differences between the times ($p \leq 0.05$). A green color indicates the end of the shelf life (day). A red color indicates the measured values of the rejected samples after the end of the shelf life.

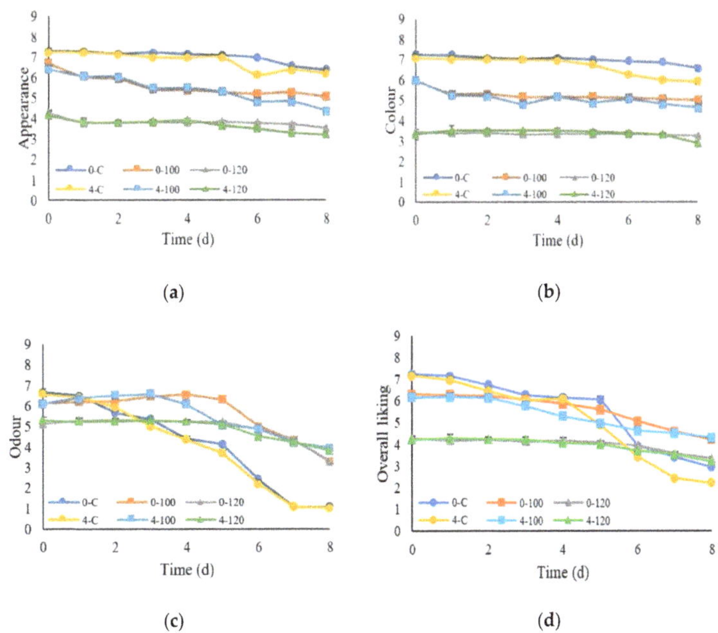

Figure 7. Means and standard errors of the sensory scores from the 9-point hedonic scores of the pork samples: (**a**) appearance, (**b**) color, (**c**) odor, and (**d**) overall liking. (0-C = nondipped sample at 0 °C, 0-100 = 100-mg/mL dipped sample at 0 °C, 0-120 = 120-mg/mL dipped sample at 0 °C, 4-C = nondipped sample at 4 °C, 4-100 = 100-mg/mL dipped sample at 4 °C, and 4-120 = 120-mg/mL dipped sample at 4 °C.)

4. Conclusions

Increasing the time of ultrasonication resulted in increasing the antioxidant activities (DPPH, ABTS, and FRAP) of the CE extracts. Increasing the ultrasonication power increased the capacity of the extracts to inhibit the growth of *S. typhimurium* and *P. aeruginosa*. The optimum ultrasonication condition was determined as ultrasonication at 80% power for 40 min. The fresh pork dipped in 100-mg/mL ultrasonicated extract could be kept for 7 days at 0 °C when compared to the control, which could be kept for only 5 days. The CE extracts by ultrasonication showed greater antioxidant and antimicrobial properties for improving the shelf life of fresh pork. In addition, the sensory evaluation of the fresh pork dipped in 100-mg/mL CE at 0 °C was also acceptable to consumers. Accordingly, these can be safety applied by the food industry to enhance the quality and extend the shelf life of meat products. More work is required to investigate the kinetic study of extraction using ultrasonication and the flavor profile of ultrasonicated CE alone and its impact on treated foods. More mechanisms of the shelf life studies of CE extracts are also needed.

Author Contributions: Conceptualization: Y.P. and P.L.; methodology, S.B.; formal analysis, S.B.; data curation, S.B.; writing—original draft preparation, Y.P. and S.B.; writing—review and editing, K.J., P.S., T.C., N.L., S.R.S., N.C., C.S.B., and J.M.R.; supervision, Y.P., C.S.B., and J.M.R.; and funding acquisition, Y.P. and P.R. All authors have read and agreed to the published version of the manuscript.

Funding: This research and the APC was funded by the Program Management Unit for Human Resources & Institutional Development, Research and Invitation, NXPO [Frontier Global Partnership for Strengthening Cutting-edge Technology and Innovations in Materials Science], grant number B16F640001. This project was also supported by Chiang Mai University.

Data Availability Statement: Not applicable.

Conflicts of Interest: All the authors declare that they have no conflict of interest.

References

1. Djekic, I.; Bozickovic, I.; Djordjevic, V.; Smetana, S.; Terjung, N.; Ilic, J.; Doroski, A.; Tomasevic, I. Can we associate environmental footprints with production and consumption using Monte Carlo simulation? Case study with pork meat. *J. Sci. Food Agric.* **2021**, *101*, 960–969. [CrossRef]
2. Perlo, F.; Fabre, R.; Bonato, P.; Jenko, C.; Tisocco, O.; Teira, G. Refrigerated storage of pork meat sprayed with rosemary extract and ascorbic acid. *Cienc. Rural* **2018**, *48*, e20170238. [CrossRef]
3. Shah, M.A.; Bosco, S.J.D.; Mir, S.A. Plant extracts as natural antioxidants in meat and meat products. *Meat Sci.* **2014**, *98*, 21–33. [CrossRef]
4. Castro, S.; Kolomeytseva, M.; Casquete, R.; Silva, J.; Queirós, R.; Saraiva, J.; Teixeira, P. Biopreservation strategies in combination with mild high pressure treatments in traditional Portuguese ready-to-eat meat sausage. *Food Biosci.* **2017**, *19*, 65–72. [CrossRef]
5. Sahraee, S.; Milani, J.M.; Regenstein, J.M.; Kafil, H.S. Protection of foods against oxidative deterioration using edible films and coatings: A review. *Food Biosci.* **2019**, *32*, 100451. [CrossRef]
6. Rocchetti, G.; Bernardo, L.; Pateiro, M.; Barba, F.J.; Munekata, P.E.; Trevisan, M.; Lorenzo, J.M.; Lucini, L. Impact of a Pitanga leaf extract to prevent lipid oxidation processes during shelf life of packaged pork burgers: An untargeted metabolomic approach. *Foods* **2020**, *9*, 1668. [CrossRef] [PubMed]
7. de Souza de Azevedo, P.O.; Converti, A.; Gierus, M.; de Souza Oliveira, R.P. Application of nisin as biopreservative of pork meat by dipping and spraying methods. *Braz. J. Microbiol.* **2019**, *50*, 523–526. [CrossRef] [PubMed]
8. Lorenzo, J.M.; Vargas, F.C.; Strozzi, I.; Pateiro, M.; Furtado, M.M.; Sant'Ana, A.S.; Rocchetti, G.; Barba, F.J.; Dominguez, R.; Lucini, L. Influence of pitanga leaf extracts on lipid and protein oxidation of pork burger during shelf life. *Food Res. Int.* **2018**, *114*, 47–54. [CrossRef]
9. Chaisuwan, W.; Manassa, A.; Phimolsiripol, Y.; Jantanasakulwong, K.; Chaiyaso, T.; Pathom-aree, W.; You, S.G.; Seesuriyachan, P. Integrated ultrasonication and microbubble-assisted enzymatic synthesis of fructooligosaccharides from brown sugar. *Foods* **2020**, *9*, 1833. [CrossRef] [PubMed]
10. Liu, Y.; Luo, X.; Lan, Z.; Tang, J.; Zhao, P.; Kan, H. Ultrasonic-assisted extraction and antioxidant capacities of flavonoids from *Camellia fascicularis* leaves. *CYTA J. Food* **2018**, *16*, 105–112. [CrossRef]
11. Machado, A.P.D.F.; Sumere, B.R.; Mekaru, C.; Martinez, J.; Bezerra, R.M.N.; Rostagno, M.A. Extraction of polyphenols and antioxidants from pomegranate peel using ultrasound: Influence of temperature, frequency and operation mode. *Int. J. Food Sci. Technol.* **2019**, *54*, 2792–2801. [CrossRef]
12. Tiwari, B.K. Ultrasound: A clean, green extraction technology. *Trends Analyt. Chem.* **2015**, *71*, 100–109. [CrossRef]

13. Sanpa, S.; Sutjarittangtham, K.; Tunkasiri, T.; Eitssayeam, S.; Chantawannakul, P. Ultrasonic extraction of Thai propolis for antimicrobial and antioxidant properties. *Adv. Mater. Res.* **2012**, *506*, 371–374. [CrossRef]
14. Balouiri, M.; Sadiki, M.; Ouedehiri, W.; Farah, A.; Abed, S.; Koraichi, S.I. Antibacterial activity of extracts from *Salvia officinalis* and *Rosmarinus officinalis* obtained by sonication and maceration methods. *Int. J. Pharm. Pharm. Sci.* **2014**, *6*, 167–170.
15. Luque-García, J.L.; Luque de Castro, M.D. Ultrasound-assisted Soxhlet extraction: An expeditive approach for solid sample treatment: Application to the extraction of total fat from oleaginous seeds. *J. Chromatogr. A* **2004**, *1034*, 237–242. [CrossRef] [PubMed]
16. Oyewole, O.; Kalejaiye, O. The antimicrobial activities of ethanolic extracts of *Basella alba* on selected microorganisms. *Sci. J. Microbiol.* **2012**, *1*, 113–118.
17. Vongsak, B.; Sithisarn, P.; Mangmool, S.; Thongpraditchote, S.; Wongkrajang, Y.; Gritsanapan, W. Maximizing total phenolics, total flavonoids contents and antioxidant activity of *Moringa oleifera* leaf extract by the appropriate extraction method. *Ind. Crops Prod.* **2013**, *44*, 566–571. [CrossRef]
18. Kumar, S.; Prasad, A.; Iyer, S.; Vaidya, S. Systematic pharmacognostical, phytochemical and pharmacological review on an ethno medicinal plant, *Basella alba* L. *J. Pharmacogn. Phytotherapy* **2013**, *5*, 53–58.
19. Kumar, B.R.; Anupam, A.; Manchikanti, P.; Rameshbabu, A.P.; Dasgupta, S.; Dhara, S. Identification and characterization of bioactive phenolic constituents, anti-proliferative, and anti-angiogenic activity of stem extracts of *Basella alba* and *rubra*. *J. Food Sci. Technol.* **2018**, *55*, 1675–1684. [CrossRef]
20. Maran, J.P.; Priya, B. Natural pigments extraction from *Basella rubra* L. fruits by ultrasound-assisted extraction combined with Box-Behnken response surface design. *Sep. Sci. Technol.* **2015**, *50*, 1532–1540. [CrossRef]
21. Adesina, A.; Adefemi, S. Lipid composition of *Basella alba* and *Basella rubra* leaves consumed in South-Western Nigeria: Nutritional implications. *Bangladesh J. Sci. Ind. Res.* **2017**, *52*, 125–134. [CrossRef]
22. Sulaiman, I.S.C.; Basri, M.; Masoumi, H.R.F.; Chee, W.J.; Ashari, S.E.; Ismail, M. Effects of temperature, time, and solvent ratio on the extraction of phenolic compounds and the anti-radical activity of *Clinacanthus nutans* Lindau leaves by response surface methodology. *Chem. Cent. J.* **2017**, *11*, 1–11. [CrossRef] [PubMed]
23. Hashemi, S.M.B.; Khaneghah, A.M.; Akbarirad, H. The effects of amplitudes ultrasound-assisted solvent extraction and pretreatment time on the yield and quality of *Pistacia khinjuk* hull oil. *J. Oleo Sci.* **2016**, *65*, 733–738. [CrossRef] [PubMed]
24. Oluwakemi, O.A.; Akindele, A.J.; Bosede, F.S. Ascorbic acid, total phenolic, flavonoid and antioxidant activity of two cultivars of *Basella alba*. *Food Sci. Technol.* **2017**, *5*, 92–96. [CrossRef]
25. Surin, S.; Surayot, U.; Seesuriyachan, P.; You, S.G.; Phimolsiripol, Y. Antioxidant and immunomodulatory activities of sulphated polysaccharides from purple glutinous rice bran (*Oryza sativa* L.). *Int. J. Food Sci.* **2018**, *53*, 994–1004. [CrossRef]
26. Chaiwong, N.; Leelapornpisid, P.; Jantanasakulwong, K.; Rachtanapun, P.; Seesuriyachan, P.; Sakdatorn, V.; Leksawasdi, N.; Phimolsiripol, Y. Antioxidant and moisturizing properties of carboxymethyl chitosan with different molecular weights. *Polymers* **2020**, *12*, 1445. [CrossRef]
27. Surin, S.; You, S.G.; Seesuriyachan, P.; Muangrat, R.; Wangtueai, S.; Jambrak, A.R.; Phongthai, S.; Jantanasakulwong, K.; Chaiyaso, T.; Phimolsiripol, Y. Optimization of ultrasonic-assisted extraction of polysaccharides from purple glutinous rice bran (*Oryza sativa* L.) and their antioxidant activities. *Sci. Rep.* **2020**, *10*, 1–10.
28. Kumar, D.; Jagarwal, P.; Shrama, R. Phytochemical screening and antimicrobial activity of *Basella alba* Linn. *Int. J. Pharm. Chem. Sci.* **2018**, *8*, 502–507.
29. Sen, A.R.; Muthukumar, M.; Naveena, B.M.; Ramanna, D.B.V. Effects on colour characteristics of buffalo meat during blooming, retail display and using vitamin C during refrigerated storage. *J. Food Sci. Technol.* **2014**, *51*, 3515–3519. [CrossRef]
30. Lekjing, S.; Venkatachalam, K. Influences of storage time and temperature on sensory and measured quality of green gram savory crackers. *LWT-Food Sci. Technol.* **2019**, *113*, 108310. [CrossRef]
31. Lu, X.; Zhang, Y.; Zhu, L.; Luo, X.; Hopkins, D.L. Effect of superchilled storage on shelf life and quality characteristics of *M. longissimus* lumborum from Chinese yellow cattle. *Meat Sci.* **2019**, *149*, 79–84. [CrossRef]
32. Kim, H.J.; Jang, A. Evaluation of the microbiological status of raw pork meat in Korea: Modification of the microbial guideline levels for meat. *Food Sci. Biotechnol.* **2018**, *27*, 1219–1225. [CrossRef] [PubMed]
33. American Meat Science Association. *Research Guidelines for Cookery, Sensory Evaluation, and Instrumental Tenderness Measurements of Meat*, 2nd ed.; American Meat Science Association: Chicago, IL, USA, 2016.
34. Vilar, E.G.; Ouyang, H.; O'Sullivan, M.G.; Kerry, J.P.; Hamill, R.M.; O'Grady, M.N.; Mohammed, H.O.; Kilcawley, K.N. Effect of salt reduction and inclusion of 1% edible seaweeds on the chemical, sensory and volatile component profile of reformulated frankfurters. *Meat Sci.* **2020**, *161*, 108001. [CrossRef] [PubMed]
35. Phimolsiripol, Y.; Siripatrawan, U.; Teekachunhatean, S.; Wangtueai, S.; Seesuriyachan, P.; Surawang, S.; Laokuldilok, T.; Regenstein, J.M.; Henry, C.J.K. Technological properties, *in vitro* starch digestibility and *in vivo* glycaemic index of bread containing crude malva nut gum. *Int. J. Food Sci. Technol.* **2017**, *52*, 1035–1041. [CrossRef]
36. Chokumnoyporn, N.; Sriwattana, S.; Phimolsiripol, Y.; Torrico, D.D.; Prinyawiwatkul, W. Soy sauce odour induces and enhances saltiness perception. *Int. J. Food Sci. Technol.* **2015**, *50*, 2215–2221. [CrossRef]
37. International Standard Organization. *Sensory Analysis: General Guidance for the Design of Test Rooms*; ISO Central Secretariat: Geneva, Switzerland, 2007.

38. Biffin, T.E.; Smith, M.A.; Bush, R.D.; Collins, D.; Hopkins, D.L. The effect of electrical stimulation and tenderstretching on colour and oxidation traits of alpaca (*Vicunga pacos*) meat. *Meat Sci.* **2019**, *156*, 125–130. [CrossRef] [PubMed]
39. Phimolsiripol, Y.; Siripatrawan, U.; Cleland, D.J. Weight loss of frozen bread dough under isothermal and fluctuating temperature storage conditions. *J. Food Eng.* **2011**, *106*, 134–143. [CrossRef]
40. Wang, X.; Wu, Q.; Wu, Y.; Chen, G.; Yue, W.; Liang, Q. Response surface optimized ultrasonic-assisted extraction of flavonoids from *Sparganii rhizoma* and evaluation of their in vitro antioxidant activities. *Molecules* **2012**, *17*, 6769–6783. [CrossRef]
41. Altemimi, A.; Watson, D.G.; Choudhary, R.; Dasari, M.R.; Lightfoot, D.A. Ultrasound assisted extraction of phenolic compounds from peaches and pumpkins. *PLoS ONE* **2016**, *11*, e0148758. [CrossRef]
42. Jerman, T.; Trebše, P.; Vodopivec, B.M. Ultrasound-assisted solid liquid extraction (USLE) of olive fruit (*Olea europaea*) phenolic compounds. *Food Chem.* **2010**, *123*, 175–182. [CrossRef]
43. Ilghami, A.; Ghanbarzadeh, S.; Hamishehkar, H. Optimization of the ultrasonic-assisted extraction of phenolic compounds, ferric reducing activity and antioxidant activity of the *Beta vulgaris* using response surface methodology. *Pharm. Sci.* **2015**, *21*, 46–50. [CrossRef]
44. Pankey, G.A.; Sabath, L. Clinical relevance of bacteriostatic versus bactericidal mechanisms of action in the treatment of Gram-positive bacterial infections. *Clin. Infect. Dis.* **2004**, *38*, 864–870. [CrossRef] [PubMed]
45. Jahouach-Rabai, W.; Trabelsi, M.; Van Hoed, V.; Adams, A.; Verhé, R.; De Kimpe, N.; Frikha, M. Influence of bleaching by ultrasound on fatty acids and minor compounds of olive oil. Qualitative and quantitative analysis of volatile compounds (by SPME coupled to GC/MS). *Ultrason. Sonochem.* **2008**, *15*, 590–597. [CrossRef]
46. Giacometti, J.; Žauhar, G.; Žuvić, M. Optimization of ultrasonic-assisted extraction of major phenolic compounds from olive leaves (*Olea europaea* L.) using response surface methodology. *Foods* **2018**, *7*, 149. [CrossRef]
47. Donsì, F.; Annunziata, M.; Vincensi, M.; Ferrari, G. Design of nanoemulsion-based delivery systems of natural antimicrobials: Effect of the emulsifier. *J. Biotechnol.* **2012**, *159*, 342–350. [CrossRef] [PubMed]
48. Weidenmaier, C.; Peschel, A. Teichoic acids and related cell-wall glycopolymers in Gram-positive physiology and host interactions. *Nat. Rev. Microbiol.* **2008**, *6*, 276–287. [CrossRef] [PubMed]
49. Sharma, G.; Sharma, S.; Sharma, P.; Chandola, D.; Dang, S.; Gupta, S.; Gabrani, R. *Escherichia coli* biofilm: Development and therapeutic strategies. *J. Appl. Microbiol.* **2016**, *121*, 309–319. [CrossRef] [PubMed]
50. Yolmeh, M.; Habibi-Najafi, M.B.; Shakouri, S.; Hosseini, F. Comparing antibacterial and antioxidant activity of annatto dye extracted by conventional and ultrasound-assisted methods. *Zahedan J. Res. Med. Sci.* **2015**, *17*, e1020. [CrossRef]
51. McKeegan, K.S.; Borges-Walmsley, M.I.; Walmsley, A.R. Microbial and viral drug resistance mechanisms. *Trends Microbiol.* **2002**, *10*, s8–s14. [CrossRef]
52. Dini, A.; Khanamani Falahati-Pour, S.; Behmaram, K.; Sedaghat, N. The kinetics of colour degradation, chlorophylls and xanthophylls loss in pistachio nuts during roasting process. *J. Food Saf.* **2019**, *3*, 251–263. [CrossRef]
53. Zhang, H.; Wu, J.; Guo, X. Effects of antimicrobial and antioxidant activities of spice extracts on raw chicken meat quality. *Food Sci. Hum. Wellness* **2016**, *5*, 39–48. [CrossRef]
54. Domínguez, R.; Pateiro, M.; Gagaoua, M.; Barba, F.J.; Zhang, W.; Lorenzo, J.M. A comprehensive review on lipid oxidation in meat and meat products. *Antioxidants* **2019**, *8*, 429. [CrossRef] [PubMed]
55. Daniloski, D.; Petkoska, A.T.; Galić, K.; Ščetar, M.; Kurek, M.; Vaskoska, R.; Kalevska, T.; Nedelkoska, D.N. The effect of barrier properties of polymeric films on the shelf life of vacuum packaged fresh pork meat. *Meat Sci.* **2019**, *158*, 107880. [CrossRef]
56. Silva, A.P.R.d.; Longhi, D.A.; Dalcanton, F.; Aragão, G.M.F.d. Modelling the growth of lactic acid bacteria at different temperatures. *Braz. Arch. Biol. Technol.* **2018**, *61*, e18160159. [CrossRef]
57. Lorenzo, J.M.; Sineiro, J.; Amado, I.R.; Franco, D. Influence of natural extracts on the shelf life of modified atmosphere-packaged pork patties. *Meat Sci.* **2014**, *96*, 526–534. [CrossRef]
58. Stefanello, F.S.; Cavalheiro, C.P.; Ludtke, F.L.; Silva, M.d.S.d.; Fries, L.L.M.; Kubota, E.H. Oxidative and microbiological stability of fresh pork sausage with added sun mushroom powder. *Ciênc. Agrotechnol.* **2015**, *39*, 381–389. [CrossRef]
59. Teuteberg, V.; Kluth, I.-K.; Ploetz, M.; Krischek, C. Effects of duration and temperature of frozen storage on the quality and food safety characteristics of pork after thawing and after storage under modified atmosphere. *Meat Sci.* **2021**, *174*, 108419. [CrossRef]
60. Akoğlu, I.; Bıyıklı, M.; Akoğlu, A.; Kurhan, Ş. Determination of the quality and shelf life of sous vide cooked Turkey cutlet stored at 4 and 12 °C. *Rev. Bras. Cienc. Avic.* **2018**, *20*, 1–8. [CrossRef]
61. Álvarez-Martínez, F.J.; Barrajón-Catalán, E.; Encinar, J.A.; Rodríguez-Díaz, J.C.; Micol, V. Antimicrobial capacity of plant polyphenols against gram-positive bacteria: A comprehensive review. *Curr. Med. Chem.* **2020**, *27*, 2576–2606. [CrossRef]
62. Viuda Martos, M.; Ciro Gómez, G.L.; Ruiz Navajas, Y.; Zapata Montoya, J.E.; Sendra, E.; Pérez-Álvarez, J.A.; Fernández-López, J. In vitro antioxidant and antibacterial activities of extracts from annatto (*Bixa orellana* L.) leaves and seeds. *J. Food Saf.* **2012**, *32*, 399–406. [CrossRef]
63. Ali, F.; Abdel-Atty, N.; Helmy, E. Improving the quality and extending the shelf life of chilled fresh sausage using natural additives and their extracts. *J. Microbiol. Biotechnol. Food Sci.* **2018**, *7*, 580–585. [CrossRef]
64. Sood, V.; Tian, W.; Narvaez-Bravo, C.; Arntfield, S.D.; González, A.R. Plant extracts effectiveness to extend bison meat shelf life. *J. Food Sci.* **2020**, *85*, 936–946. [CrossRef]

65. Ranucci, D.; Roila, R.; Andoni, E.; Braconi, P.; Branciari, R. *Punica granatum* and *Citrus spp.* extract mix affects spoilage microorganisms growth rate in vacuum-packaged cooked sausages made from pork meat, emmer wheat (*Triticum dicoccum* Schübler), almond (*Prunus dulcis* Mill.) and hazelnut (*Corylus avellana* L.). *Foods* **2019**, *8*, 664. [CrossRef] [PubMed]
66. Ramírez-Rojo, M.I.; Vargas-Sánchez, R.D.; Torres-Martínez, B.d.M.; Torrescano-Urrutia, G.R.; Lorenzo, J.M.; Sánchez-Escalante, A. Inclusion of ethanol extract of Mesquite leaves to enhance the oxidative stability of pork patties. *Foods* **2019**, *8*, 631. [CrossRef] [PubMed]

 *foods* 

Article

Assessing the Potential of Algae Extracts for Extending the Shelf Life of Rainbow Trout (*Oncorhynchus mykiss*) Fillets

María I. Sáez, María D. Suárez, Francisco J. Alarcón and Tomás F. Martínez *

Departamento de Biología y Geología, Campus de Excelencia Internacional del Mar CEIMAR, Universidad de Almería, 04120 Almería, Spain; msc880@ual.es (M.I.S.); dsuarez@ual.es (M.D.S.); falarcon@ual.es (F.J.A.)
* Correspondence: tomas@ual.es; Tel.: +34-950015267

Abstract: This study evaluates the potential of different algae extracts (*Crassiphycus corneus, Cc; Ulva ohnoi, Uo; Arthrospira platensis, Ap; Haematococcus pluvialis, Hp*) as additives for the preservation of rainbow trout fillets. The extracts were prepared with different water to ethanol ratios from the four algae species. The highest ferric reducing antioxidant power (FRAP) was observed in *Uo* extracted in 80% ethanol. *Ap* aqueous extract also had considerable FRAP activity, in agreement with a high total phenolic content. Radical scavenging activity (DPPH) was higher in *Cc* 80% ethanol extract, in agreement with a high total carotenoid content. In fact, when the algae aqueous extracts were assayed on the fish fillets, their antioxidant activity exceeded that of ascorbic acid (ASC). All algae extracts delayed microbial growth and lipid oxidation processes in trout fillets throughout the cold storage period compared to controls, and also improved textural parameters, these effects being more evident for *Ap* and *Hp*. With respect to the color parameters, the *Hp* extract prevented the a* values (redness) from decreasing throughout cold storage, a key point when it comes to colored species, not least salmonids. On the other hand, the *Ap* extract was not as effective as the rest of treatments in avoiding a* and b* decrease throughout the storage period, and thereby the color parameters were impaired. The results obtained, together with the natural origin and the viability for large-scale cultivation, make algae extracts interesting fish preservative agents for the food industry.

Keywords: algae extracts; antioxidants; fish preservatives; total carotenoids; total phenolics; trout fillets

Citation: Sáez, M.I.; Suárez, M.D.; Alarcón, F.J.; Martínez, T.F. Assessing the Potential of Algae Extracts for Extending the Shelf Life of Rainbow Trout (*Oncorhynchus mykiss*) Fillets. *Foods* **2021**, *10*, 910. https://doi.org/10.3390/foods10050910

Academic Editors: Matteo Alessandro Del Nobile and Amalia Conte

Received: 23 February 2021
Accepted: 14 April 2021
Published: 21 April 2021

1. Introduction

The high polyunsaturated fatty acids content in fish fillets contribute to an increased susceptibility to oxidative processes, which leads to decreased shelf life and sensorial quality. Therefore, the prevention of lipid oxidation with natural additives represents a major challenge for the seafood industry [1]. The interest in algae, both macro and microalgae, as a valuable source of potential additives for the food industry has increased considerably in the last few years. They are acknowledged for being rich in a wide variety of bioactive compounds, such as polyphenols and carotenoids, linked to remarkable antioxidant and antimicrobial activities [2]. Thus, some studies have broached the application of algae solutions as natural preservatives for fish products [3,4].

Crassiphycus sp. and *Ulva* sp. are two of the most extensively assessed genera of edible macroalgae, owing to their easiness for cultivation, together with their richness in bioactive compounds [5]. On the other hand, the interest in microalgae as natural sources of bioactive compounds is increasing [6], not least taking into consideration that they can be grown at large scale in bioreactors. Their cultivation under controlled conditions not only enables continuous supply, but also guarantees a production free from the eventual bioaccumulation of toxic substances [7]. Among the vast variety of microalgae species, *Haematococcus pluvialis* stands out as a source of astaxanthin [8], a pigment included routinely in commercial feeds for salmonids. With respect to *Arthrospira*, this is one of the most widely assessed cyanobacteria genera in aquaculture, not only owing to its high nutritional value, but also to its acknowledged antioxidant activity [9,10].

In view of the above, this study was aimed at assessing the potential preservative effects of four algae species (*Crassiphycus corneus*, *Ulva ohnoi*, *Arthrospira platensis*, and *Haematococcus pluvialis*) on the shelf life and quality attributes of rainbow trout (*Oncorhynchus mykiss*) fillets kept under cold storage. The phenolic and carotenoid contents, as well as the antioxidant activity of algae extracts prepared with different water to ethanol ratios, were also studied.

2. Materials and Methods

2.1. Materials

Freeze-dried biomass of two marine macroalgae, *Ulva ohnoi* (*Uo*) and *Crassiphycus corneus* (*Cc*) were provided by LifeBioencapsulation S.L. (Almería, Spain). SABANA project (H2020 EU research program) supplied cyanobacteria *Arthrospira platensis* (*Ap*) and microalgae *Haematococcus pluvialis* (*Hp*) freeze-dried powder. Ascorbic acid was purchased from Sigma-Aldrich (Madrid, Spain).

2.2. Preparation of Extracts and Antioxidant Activity

Algal extracts used in antioxidant activity determinations were prepared by mixing 10 g of lyophilized powder with 250 mL ethanol in distilled water at different concentrations (0%, 30%, 50% or 80% *v/v*), according to the procedure described by Santoso et al. [11]. The mixtures were homogenized by vigorous shaking (2 min) and then agitated for 24 h at room temperature (22 °C) in darkness with a magnetic stirrer. Then, the mixtures were centrifuged (8000× *g*, 20 min), and supernatants filtered through Whatman #1 paper. Extractions were carried out in triplicate. Extracts were stored at 4 °C until further use within the next 24 h.

2.2.1. Total Phenolic and Carotenoid Contents

Folin–Ciocalteu spectrophotometric procedure was carried out as described by Singh et al. [12]. A gallic acid standard was prepared (0 to 200 μg mL^{-1}) and the results for total phenolic content were expressed as mg of gallic acid equivalents g^{-1}. Total carotenoid content of extracts was estimated spectrophotometrically at 470 nm according to the equations proposed by Lichtenthaler and Buschmann [13]. For both parameters, results were expressed as mg g^{-1}.

2.2.2. Ferric Reducing Antioxidant Power (FRAP)

The antioxidant capacity of algae extracts was estimated according to the methodology described by Hajimahmoodi et al. [14]. Working solutions were prepared by mixing 100 μL of the samples or standards with 3 mL of FRAP reagent (50 mL of 0.3 M acetate buffer, pH 3.6, 5 mL of 10 mM tripyridyltriazine prepared in 40 mM HCl, and 5 mL of 20 mM $FeCl_3$), and kept in the dark for 20 min at room temperature. Absorbance was then measured at 593 nm. Standards were prepared with ethanolic solutions of 6-hydroxy-2,5,7,8-tetramethylchroman-2-carboxylic acid (Trolox; Sigma Aldrich), and the results were expressed as μmol Trolox equivalents g^{-1}.

2.2.3. Radical Scavenging Activity Determination (DPPH)

This activity was measured according to the method described by Brand-Williams [15]. The reaction mixtures were prepared by adding 75 μL of the algae extracts into 150 μL of 100 μg mL^{-1} 2,2-diphenyl-1-picryl-hydrazyl-hydrate (DPPH) solution, and then incubating at room temperature in the dark for 24 h. The transformation of DPPH from oxidized to reduced form was determined spectrophotometrically at 515 nm. Standards were prepared with ethanolic solutions of Trolox. Results were expressed as μmol Trolox equivalents g^{-1}.

2.3. *Treatment of Fish Fillets with Algae Extracts*

2.3.1. Preparation of the Dipping Solutions

Only aqueous algae extracts were used for dipping the fillets. The solutions were prepared by mixing 1.5 g of the respective lyophilized algae with autoclaved distilled water up to a final volume of 500 mL (0.3% *w/v* final concentration). The mixtures were homogenized by vigorous shaking (2 min), then agitated with a magnetic stirrer for 24 h at room temperature (22 °C) in darkness, and filtered through Whatman #1 paper. Ascorbic acid solution (0.3% *w/v*) was also prepared for comparative purposes. The six solutions were designed as follows: (i) control (CONTROL, distilled water); (ii) 0.3% ascorbic acid solution (ASC); (iii) 0.3% *Crassiphycus corneus* extract solution (*Cc*); (iv) 0.3% *Ulva ohnoi* extract solution (*Uo*); (v), 0.3% *Arthrospira platensis* extract solution (*Ap*); and (vi) 0.3% *Haematococcus pluvialis* extract solution (*Hp*). Values of pH and buffering capacity (mEq g^{-1} biomass) of the experimental extracts were determined by acid–base neutralization in order to estimate their possible influence on fillet pH determinations.

2.3.2. Fillet Treatment and Sampling

A total of 60 rainbow trout (500 $\pm$ 25 g) were provided by a commercial fish farm (Piscifactorías Andaluzas, Granada, Spain). Immediately after slaughtering, fish were mechanically filleted at the farm, then washed, dried, and kept on ice. Once in the laboratory (less than 2 h after slaughtering), fillets (210 $\pm$ 22 g) were distributed in 6 experimental lots of 20 units each (5 sampling points $\times$ 4 fillets). The 20 fillets of each treatment group were dipped into the experimental solutions. Autoclaved water was used for dipping control batch. Fillets were carefully placed in sterile trays where the solutions were previously added (500 mL solution per each 4 fillets), kept for 30 s, then removed and placed in a grid for drying during 10 min. Next, fillets were stored in groups of 4 in sterile polyethylene bags, and stored in a cold-room (4 $\pm$ 1 °C). Samples were withdrawn from each lot at 1, 4, 6, 8, and 12 days postmortem (dpm) for the experimental determinations.

2.3.3. Microbial Counts

Total viable psychrophilic bacteria counts were carried out according to Sáez et al. [16]. Muscle pieces (1 g) were introduced into aseptic tubes with 10 mL of 0.1% (*w/v*) peptone water (Cultimed SL, Murcia, Spain) and homogenized for 60 s. Four 1-g pieces were withdrawn from each fillet. Appropriate dilutions were serially prepared (12 serial dilutions by mixing 100 µL with 900 µL of peptone water) and then 0.1 mL of each were spread onto plate count agar media. The prepared plates were incubated at 4 °C for 120 h. Microbiological loads were expressed as logarithm of colony-forming units (cfu) g^{-1} tissue.

2.3.4. Determination of Thiobarbituric Acid-Reactive Substances (TBARS)

TBARS were determined according to the method of Buege and Aust [17], as detailed in Molina et al. [18]. Extracts were prepared in triplicate. TBARS value was expressed as mg of malonyldialdehyde (MDA) kg^{-1} fresh tissue.

2.3.5. pH and Water Holding Capacity (WHC)

Fillet pH was determined by means of a penetration electrode as described in [19]. WHC was estimated by weighing 1 cm^3 samples before and after centrifugation, as described in Suárez et al. [19].

2.3.6. Texture Profile Analysis (TPA)

Texture was measured by compression using a Texture Analyser (TXT2 plus, Stable Micro Systems, Surrey, England, UK), with a load cell of 5 kN. A cylindrical probe (20 mm diameter) was used for pressing downwards into the fillet at a constant speed of 1 mm s^{-1}. The determinations were made in all fillets at two adjacent points at the front dorsal muscle. Results are the average of these two values. The probe was placed parallel to the muscle fibers. The parameters hardness, springiness, cohesiveness, gumminess, chewiness, and resilience were calculated as described in Bourne [20].

2.3.7. Fillet Color Determination

Flesh color was measured by L*, a*, and b* system [21], using a Minolta Chroma meter CR400 device (Minolta, Osaka, Japan). The determinations were made at two adjacent points on the front dorsal part of each fillet. The parameters L* (lightness, on a 0–100 point scale from black to white), a* (estimates the position between red, positive values, and green, negative values), and b* (estimates the position between yellow, positive values, and blue, negative values)) were recorded.

2.4. Statistics

Determinations on antioxidant activity of extracts were carried out in triplicate, and the effect of the categorical variables "algae species" and "water/ethanol ratio" were determined by analysis of variance (ANOVA). With regard to fillet determinations, the effect of the categorical variables "algae treatment" and "storage time", as well as their interactions, were determined for each numeric parameter studied by fitting a generalized lineal statistical model (GLM). Least squares means were tested for differences using Fisher's least significant difference (LSD) procedure. A significance level of 95% was considered to indicate statistical difference. Specific statistical software (SPSS 25, IBM Corporation Inc., Armonk, NY, USA) was used.

3. Results and Discussion

3.1. Total Polyphenol (TPC) and Carotenoid (TCC) Contents and Antioxidant Activity

The influence of solvent water to ethanol ratio on the characteristics of the algae extracts is shown in Table 1. The *Ap* water extract yielded the highest TPC value among the algae studied, followed by *Hp* and *Cc*, whereas *Uo* showed the lowest value. Previous studies have reported a wide variability in TPC for *Arthrospira* sp., likely as a result of both cultivation factors and extraction processes [22]. With regard to *H. pluvialis*, disparate TPC values have also been reported, attributable to different aspects, such as the culture growth phase and the extraction solvent used [23]. Similar TCP figures were obtained for both seaweeds (*Crassiphycus corneus* and *Ulva ohnoi*), which are within the range reported by other authors [24].

The efficiency of polyphenol extraction depended on the solvent water to ethanol ratio considered (Table 1). Roughly, water extraction was more efficient than the rest of mixtures, in agreement with previous studies [14,23], a fact that might be attributed to the partially polar nature of the phenolic compounds. However, opposite results were reported by Mazumder [22] for *A. platensis*, who observed higher phenolic yield, as well as higher antioxidant activity, after organic solvent extraction, such as 60% ethanol, hexane, or methanol.

With respect to carotenoids (TCC) in microalgae, aqueous extracts yielded higher values for *H. pluvialis* than for *A. platensis*, in line with previous studies [25]. Nevertheless, comparison among different studies should be made cautiously, given that microalgae composition could be influenced by several cultivation factors, mostly nutrient availability and light intensity. With the exception of *Ap*, increasing the proportion of ethanol in the extraction solutions yielded higher TCC in extracts.

With regard to the antioxidant activity of algae extracts, the relative contribution of TPC and TCC to antioxidant capacity has not been well established yet. In our study, FRAP activity was remarkable in *A. platensis* aqueous extracts, in agreement with their high phenolic and carotenoid contents. For seaweed extracts, those with superior TCC values have showed outstanding antioxidant activity (not least DPPH in *Cc* and FRAP in *Uo*). Other authors have reported that phenolics and carotenoids contributed similarly to the antioxidant activity in several microalgae species [23]. The study by Yarnpakdee et al. [26] found a correlation between phenolic content in extracts and FRAP and DPPH values in the seaweed *Cladophora glomerata*. In our study, the aqueous extracts showed a correlation between phenolic contents and FRAP activity (R^2 = 0.727), as well as between total carotenoid and DPPH activity (R^2 = 0.990) considering all the algal extracts as a whole. These differences suggest a dissimilar mechanism of action for both groups of compounds.

Since no extensive but only preliminary research has been carried out in this respect in our work, further studies are required to optimize the extraction process of antioxidants for food use, not only in these, but also in many other microalgae species.

Even if the extraction efficiency was higher by using organic solvents in some of the algae species in this study (not least in macroalgae), when it comes to possible practical application, it should be borne in mind that certain organic solvents might not be considered as safe for direct food use, and thus, water extraction might well represent a clear advantage. Moreover, it is reasonable to presume that aqueous extracts would cause a lesser impact on fish fillet sensorial parameters than organic solvent-based extracts. Keeping in mind all the above, only aqueous extracts were considered in the following assays on fillets.

Table 1. Total phenolic and carotenoid contents and antioxidant activity of algae extracts prepared with different water to ethanol ratios.

Algae Species	W/EtOH	TPC	TCC	FRAP	DPPH
Crassiphycus corneus	100/0	1.38 ± 0.01 ^{B,b}	1.63 ± 0.08 ^{B,a}	3.90 ± 0.49 ^{A,a}	9.4 ± 0.61 ^{B,a}
	70/30	0.98 ± 0.03 ^{A,a}	2.08 ± 001 ^{B,b}	3.51 ± 0.06 ^{A,a}	21.55 ± 0.05 ^{C,c}
	50/50	0.97 ± 0.03 ^{A,a}	2.49 ± 0.03 ^{C,c}	4.34 ± 0.42 ^{A,b}	19.02 ± 0.71 ^{C,b}
	20/80	1.12 ± 0.02 ^{A,b}	3.46 ± 0.08 ^{D,d}	4.14 ± 0.29 ^{A,b}	23.5 ± 0.9 ^{D,d}
Ulva ohnoi	100/0	1.18 ± 0.18 ^{A,b}	1.16 ± 0.09 ^{A,a}	8.37 ± 0.06 ^{C,a}	4.14 ± 0.16 ^{A,c}
	70/30	1.08 ± 0.02 ^{B,a}	1.41 ± 0.09 ^{A,b}	8.83 ± 0.13 ^{B,b}	4.56 ± 0.34 ^{A,c}
	50/50	1.08 ± 0.03 ^{B,a}	1.68 ± 0.06 ^{B,c}	8.62 ± 0.06 ^{C,b}	3.56 ± 0.30 ^{A,b}
	20/80	1.42 ± 0.15 ^{B,b}	1.86 ± 0.15 ^{B,c}	16.52 ± 0.18 ^{C,c}	3.05 ± 0.08 ^{A,a}
Artrhospira platensis	100/0	3.22 ± 0.07 ^{D,c}	1.62 ± 0.36 ^{B}	13.93 ± 0.02 ^{D,b}	9.82 ± 0.40 ^{B,c}
	70/30	1.85 ± 0.02 ^{C,b}	1.32 ± 0.03 ^{A}	13.13 ± 0.01 ^{C,b}	8.43 ± 0.29 ^{B,b}
	50/50	1.37 ± 0.01 ^{C,a}	1.50 ± 0.01 ^{A}	6.13 ± 0.66 ^{B,a}	5.37 ± 0.05 ^{B,a}
	20/80	1.40 ± 0.02 ^{B,a}	1.57 ± 0.04 ^{A}	6.55 ± 0.61 ^{B,a}	5.16 ± 0.04 ^{B,a}
Haematococcus pluvialis	100/0	1.64 ± 0.03 ^{C}	2.06 ± 0.02 ^{C,a}	6.40 ± 0.18 ^{B,a}	13.66 ± 0.15 ^{C,a}
	70/30	1.62 ± 0.02 ^{C}	2.23 ± 0.21 ^{C,b}	6.53 ± 0.22 ^{B}	13.58 ± 0.17 ^{C}
	50/50	1.67 ± 0.01 ^{C}	2.44 ± 0.15 ^{C,c}	6.64 ± 0.34 ^{B}	13.22 ± 0.31 ^{C}
	20/80	1.69 ± 0.07 ^{C}	2.52 ± 0.03 ^{C,b}	6.65 ± 0.22 ^{B,b}	14.61 ± 0.11 ^{C,b}

W/EtOH: water to ethanol ratio in solvents. TPC: total phenolic content of extracts, expressed as mg of gallic acid equivalents per g. TCC: total carotenoid content of extracts, expressed as μmol equivalents Trolox per g. FRAP: ferric reducing antioxidant power, expressed as μmol equivalent of Trolox per g. DPPH: radical scavenging activity, expressed as μmol equivalents Trolox per g. Superscript uppercase letters indicate differences attributable to algae species for the same water/ethanol proportion. Superscript lowercase letters indicate differences attributable to water/ethanol proportion for the same algal species. Values are mean ± sd on dry weight basis.

3.2. *Effects of Extracts on Fillet Quality Parameters*

3.2.1. Total Viable Counts

Initial psychrophilic bacterial count in control fillets was 2.2 log CFU g^{-1}, and increased over storage time, exceeding the maximum acceptable limit for fish (6 log CFU g^{-1}; ICMSF [27]) at 12 dpm. Compared to controls, all treatments inhibited bacterial growth in fillets to a greater or lesser degree (Figure 1, Table S1). ASC markedly delayed bacterial growth in trout fillets, in agreement with previous studies [28,29]. Indeed, this compound with acknowledged antimicrobial activity is, up until now, one of the very few additives authorized for unprocessed fresh fish in the EU (Regulation 1333/2008/EC) [30].

In this regard, it is remarkable that the inhibition of microbial growth caused by all the aqueous algae extracts was significantly more intense than that observed for ASC, not least during the first six days of the storage period. The effects of the *H. pluvialis* and *A. platensis* solutions outweighed the other extracts at the initial stages of cold storage. These results agree with the antimicrobial efficiency reported for the seaweed *Fucus spiralis* on refrigerated hake [31].

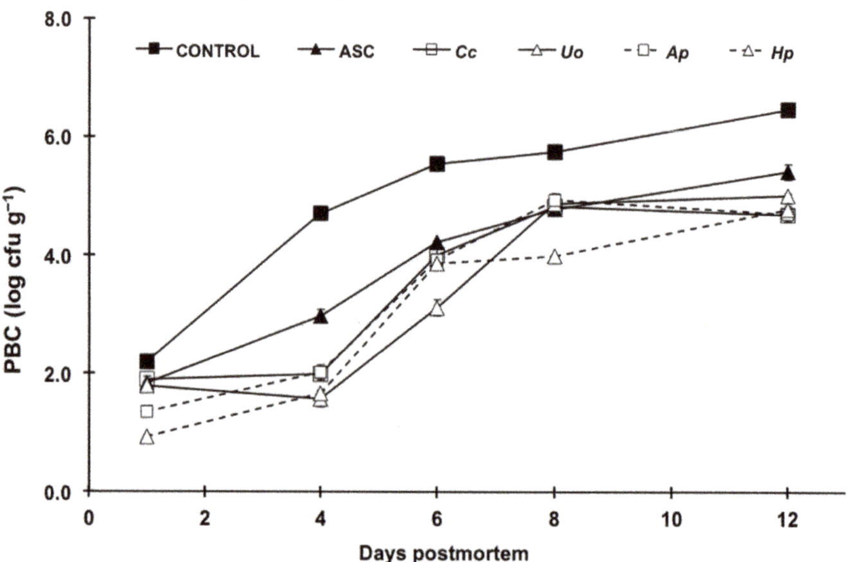

Figure 1. Changes in psychrophilic bacterial counts (PBC) in rainbow trout fillets treated with distilled water (CONTROL); ascorbic acid (ASC) and aqueous extracts of *Crassiphycus corneus* (*Cc*), *Ulva ohnoi* (*Uo*), *Arthrospira platensis* (*Ap*) and *Haematococcus pluvialis* (*Hp*) during a 12-day cold storage (4 °C) period. Values are expressed as mean ± sd. CFU stands for colony-forming units.

3.2.2. Lipid Oxidation (TBARS)

The initial TBARS content (Figure 2) in fillets was within the values established for good quality fish products (1–2 mg MDA kg^{-1}) [32]. The values increased during storage in all treatments, but ASC showed remarkably lower values than controls, confirming its powerful antioxidant effects [29,33]. Interestingly, the seaweed extracts assessed also displayed outstanding effectiveness in protecting muscle lipids from oxidative processes (Figure 2, Table S2), which is the same as described in previous studies on different fish products (*Durvillaea antartica*, *Ulva lactuca*, and *Pyropia columbina* on canned salmon, [3]; *Fucus spiralis* on fresh hake, [30]; *Cladophora glomerata* on sliced tuna, [26]). Moreover, it should be pointed out that the algae extracts not only exceeded the antioxidant capacity of ASC, but also the effects persisted longer than those of ASC up to the end of the storage period.

Surprisingly, despite their potent antioxidant activity, research on the use of microalgae with the purpose of preserving fish products is not extensive. Takyar et al. [34] found that ethanolic extracts of *Chlorella vulgaris* and *A. platensis* yielded significant antioxidant activity on rainbow trout fillets. Studies on other food products also reported remarkable antioxidant effects (*A. platensis* in olive oil, [35]; *H. pluvialis* in ground pork [36]).

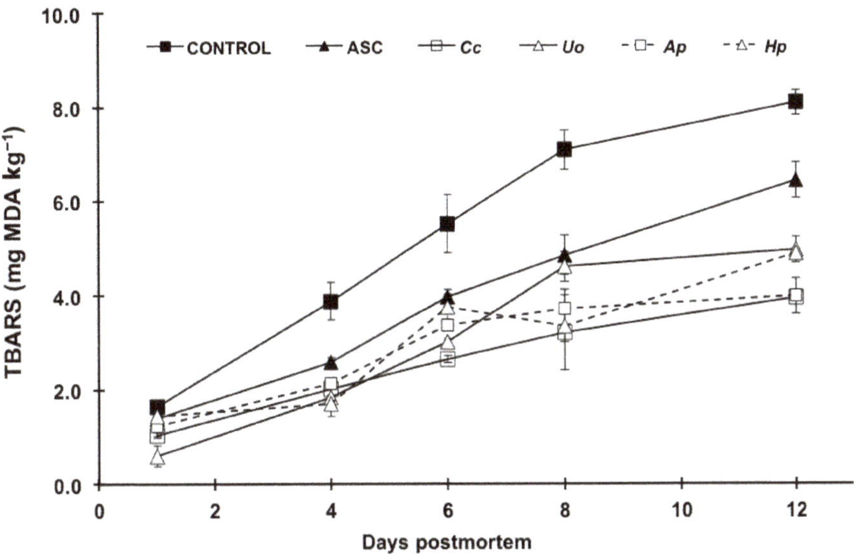

Figure 2. Time course of lipid oxidation in rainbow trout fillets treated with distilled water (CONTROL); ascorbic acid (ASC) and aqueous extracts of *Crassiphycus corneus* (*Cc*), *Ulva ohnoi* (*Uo*), *Arthrospira platensis* (*Ap*) and *Haematococcus pluvialis* (*Hp*) during a 12-day cold storage (4 °C) period. Values are given as mean ± sd. TBARS stands for thiobarbituric acid reactive substances, expressed as mg malonyldialdehyde (MDA) kg^{-1}.

3.2.3. pH and Water Holding Capacity (WHC)

The postmortem changes in pH and WHC are shown in Figure 3 and Table S3. The values increased significantly throughout the 12-day trial for pH in all experimental batches, likely owing to the emergence of alkaline compounds from protein bacterial degradation [37]. The significantly lower pH values observed in all treatments compared to the control fillets (being the lowest those caused by *Ap* and *Hp* extracts) might well be attributed to the antimicrobial effects also observed (Figure 1), which is same as found in previous studies [29]. Nevertheless, doubts can arise regarding the possible influence on fillet pH of the dipping solutions themselves, and therefore, this was also assessed. Overall, the extracts prepared at the concentration assayed (3 g L^{-1}) yielded pH values within half a pH unit from neutrality (Table 2), except *H. pluvialis* and, especially, ASC, which showed a clearly acidic pH. Not only the pH value, but also the buffering capacity was estimated, and the results indicated that, with the exception of ASC, this parameter can be considered as negligible.

Table 2. Values of pH and buffering capacity measured for the aqueous dipping solutions (0.3% *w/v*).

Dipping Solutions	pH	Buffering Capacity (mEq L^{-1}) *
ASC	3.14 ± 0.03	14.67 ± 0.08
Ulva ohnoi	6.93 ± 0.02	0.03 ± 0.01
Crassiphycus corneus	6.64 ± 0.02	0.12 ± 0.02
Artrhospira platensis	4.62 ± 0.01	0.47 ± 0.03
Haematococcus pluvialis	7.12 ± 0.03	0.31 ± 0.05

* Acidic extracts (ASC, *U. ohnoi*, *C. corneus*, *H. pluvialis*) were neutralized with 0.1 N NaOH, whereas alkaline extracts (*A. platensis*) were neutralized with 0.1 N HCl.

No evidence pointing to any effect of dipping solutions on fillet pH was observed, not even for ASC, likely owing to the fact that pH was not measured on fillet surface, but in dorsal muscle depth by means of a penetration electrode.

With regard to WHC, this parameter decreased throughout storage time, with the remaining values being statistically higher for *Hp*-treated fillets during the entire trial compared to the rest of treatments. This fact could indicate significant contribution of this extract in maintaining muscle mechanical properties, which is in agreement with the improvement observed in the textural parameters as well.

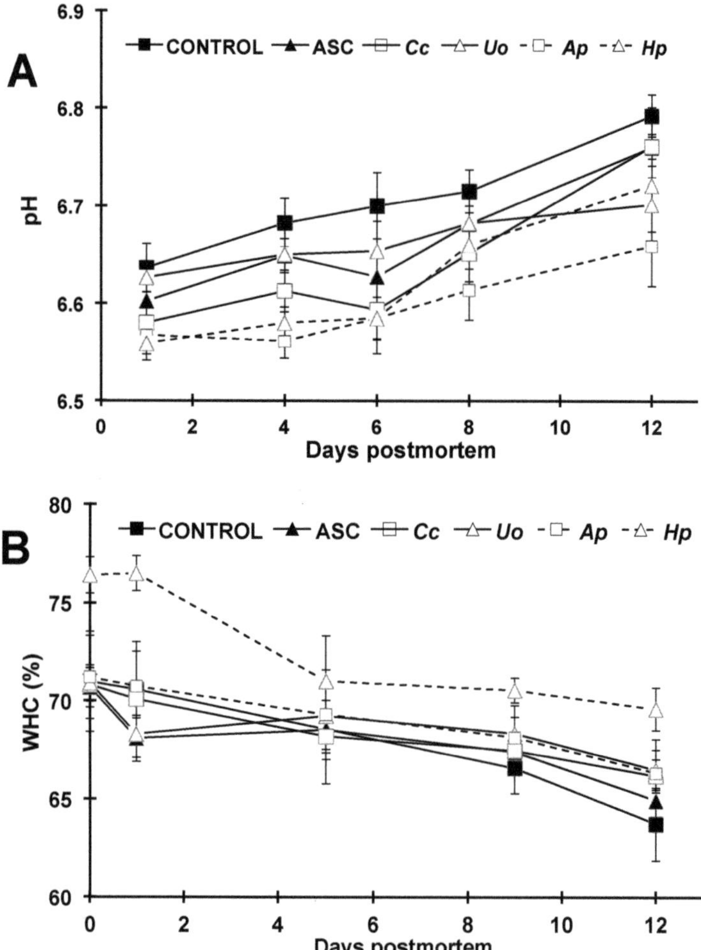

Figure 3. Postmortem changes in pH (**A**) and water holding capacity (WHC; **B**) of rainbow trout fillets treated with distilled water (CONTROL); ascorbic acid (ASC) and aqueous extracts of *Crassiphycus corneus* (*Cc*), *Ulva ohnoi* (*Uo*), *Arthrospira platensis* (*Ap*) and *Haematococcus pluvialis* (*Hp*) during a 12-day cold storage (4 °C) period. Values are expressed as mean ± sd.

3.2.4. Texture Profile Analysis (TPA)

The TPA parameters are shown in Table 3. The hardness decreased in all treatments during cold storage, although all treatments yielded consistently higher values for this parameter than the control fillets at any sampling time from day one onwards. The postmortem deterioration of the textural parameters (not least muscle softening) is caused by myofibrillar and connective tissue proteolysis, which leads to a relaxation of the muscle structure [38]. Not a single factor, but a complex constellation of them (both biochemical and microbiological) is responsible for the alterations of the muscle structures throughout storage time, this ending in the unacceptable softening of the fresh fish. Given that muscle hardness is crucial in terms of purchasing decision, any strategy aimed at preserving this parameter deserves attention.

Table 3. Changes in texture profile analysis parameters of rainbow trout fillets treated with algae aqueous extracts throughout a 12-day cold storage (4 °C) period.

	dpm	C	ASC	*Cc*	*Uo*	*Ap*	*Hp*	*p*
Hardness (N)	1	18.6 ± 1.2^{B}	19.9 ± 1.4^{B}	18.7 ± 0.6	19.8 ± 2.4^{B}	19.8 ± 1.2^{B}	20.6 ± 1.3^{B}	0.133
	4	$16.4 \pm 1.2^{B,a}$	$19.8 \pm 3.3^{B,ab}$	17.6 ± 2.1^{ab}	19.4 ± 1.4^{Bab}	21.1 ± 1.2^{Bb}	$22.1 \pm 1.3^{B,b}$	0.004
	6	$16.7 \pm 1.5^{B,a}$	$18.2 \pm 1.5^{AB,a}$	18.1 ± 1.5^{a}	$18.7 \pm 1.2^{BC,a}$	$20.9 \pm 1.7^{B,b}$	$22.2 \pm 1.7^{B,b}$	<0.001
	8	$16.6 \pm 1.4^{B,a}$	$17.4 \pm 1.7^{AB,a}$	$17.4 \pm 1.6^{B,a}$	$18.2 \pm 1.3^{B,b}$	$17.2 \pm 1.5^{B,b}$	$19.0 \pm 1.8^{B,c}$	0.005
	12	$13.7 \pm 1.9^{A,a}$	$16.4 \pm 1.2^{A,b}$	16.2 ± 1.6^{b}	$14.8 \pm 1.2^{A,ab}$	$16.4 \pm 1.9^{A,b}$	$15.9 \pm 2.1^{A,b}$	0.007
	p	0.001	0.001	0.061	<0.001	0.002	<0.001	
Springiness (mm)	1	0.72 ± 0.05^{A}	0.76 ± 0.04	0.75 ± 0.07^{A}	0.80 ± 0.07^{A}	0.73 ± 0.05^{A}	0.78 ± 0.06^{AB}	0.097
	4	$0.76 \pm 0.05^{A,a}$	0.78 ± 0.04^{a}	$0.89 \pm 0.05^{B,b}$	$0.85 \pm 0.05^{B,b}$	$0.90 \pm 0.05^{C,b}$	$0.91 \pm 0.03^{C,b}$	<0.001
	6	$0.82 \pm 0.04^{B,a}$	0.80 ± 0.07^{a}	$0.84 \pm 0.06^{B,b}$	$0.79 \pm 0.05^{A,a}$	$0.89 \pm 0.07^{C,c}$	$0.87 \pm 0.05^{C,c}$	0.003
	8	$0.76 \pm 0.06^{A,a}$	0.74 ± 0.06^{a}	$0.76 \pm 0.04^{A,a}$	$0.78 \pm 0.03^{A,ab}$	$0.83 \pm 0.05^{B,b}$	$0.81 \pm 0.02^{B,b}$	0.006
	12	$0.73 \pm 0.09^{A,a}$	0.74 ± 0.08^{ab}	$0.75 \pm 0.06^{A,b}$	$0.74 \pm 0.08^{A,ab}$	$0.76 \pm 0.05^{A,b}$	$0.76 \pm 0.05^{A,b}$	0.018
	p	*0.020*	*0.099*	*<0.001*	*0.011*	*<0.001*	*<0.001*	
Cohesiveness	1	0.16 ± 0.03^{b}	0.16 ± 0.01^{b}	$0.13 \pm 0.03^{A,a}$	$0.17 \pm 0.03^{B,b}$	0.15 ± 0.03^{b}	0.16 ± 0.03^{b}	0.010
	4	0.15 ± 0.02^{a}	0.15 ± 0.02^{a}	0.16 ± 0.02^{C}	0.14 ± 0.01^{Aa}	0.17 ± 0.02^{b}	0.17 ± 0.02^{b}	0.018
	6	0.14 ± 0.01^{a}	0.15 ± 0.01^{a}	$0.16 \pm 0.02^{C,b}$	0.13 ± 0.01^{Aa}	0.17 ± 0.01^{bc}	0.18 ± 0.03^{c}	0.009
	8	0.15 ± 0.01	0.15 ± 0.03	0.14 ± 0.02^{B}	0.17 ± 0.02^{B}	0.15 ± 0.01	0.15 ± 0.03	0.932
	12	0.16 ± 0.02	0.16 ± 0.03	0.14 ± 0.02^{B}	0.15 ± 0.03^{AB}	0.15 ± 0.03	0.15 ± 0.02	0.557
	p	0.083	0.511	0.012	0.016	0.063	0.107	
Gumminess ($N\ mm^{-2}$)	1	2.8 ± 0.2^{C}	2.7 ± 0.3	2.7 ± 0.3^{B}	2.7 ± 0.4^{B}	2.9 ± 0.3^{B}	2.9 ± 0.3^{B}	0.538
	4	$2.5 \pm 0.3^{BC,a}$	3.1 ± 0.7^{ab}	$3.5 \pm 0.4^{D,b}$	$2.7 \pm 0.3^{B,ab}$	$3.5 \pm 0.5^{C,b}$	$3.3 \pm 0.4^{C,ab}$	0.005
	6	$2.5 \pm 0.6^{BC,a}$	2.6 ± 0.4^{a}	$3.0 \pm 0.5^{C,c}$	$2.8 \pm 0.3^{BC,b}$	$3.1 \pm 0.4^{B,c}$	$3.3 \pm 0.4^{C,c}$	0.004
	8	$2.3 \pm 0.2^{A,a}$	2.6 ± 0.3^{b}	2.2 ± 0.4^{Aa}	$3.1 \pm 0.3^{C,c}$	$2.8 \pm 0.2^{AB,b}$	$2.8 \pm 0.2^{B,b}$	0.001
	12	$2.2 \pm 0.2^{A,a}$	2.6 ± 0.6^{c}	$2.4 \pm 0.4^{AB,bc}$	$1.9 \pm 0.3^{A,a}$	$2.5 \pm 0.4^{A,c}$	2.2 ± 0.4^{A}	0.004
	p	0.022	0.261	<0.001	<0.001	0.001	<0.001	
Chewiness ($N\ mm^{-1}$)	1	1.9 ± 0.3^{a}	$2.2 \pm 0.5^{BC,ab}$	$2.2 \pm 0.4^{AB,ab}$	$2.6 \pm 0.4^{BC,b}$	2.3 ± 0.3^{AB}	2.3 ± 0.3^{B}	0.004
	4	2.1 ± 0.4^{a}	$2.5 \pm 0.3^{D,ab}$	$2.8 \pm 0.4^{B,c}$	$2.3 \pm 0.3^{B,ab}$	$3.1 \pm 0.4^{B,c}$	$2.7 \pm 0.3^{BC,bc}$	0.003
	6	1.9 ± 0.5^{a}	$2.3 \pm 0.4^{BC,ab}$	$2.7 \pm 0.4^{B,b}$	$2.1 \pm 0.4^{B,a}$	$2.7 \pm 0.4^{B,b}$	$3.3 \pm 0.3^{C,c}$	<0.001
	8	1.9 ± 0.3^{a}	$1.8 \pm 0.4^{A,a}$	1.8 ± 0.2^{A}	$2.5 \pm 0.4^{BC,b}$	$1.8 \pm 0.1^{A,a}$	$2.3 \pm 0.4^{B,b}$	0.010
	12	1.6 ± 0.4^{b}	$1.9 \pm 0.5^{AB,c}$	$1.9 \pm 0.6^{A,c}$	$1.3 \pm 0.4^{A,a}$	$1.9 \pm 0.3^{A,c}$	1.6 ± 0.3^{Ab}	0.001
	p	0.149	0.008	<0.001	<0.001	<0.001	<0.001	
Resilience ($N\ mm^{-1}$)	1	0.16 ± 0.03^{b}	0.17 ± 0.01^{c}	$0.13 \pm 0.03^{A,a}$	$0.17 \pm 0.03^{B,c}$	0.15 ± 0.03^{ab}	0.16 ± 0.03^{b}	0.021
	4	0.15 ± 0.02	0.15 ± 0.03	$0.16 \pm 0.02^{B,b}$	0.14 ± 0.01^{A}	0.17 ± 0.02	0.17 ± 0.02	0.118
	6	0.14 ± 0.01^{a}	0.16 ± 0.02^{b}	$0.16 \pm 0.02^{A,b}$	$0.13 \pm 0.01^{A,a}$	0.16 ± 0.02^{b}	0.18 ± 0.03^{c}	0.009
	8	0.15 ± 0.01	0.15 ± 0.03	0.14 ± 0.02^{A}	0.15 ± 0.02^{A}	0.15 ± 0.01	0.15 ± 0.02	0.932
	12	0.16 ± 0.02	0.16 ± 0.03	0.14 ± 0.02^{A}	0.15 ± 0.03^{A}	0.15 ± 0.03	0.15 ± 0.03	0.557
	p	0.083	0.511	<0.001	0.017	0.063	0.107	

dpm: days postmortem. C: control (distilled water). ASC: ascorbic acid. *Cc*: *Crassiphycus corneus*. *Uo*: *Ulva ohnoi*. *Ap*: *Arthrospira platensis*. *Hp*: *Haematococcus pluvialis*. Values are mean ± sd. Superscript uppercase letters indicate differences ($p < 0.05$) attributable to storage time within each additive treatment. Superscript lowercase letters indicate differences ($p < 0.05$) attributable to treatments within each storage time.

The delayed softening in all treated fillets might well have been linked to lower microbial counts (Figure 1), which is also in agreement with a recent study on rainbow trout fillets [29]. In addition, the inhibition of muscle endogenous protease activity owing to the antioxidant effect of the additives might also have occurred [39]. Although all the algae treatments kept a firmness above that of control fillets, it is remarkable that *Ap* and *Hp* were especially effective in this regard, outweighing even the effects of ASC from day six onwards. This might be related to the higher antioxidant activity observed for those aqueous extracts (Table 1). On the other hand, no consistent trend was observed for the rest of the texture parameters measured (gumminess, chewiness, cohesiveness and resilience) neither regarding storage time nor additive treatment.

3.2.5. Instrumental Color

The influence of the experimental treatments on fillet color parameters is shown in Table 4. The lightness (L*) values increased over time in all treatments. Such an increase has been attributed in salmonids to lipid oxidation and protein denaturation, both effects together leading to higher light refraction on the fillet surface [40]. Color-related quality loss in trout fillets during cold storage results from the combination of increased L*, and decreased a* and b* parameters, together jeopardizing their market value [41]. Such loss of pigmentation has also been partly attributed to astaxanthin degradation [42], included routinely in finishing diets for farmed salmonids, and responsible for their distinctive color.

The most evident effect of algae extracts on L* was the significant decrease in this parameter up to 12 dpm caused by *Ap*, compared to the controls and to the rest of the lots, which, roughly, were similar. The *Ap* extract also caused a clear detrimental effect on the a* and b* parameters (as also found by Takyar et al. [34]). *Uo* also decreased fillet redness throughout the complete storage period, and it is likely that this effect can be attributed to the richness of *Uo* and *Ap* in chlorophylls and phycocyanins, respectively.

At the other end, the *Hp* extract accounted not only for yielding the highest a* and b* values among treatments, but also for causing the most persistent effects during the entire storage period (Table 4), a very desirable effect from the point of view of market acceptability. It is likely that the outstanding astaxanthin content in *Hp* was responsible not only for these favorable effects on color properties, but also for the remarkably positive antimicrobial, antioxidant, and textural effects found in this work, which makes this microalga a promising candidate as a fish preservative in colored muscle fillets.

However, when it comes to practical utilization, there is a clear incompatibility between the intense antioxidant and antimicrobial activities of *Uo* and *Ap* extracts and their coloring properties, which could limit their applicability on fresh fillets of white muscle fish species.

Table 4. Changes in color parameters of rainbow trout fillets treated with algae aqueous extracts throughout a 12-day cold storage (4 °C) period.

	dpm	C	ASC	*Cc*	*Uo*	*Ap*	*Hp*	*p*
L*	1	$45.7 \pm 1.3^{A,b}$	$45.7 \pm 0.7^{A,b}$	$44.5 \pm 0.7^{A,b}$	$46.7 \pm 0.9^{AB,c}$	$43.3 \pm 0.8^{A,a}$	$45.1 \pm 0.5^{A,b}$	0.012
	4	$45.5 \pm 1.2^{A,b}$	$48.7 \pm 1.2^{B,c}$	$45.5 \pm 0.9^{A,b}$	$45.0 \pm 0.6^{A,b}$	$43.0 \pm 0.8^{A,a}$	$46.1 \pm 0.9^{AB,b}$	<0.001
	8	$47.8 \pm 1.4^{B,b}$	$49.0 \pm 0.7^{B,b}$	$48.9 \pm 0.8^{B,b}$	$48.3 \pm 0.7^{B,b}$	$45.2 \pm 0.7^{B,a}$	$47.3 \pm 1.1^{B,b}$	<0.001
	12	$48.4 \pm 1.5^{B,a}$	$51.8 \pm 1.2^{C,bc}$	$52.8 \pm 0.9^{C,c}$	$48.5 \pm 0.9^{C,a}$	$48.1 \pm 0.7^{C,a}$	$50.3 \pm 1.3^{C,b}$	0.004
	p	0.001	<0.001	<0.001	0.001	<0.001	<0.001	
a*	1	$12.2 \pm 0.4^{B,c}$	$9.7 \pm 0.4^{A,b}$	11.9 ± 0.6^{c}	$9.9 \pm 0.4^{B,b}$	$8.7 \pm 0.4^{B,a}$	12.4 ± 0.5^{c}	<0.001
	4	$12.5 \pm 0.1^{B,c}$	$11.1 \pm 0.4^{B,b}$	11.7 ± 0.2^{bc}	$10.5 \pm 0.3^{B,bc}$	$9.1 \pm 0.2^{B,a}$	12.1 ± 0.4^{bc}	<0.001
	8	$11.5 \pm 0.3^{AB,d}$	$11.2 \pm 0.5^{B,d}$	10.9 ± 0.3^{c}	$8.5 \pm 0.3^{A,b}$	$6.3 \pm 0.8^{A,a}$	12.5 ± 0.4^{e}	<0.001
	12	$10.3 \pm 0.3^{A,c}$	$10.9 \pm 0.3^{B,d}$	10.3 ± 0.3^{d}	$9.4 \pm 0.4^{A,b}$	$6.7 \pm 0.1^{A,a}$	12.5 ± 0.4^{e}	<0.001
	p	0.001	0.002	0.098	<0.001	0.001	0.143	
b*	1	$13.2 \pm 0.2^{B,c}$	$9.6 \pm 0.4^{A,a}$	$11.5 \pm 0.4^{B,b}$	$11.4 \pm 0.4^{AB,b}$	10.3 ± 0.6^{a}	$13.5 \pm 0.9^{AB,c}$	0.001
	4	$12.6 \pm 0.7^{AB,b}$	$11.9 \pm 1.1^{B,b}$	$10.8 \pm 1.1^{A,ab}$	$12.0 \pm 1.2^{AB,b}$	9.8 ± 1.2^{a}	$13.1 \pm 1.0^{A,b}$	0.013
	8	$11.4 \pm 0.5^{A,b}$	$11.4 \pm 1.1^{AB,b}$	$10.2 \pm 0.7^{A,ab}$	$10.7 \pm 0.7^{A,ab}$	9.5 ± 0.6^{a}	$13.7 \pm 0.4^{AB,c}$	0.008
	12	$11.0 \pm 0.4^{A,ab}$	$13.0 \pm 0.7^{B,bc}$	$11.6 \pm 0.8^{B,ab}$	$12.8 \pm 0.8^{B,bc}$	9.0 ± 0.9^{a}	$14.1 \pm 0.5^{B,c}$	<0.001
	p	0.024	0.006	<0.001	0.011	0.376	<0.001	

dpm: days postmortem. C: control (distilled water). ASC: ascorbic acid. *Cc*: *Crassiphycus corneus*. *Uo*: *Ulva ohnoi*. *Ap*: *Arthrospira platensis*. *Hp*: *Haematococcus pluvialis*. Values are mean ± sd. Superscript uppercase letters indicate differences ($p < 0.05$) attributable to storage time within each additive treatment. Superscript lowercase letters indicate differences ($p < 0.05$) attributable to treatments within each storage time.

4. Conclusions

The results indicate that all the algae species tested showed valuable antioxidant and antimicrobial effects, which might well be linked to their richness in both carotenoid and phenolic compounds. Outstanding FRAP and DPPH activities were found in *A. platensis* aqueous extract, in agreement with the highest phenolic content. Furthermore, the seaweed extracts with superior carotenoid contents yielded the highest antioxidant capacity (*C. corneus* and *U. ohnoi*).

Not only the antioxidant, but also the antimicrobial effects of the aqueous extracts of all the algae species tested were noteworthy, outweighing even those caused by ascorbic acid, the most widely used authorized additive for fresh fish. Together with delayed firmness loss, all the algae extracts assayed were able to extend the shelf life of the trout fillets compared to untreated controls.

When the impact on trout fillet color is also included in the overall assessment, *H. pluvialis* exceeded the effects of the other species, given that this extract lacks the detrimental effects observed for *A. platensis* on color parameters, despite its powerful antioxidant and antimicrobial activity. In view of the results, extracts of the algae species tested could represent valuable alternative sources of natural food additives with remarkable effects on fresh fish objective quality parameters. In addition to favorable antioxidant and antimicrobial properties, the extracts did not cause negative impact on trout flesh color parameters, which are considered crucial quality attributes for this species.

Supplementary Materials: The following are available online at https://www.mdpi.com/article/10.3390/foods10050910/s1, Table S1: Changes in psychrophilic bacterial counts (PBC, log cfu g^{-1}) in rainbow trout fillets treated with distilled water (C); ascorbic acid (ASC) and aqueous extracts of *Crassiphycus corneus* (*Cc*), *Ulva ohnoi* (*Uo*), *Arthrospira platensis* (*Ap*) and *Haematococcus pluvialis* (*Hp*) during a 12-day cold storage (4 °C) period. Values are given as mean ± standard deviation (n = 4 fillets), Table S2: Changes in lipid oxidation (estimated as mg MDA kg^{-1} content) in rainbow trout fillets treated with distilled water (C); ascorbic acid (ASC) and aqueous extracts of *Crassiphycus corneus* (*Cc*), *Ulva ohnoi* (*Uo*), *Arthrospira platensis* (*Ap*) and *Haematococcus pluvialis* (*Hp*) during a 12-day cold storage (4 °C) period. Values are given as mean ± standard deviation (n = 4 fillets), Table S3: Changes in pH and water holding capacity (WHC) in rainbow trout fillets treated with distilled water (C); ascorbic acid (ASC) and aqueous extracts of *Crassiphycus corneus* (*Cc*), *Ulva ohnoi* (*Uo*), *Arthrospira platensis* (*Ap*) and *Haematococcus pluvialis* (*Hp*) during a 12-day cold storage (4 °C) period. Values are given as mean ± standard deviation (n = 4 fillets).

Author Contributions: Conceptualization: T.F.M., M.I.S., M.D.S., F.J.A. Investigation: T.F.M., M.I.S., M.D.S., F.J.A. Formal analysis: M.I.S. and M.D.S. Data curation: M.I.S. and M.D.S. Writing—original draft preparation: M.D.S. Writing—review and editing T.F.M. Supervision, funding acquisition and project administration: T.F.M. All authors have read and agreed to the published version of the manuscript.

Funding: This research was funded by SABANA project (the European Union's Horizon 2020 Research and Innovation program, grant # 727874), the Spanish Government's MCIU RTI2018-096625-B-C31; and AquaTech4Feed (grant # PCI2020-112204) granted by AEI within the ERA-NET BioBlue COFUND. APC was co-funded by MCIU RTI2018-096625-B-C31 and Universidad de Almería (Spain).

Institutional Review Board Statement: Not applicable.

Informed Consent Statement: Not applicable.

Data Availability Statement: Data available on request.

Acknowledgments: The authors thank Pablo Medina, from Piscifactorías Andaluzas, S.L. (Granada, Spain) for kindly providing the fish used in the study and LifeBioencapsulation S.L. (Almería, Spain).

Conflicts of Interest: The authors declare no conflict of interest.

References

1. Domínguez, R.; Pateiro, M.; Gagaoua, M.; Barba, F.J.; Zhang, W.; Lorenzo, J.M. A Comprehensive Review on Lipid Oxidation in Meat and Meat Products. *Antioxidants* **2019**, *8*, 429. [CrossRef]
2. Fernando, I.S.; Kim, M.; Son, K.-T.; Jeong, Y.; Jeon, Y.-J. Antioxidant Activity of Marine Algal Polyphenolic Compounds: A Mechanistic Approach. *J. Med. Food* **2016**, *19*, 615–628. [CrossRef]
3. Ortiz, J.; Vivanco, J.P.; Aubourg, S.P. Lipid and sensory quality of canned Atlantic salmon (Salmo salar): Effect of the use of different seaweed extracts as covering liquids. *Eur. J. Lipid Sci. Technol.* **2014**, *116*, 596–605. [CrossRef]
4. Halldorsdottir, S.M.; Sveinsdottir, H.; Gudmundsdottir, A.; Thorkelsson, G.; Kristinsson, H.G. High quality fish protein hydrolysates prepared from by-product material with Fucus vesiculosus extract. *J. Funct. Foods* **2014**, *9*, 10–17. [CrossRef]
5. Holdt, S.L.; Kraan, S. Bioactive compounds in seaweed: Functional food applications and legislation. *J. Appl. Phycol.* **2011**, *23*, 543–597. [CrossRef]
6. Begum, H.; Yusoff, F.M.; Banerjee, S.; Khatoon, H.; Shariff, M. Availability and Utilization of Pigments from Microalgae. *Crit. Rev. Food Sci. Nutr.* **2016**, *56*, 2209–2222. [CrossRef]
7. Gomez-Zavaglia, A.; Lage, M.A.P.; Jimenez-Lopez, C.; Mejuto, J.C.; Simal-Gandara, J. The Potential of Seaweeds as a Source of Functional Ingredients of Prebiotic and Antioxidant Value. *Antioxidants* **2019**, *8*, 406. [CrossRef] [PubMed]
8. Kobayashi, M.; Kakizono, T.; Nagai, S. Enhanced Carotenoid Biosynthesis by Oxidative Stress in Acetate-Induced Cyst Cells of a Green Unicellular Alga, Haematococcus pluvialis. *Appl. Environ. Microbiol.* **1993**, *59*, 867–873. [CrossRef]
9. Chu, W.-L.; Lim, Y.-W.; Radhakrishnan, A.K.; Lim, P.-E. Protective effect of aqueous extract from Spirulina platensis against cell death induced by free radicals. *BMC Complement. Altern. Med.* **2010**, *10*, 53. [CrossRef]
10. Galafat, A.; Vizcaíno, A.J.; Sáez, M.I.; Martínez, T.F.; Jerez-Cepa, I.; Mancera, J.M.; Alarcón, F.J. Evaluation of Arthrospira sp. enzyme hydrolysate as dietary additive in gilthead seabream (Sparus aurata) juveniles. *Environ. Boil. Fishes* **2020**, *32*, 3089–3100. [CrossRef]
11. Santoso, J.; Yoshie-Stark, Y.; Suzuki, T.; Gilmore, J.; Starr, D. Anti-oxidant activity of methanol extracts from Indonesian seaweeds in an oil emulsion model. *Fish. Sci.* **2004**, *70*, 183–188. [CrossRef]
12. Singh, R.P.; Murthy, K.N.C.; Jayaprakasha, G.K. Studies on the Antioxidant Activity of Pomegranate (Punicagranatum) Peel and Seed Extracts Using in Vitro Models. *J. Agric. Food Chem.* **2002**, *50*, 81–86. [CrossRef]
13. Lichtenthaler, H.K.; Buschmann, C. Chlorophylls and Carotenoids: Measurement and Characterization by UV-VIS Spectroscopy. *Curr. Protoc. Food Anal. Chem.* **2001**, *1*, F4.3.1–F4.3.8. [CrossRef]
14. Hajimahmoodi, M.; Faramarzi, M.A.; Mohammadi, N.; Soltani, N.; Oveisi, M.R.; Nafissi-Varcheh, N. Evaluation of antioxidant properties and total phenolic contents of some strains of microalgae. *Environ. Boil. Fishes* **2010**, *22*, 43–50. [CrossRef]
15. Brand-Williams, W.; Cuvelier, M.; Berset, C. Use of a free radical method to evaluate antioxidant activity. *LWT* **1995**, *28*, 25–30. [CrossRef]
16. Sáez, M.I.; Martínez, T.F.; Cárdenas, S.; Suárez, M.D. Effects of Different Preservation Strategies on Microbiological Counts, Lipid Oxidation and Color of Cultured Meagre (*Argyrosomus regius*, L.) Fillets. *J. Food Process. Preserv.* **2015**, *39*, 768–775. [CrossRef]
17. Buege, J.A.; Aust, S.D. Microsomal lipid peroxidation. *Methods Enzymol.* **1978**, *52*, 302–310. [CrossRef]
18. Molina, B.; Sáez, M.; Martínez, T.; Guil-Guerrero, J.; Suárez, M. Effect of ultraviolet light treatment on microbial contamination, some textural and organoleptic parameters of cultured sea bass fillets (Dicentrarchus labrax). *Innov. Food Sci. Emerg. Technol.* **2014**, *26*, 205–213. [CrossRef]
19. Suárez, M.; Martínez, T.; Sáez, M.; Morales, A.; García-Gallego, M.; Moya, T.F.M. Effects of dietary restriction on post-mortem changes in white muscle of sea bream (Sparus aurata). *Aquaculture* **2010**, *307*, 49–55. [CrossRef]
20. Bourne, M.C. Texture profile analysis. *Food Tech.* **1978**, *32*, 62–72.
21. Commission Internationale de l'Eclairage (CIE). *Colorimetry*, 2nd ed.; Publication CIE No. 15.2; Commission Internationale de l'Eclairage: Vienna, Austria, 1986.
22. Mazumder, A.; Prabuthas, P.; Mishra, H.N. Optimization of ultrasound-assisted solvent extraction of phycocyanin and phe-nolics from *Arthrospira platensis* var. 'lonor' biomass. *Nutrafoods* **2017**, *16*, 231–239. [CrossRef]
23. Goiris, K.; Muylaert, K.; Fraeye, I.; Foubert, I.; De Brabanter, J.; De Cooman, L. Antioxidant potential of microalgae in relation to their phenolic and carotenoid content. *Environ. Boil. Fishes* **2012**, *24*, 1477–1486. [CrossRef]
24. Farvin, K.S.; Jacobsen, C. Phenolic compounds and antioxidant activities of selected species of seaweeds from Danish coast. *Food Chem.* **2013**, *138*, 1670–1681. [CrossRef]
25. Mendes-Pinto, M.; Raposo, M.; Bowen, J.; Young, A.; Morais, R. Evaluation of different cell disruption processes on encysted cells of Haematococcus pluvialis: Effects on astaxanthin recovery and implications for bio-availability. *Environ. Boil. Fishes* **2001**, *13*, 19–24. [CrossRef]
26. Yarnpakdee, S.; Benjakul, S.; Senphan, T. Antioxidant activity of the extracts from freshwater macroalgae (Cladophora glom-erata) grown in northern Thailand and its preventive effect against lipid oxidation of refrigerated eastern little tuna slice. *Turk. J. Fish. Aquat. Sci.* **2018**, *19*, 209–219. [CrossRef]
27. International Commission on Microbial Specifications of Foods (ICMSF). Sampling plans for fish and shellfish. In *Microorganisms in Foods Sampling for Microbiological Analysis: Principles and Scientific Applications*; University of Toronto Press: Toronto, ON, Canada, 1986; pp. 181–196.

28. Monirul, I.; Yang, F.; Niaz, M.; Qixing, J.; Wenshui, X. Effectiveness of combined acetic acid and ascorbic acid spray on fresh silver carp (Hypophthalmichthys molitrix) fish to increase shelf-life at refrigerated temperature. *Curr. Res. Nutr. Food Sci. J.* **2019**, *7*, 415–426. [CrossRef]
29. Sáez, M.; Suárez, M.; Martínez, T. Effects of alginate coating enriched with tannins on shelf life of cultured rainbow trout (*Oncorhynchus mykiss*) fillets. *LWT* **2020**, *118*, 108767. [CrossRef]
30. Regulation 1333/2008/EC of the European Parliament and of the Council of 16 December 2008, on Food Additives. (Official Journal, Series L 354, 31.12.2008, p. 16). Available online: https://eur-lex.europa.eu/legal-content/EN/TXT/?uri=celex%3A32008R1333 (accessed on 12 April 2021).
31. Barros-Velázquez, J.; Miranda, J.M.; Ezquerra-Brauer, J.M.; Aubourg, S.P. Impact of icing systems with aqueous, ethanolic and ethanolic-aqueous extracts of alga Fucus spiralis on microbial and biochemical quality of chilled hake (Merluccius merluccius). *Int. J. Food Sci. Technol.* **2016**, *51*, 2081–2089. [CrossRef]
32. Gomes, H.D.A.; Da Silva, E.N.; Nascimento, M.R.L.D.; Fukuma, H.T. Evaluation of the 2-thiobarbituric acid method for the measurement of lipid oxidation in mechanically deboned gamma irradiated chicken meat. *Food Chem.* **2003**, *80*, 433–437. [CrossRef]
33. Van Nguyen, M.; Thorarinsdottir, K.A.; Thorkelsson, G.; Gudmundsdóttir, Á.; Arason, S. Influences of potassium ferrocyanide on lipid oxidation of salted cod (Gadus morhua) during processing, storage and rehydration. *Food Chem.* **2012**, *131*, 1322–1331. [CrossRef]
34. Takyar, M.B.T.; Khajavi, S.H.; Safari, R. Evaluation of antioxidant properties of Chlorella vulgaris and Spirulina platensis and their application in order to extend the shelf life of rainbow trout (*Oncorhynchus mykiss*) fillets during refrigerated storage. *LWT* **2019**, *100*, 244–249. [CrossRef]
35. Alavi, N.; Golmakani, M.-T. Evaluation of synergistic effect of Chlorella vulgaris and citric acid on oxidative stability of virgin olive oil. *Food Sci. Biotechnol.* **2017**, *26*, 901–910. [CrossRef]
36. Pogorzelska, E.; Godziszewska, J.; Brodowska, M.; Wierzbicka, A. Antioxidant potential of Haematococcus pluvialis extract rich in astaxanthin on color and oxidative stability of raw ground pork meat during refrigerated storage. *Meat Sci.* **2018**, *135*, 54–61. [CrossRef]
37. Masniyom, P.; Benjakul, S.; Visessanguan, W. Shelf-life extension of refrigerated seabass slices under modified atmosphere packaging. *J. Sci. Food Agric.* **2002**, *82*, 873–880. [CrossRef]
38. Delbarre-Ladrat, C.; Chéret, R.; Taylor, R.; Verrez-Bagnis, V. Trends in Postmortem Aging in Fish: Understanding of Proteolysis and Disorganization of the Myofibrillar Structure. *Crit. Rev. Food Sci. Nutr.* **2006**, *46*, 409–421. [CrossRef]
39. Sriket, C. Proteases in fish and shellfish: Role on muscle softening and prevention. *Int. Food Res. J.* **2014**, *21*, 433–445.
40. Ozbay, G.; Spencer, K.; Gill, T.A. Investigation of protein denaturation and pigment fading in farmed steelhead (Onchorhychus mykiss) fillets during frozen storage. *J. Food Process. Preserv.* **2006**, *30*, 208–230. [CrossRef]
41. Fellenberg, M.A.; Carlos, F.; Peña, I.; Ibáñez, R.A.; Vargas-Bello-Pérez, E. Oxidative quality and color variation during refrigeration (4 °C) of rainbow trout fillets marinated with different natural antioxidants from oregano, quillaia and rosemary. *Agric. Food Sci.* **2020**, *29*, 43–54. [CrossRef]
42. Ortiz, J.; Palma, Ó.; González, N.; Aubourg, S.P. Lipid damage in farmed rainbow trout (*Oncorhynchus mykiss*) after slaughtering and chilled storage. *Eur. J. Lipid Sci. Technol.* **2008**, *110*, 1127–1135. [CrossRef]

Article

Effect of Cross-Linked Alginate/Oil Nanoemulsion Coating on Cracking and Quality Parameters of Sweet Cherries

Camilo Gutiérrez-Jara [1], Cristina Bilbao-Sainz [2], Tara McHugh [2], Bor-Sen Chiou [2], Tina Williams [2] and Ricardo Villalobos-Carvajal [1,*]

1 Food Engineering Department, Universidad del Bío-Bío, P.O. Box 447, Av. Andrés Bello 720, Chillán 3800708, Chile; camilo.gtz.j@gmail.com

2 Western Regional Research Center, United States Department of Agriculture, Albany, CA 94710, USA; cristina.bilbao@usda.gov (C.B.-S.); tara.mchugh@ars.usda.gov (T.M.); bor-sen.chiou@usda.gov (B.-S.C.); tina.williams@ars.usda.gov (T.W.)

* Correspondence: r.villalobos@ubiobio.cl

Citation: Gutiérrez-Jara, C.; Bilbao-Sainz, C.; McHugh, T.; Chiou, B.-S.; Williams, T.; Villalobos-Carvajal, R. Effect of Cross-Linked Alginate/Oil Nanoemulsion Coating on Cracking and Quality Parameters of Sweet Cherries. *Foods* **2021**, *10*, 449. https://doi.org/10.3390/foods10020449

Academic Editors: Matteo Alessandro Del Nobile and Amalia Conte

Received: 21 January 2021
Accepted: 9 February 2021
Published: 18 February 2021

Abstract: The cracking of sweet cherries causes significant crop losses. Sweet cherries (cv. Bing) were coated by electro-spraying with an edible nanoemulsion (NE) of alginate and soybean oil with or without a $CaCl_2$ cross-linker to reduce cracking. Coated sweet cherries were stored at 4 °C for 28 d. The barrier and fruit quality properties and nutritional values of the coated cherries were evaluated and compared with those of uncoated sweet cherries. Sweet cherries coated with NE + $CaCl_2$ increased cracking tolerance by 53% and increased firmness. However, coated sweet cherries exhibited a 10% increase in water loss after 28 d due to decreased resistance to water vapor transfer. Coated sweet cherries showed a higher soluble solid content, titratable acidity, antioxidant capacity, and total soluble phenolic content compared with uncoated sweet cherries. Therefore, the use of the NE + $CaCl_2$ coating on sweet cherries can help reduce cracking and maintain their postharvest quality.

Keywords: sweet cherry; nanoemulsion coating; cracking; fruit quality, nutraceutical value; crosslinking

1. Introduction

The sweet cherry fruit has a high nutritional value, mainly due to its high antioxidant capacity associated with ascorbic acid, carotenoids, and phenolic compounds [1]. The phenolic compounds in sweet cherries play a protective role against oxidative stress, ultraviolet radiation, and free radical damage [2]. However, the rapid deterioration of sweet cherries after harvest often leads to quality loss. More research is needed to develop novel strategies to prevent or reduce postharvest deterioration [3].

The cracking of sweet cherries caused by rain during the harvest period is the most important source of crop loss in the industry [4]. Rainfall, high humidity, high temperature, rootstock type, crop load, soil moisture levels, and irrigation management are some of the main factors that affect sweet cherry cracking [5]. As for the development of cracks on the skin of the fruits, the main mechanism proposed is related to the increase in turgor pressure caused by water absorption during and after rain. The two main routes of water absorption occur through the fruit surface [6] and/or the roots of the tree [7]. Various strategies have been used to reduce cracking. These include the use of plastic rain shields; adequate irrigation management; the application of calcium salts; and, more recently, the use of protective waterproofing agents [5]. The use of these technologies can reduce the severity of the damage. However, the degree of effectiveness widely varies among seasons, cultivars, and geographic locations. Some strategies, such as plastic rain shields, have high implementation costs [8].

The use of edible coatings is an alternative strategy that has emerged in recent decades [9,10]. This approach could be used to decrease the cracking phenomena of sweet cherries and extend their postharvest shelf life. Some of these edible coatings include chitosan [11], Aloe vera gel [12], and Semperfresh [13,14].

Alginate is another film-forming material that has been used as a thickening agent, gelling agent, and stabilizer in a variety of food emulsions [15]. It is a natural polysaccharide that is extracted from brown seaweed (Phaeophyceae) and comprises the two uronic acids, β-D-manuronic and α-L-guluronic. Sodium alginate consists of block polymers of sodium poly-L-guluronate, sodium poly-D-mannuronate, and alternate sequences of both sugars [16]. It has been effective in maintaining the postharvest quality of tomatoes [17], peaches [18], and sweet cherries [1]. However, to the best of our knowledge there are no previous studies on using alginate-based coatings to prevent sweet cherry cracking.

The cross-linking of the alginate film surface can be used to improve its mechanical and barrier properties because the film disintegrates when subjected to high humidity conditions due to its hydrophilic nature [19]. Cross-linking methods usually include drying, heating, ultraviolet (UV) irradiation, and chemical methods [20]. The chemical cross-linking method for sodium alginate involves the ionic interaction between polymer chains and multivalent ions to form ionomers. This improves their water barrier properties, mechanical strength, cohesiveness, and rigidity [10,21].

Another way to improve the water vapor barrier properties of the film involves adding lipids and nanofillers to form composite or nanocomposite films [22–25]. Smaller lipid globules and a more homogeneous distribution of the oil droplets in the films generally lead to better water vapor barrier properties [26].

Therefore, the aim of this study was to apply a nanoemulsion (soybean oil with alginate solution) on sweet cherries with additional ionic cross-linking and evaluate its effect on the water barrier properties of the sweet cherry cuticle, such as fruit cracking, postharvest qualities, and nutraceutical values.

2. Materials and Methods

2.1. Plant Material and Experimental Design

Sweet cherries (*Prunus avium* L.) cv. Bing were randomly collected at the commercial maturity stage from the midsection of 10 trees grown under standard commercial practices on the same commercial farm located in Brentwood (CA, USA). Cultural practices were regularly implemented for all trees equally. Fruits were transported immediately to the laboratory and selected by color, size, and the absence of physical defects or decay. Subsequently, the fruits were randomly distributed in 3 batches of 111 fruits each prior to treatment. The first batch was treated with a nanoemulsion (NE). The second batch was treated with a nanoemulsion and $CaCl_2$ solution (NE + $CaCl_2$). The third batch was used as a control. The barrier properties were determined immediately after each coating treatment. The quality parameters of the fruits stored at 4 °C were evaluated weekly for 28 d.

2.2. Nanoemulsion Preparation

The alginate solution (1.0%, *w*/*v*) was prepared by dissolving sodium alginate in a 2.5% ethanol solution. Ethanol was used to decrease the surface tension of nanoemulsions [27]. Glycerol was added at 15% alginate mass and the alginate solution was stirred for 30 min. Tween 80 (1.0% *v*/*v*) and soybean oil (0.5% *v*/*v*) were added to the solution and homogenized at 11,000 rpm for 2 min with a rotor-stator homogenizer (Polytron 3000, Kinematica, Littau, Switzerland). These coarse emulsions were passed six times through a microfluidizer processor (model 110T, Microfluidics, Asheville, NC, USA) at 200 MPa to obtain the nanoemulsions. The composition of the nanoemulsion and the process variables were selected on the basis of a prior study [28].

Particle Size and Polydispersity Index (PdI) of Nanoemulsions

The average nanoemulsion particle size and polydispersity index (PdI) were determined by dynamic light scattering (DLS) with a Zetasizer Nano ZS laser diffractometer (Malvern Instruments Ltd., Westborough, MA, USA). Emulsion samples were diluted in ultrapure water to 10% of the original concentration, placed in a cuvette, and analyzed at 25 °C. The average particle size (z-average) value and PdI were recorded.

2.3. Coating Application

The nanoemulsion (NE) was sprayed for 30 s at 30 cm from the surface of the sweet cherries with a cordless 85 kV vector solo waterborne electrostatic gun applicator (ITW Ransburg, Toledo, OH, USA). A 3.0% calcium chloride solution was applied using the same spray system after a coating was formed on the sweet cherries.

2.4. Microstructure

The microstructure of sweet cherry cross-sections was observed with a JEOL 7900F field emission scanning electron microscope (SEM) (JEOL, Kyoto, Japan) with a Quorum PP3010T cryo-system. First, the sweet cherries were cut parallel to the longitudinal axis with a scalpel. The sample was placed in the SEM sample holder and plunged into subcooled nitrogen (−210 °C). Afterward, the frozen sample was transferred to the cryo stage and freeze-fractured and gold-coated. The samples were viewed and photographed at 5 kV in the SEM.

2.5. Barrier Properties

2.5.1. Cracking Index (CI)

The CI was determined using the method reported by Christensen [29]. For this purpose, sweet cherries harvested on the same day were selected based on size (22.3–24.9 mm), total soluble solids (20.03–20.24° Bx), firmness (3.70–3.78 N), water activity (0.964–0.978), and color (a * = 10.16–13.83). Thirty fruits with stems from each batch were submerged in distilled water at 20 °C for 5 h to induce cracking. The number of cracked fruits was counted at 1 h intervals. The CI was calculated using the formula expressed in Equation (1).

$$\mathrm{CI} = ((5a + 4b + 3c + 2d + 1e)/(\mathrm{MPV})) \times 100, \quad (1)$$

where a, b, c, d, and e represent the number of cracked samples at each time interval and MPV is the maximum possible value (30 fruits × 5 h = 150).

2.5.2. Resistance to Water Vapor Transfer (RWVT)

During the same day of harvest, coated and uncoated sweet cherries were placed in a desiccator at 75.65% relative humidity using a saturated sodium chloride solution. Fans were used to ensure a uniform relative humidity throughout the desiccator. The desiccator was placed in a thermostatic chamber maintained at 4 ± 1 °C. Sweet cherries were weighed at 2 h intervals at 0.0001 g accuracy. The Resistance to Water Vapor Transfer (RWVT) was estimated by the equation of the first modified Fick law as established by different authors [30,31]. Weight loss data were used under stationary conditions. The RWVT was calculated by Equation (2).

$$\mathrm{RWVT} = [((a_w - \%\mathrm{RH}/100)\ \mathrm{PWV})/\mathrm{RT}] \times (A/J), \quad (2)$$

where RWVT is the resistance to water vapor transfer (s/cm), a_w is the sweet cherry water activity determined with a water activity meter (Aqua LAB 4TE, Pullman, WA, USA), %RH is the relative humidity inside the desiccator, PWV is the water vapor pressure at the chamber temperature (mm Hg), R is the universal gas constant (3,464,629 mm Hg cm^3/g K), T is the storage chamber temperature (K), A is the sweet cherry surface area at the beginning of the test (cm^2), and J is the slope of the weight loss curve under stationary conditions (g/s).

2.6. Fruit Quality Parameters

2.6.1. Weight Loss

Fruit weight loss was evaluated with a digital balance (Precisa XB 320M, Dietikon, Switzerland). Sweet cherries were individually weighed at the beginning of the experiment and on each sampling day (7, 14, 21, and 28). Weight loss was expressed as a percentage of the initial weight and calculated by Equation (3).

$$\text{Weight loss } (\%) = ((W_o - W_f)/W_o) \times 100 \quad (3)$$

where W_o is the initial weight and W_f is the weight on the sampling day.

2.6.2. Optical Properties

Color measurements were performed with a CR-300 colorimeter (Minolta Camera Co., Ltd., Osaka, Japan). The CIELAB parameters a^*, b^*, and L^* were obtained with the D65 light source and an observation angle of 10° using the reflectance specular mode. The L^* coordinate represented the lightness of the color ($L^* = 0$ denoted black and $L^* = 100$ denoted white), a^* indicated the position between green and red (a^* varied from −80 to +100), and b^* was the extent of blueness/yellowness (b^* varied from −50 to +70).

The hue angle (h°) was calculated by Equation (4) as:

$$\text{Hue angle} = \arctan (b^*/a^*). \quad (4)$$

2.6.3. Fruit Firmness

Mechanical tests were performed with a Texture Analyzer (TA-XT2i, Stable Microsystems Ltd., Surrey, UK) at room temperature using a puncture test. A probe (3 mm diameter stainless steel cylinder) with a trigger force of 5 N penetrated the sample to a depth of 8 mm at a speed of 1 mm s^{-1}. Fruit firmness was measured as the maximum penetration force, and the results were expressed in newtons.

2.6.4. Determination of Total Soluble Solids (TSS) and Titratable Acidity (TA)

For the TSS and TA tests, 5 g of sweet cherry tissue was homogenized in 25 mL of distilled water and filtered. The TSS was determined in the juice at 20 °C with a temperature-compensated LR-01 laboratory refractometer (Maselli Measurements Inc., Stockton, CA, USA). The TA was determined by titrating with 0.025 N of NaOH to pH 8.2 with a semi-automated titrator (Hanna Instruments, Woonsocket, RI, USA).

2.6.5. Total Soluble Phenolic (TSP) Content

The TSP content was determined by the Folin–Ciocalteu method as described by Bilbao-Sainz et al. [32] with slight modifications. Samples (5 g) were homogenized with 20 mL of methanol with a Waring Laboratory Blender (Waring Commercial, Torrington, CT, USA) surrounded with dry ice for 1 min at medium speed. Samples were placed in tubes and stored for 20 to 72 h at 4 °C. Homogenates were centrifuged (rotor SA-600, Sorvall RC 5C Plus, Kendro Laboratory Products, Newtown, CT, USA) at 29,000× g for 15 min at 4 °C. Duplicate samples from each extract were used for the final analysis. A 150 µL aliquot of methanol extract was taken from the clear supernatant, diluted with 2400 µL of ultrapure water and 150 µL 0.125 mol L^{-1} Folin–Ciocalteu reagent, and incubated for 3 min at room temperature. The reaction was stopped by adding 300 µL of 0.5 mol L^{-1} Na_2CO_3 and the mixture was incubated for 25 min. Absorbance readings at 725 nm of clear supernatant samples were measured with a Shimadzu PharmaSpec UV-1700 spectrophotometer (Shimadzu Scientific Instruments, Inc., Columbia, MD, USA). A blank sample prepared with methanol was used as a control. The total amount of phenols was determined using a gallic acid standard curve and the results were expressed as milligrams of gallic acid equivalent (GAE) per gram of fresh weight (FW) of cherry purée. Three replicates were performed for each sample.

2.6.6. Antioxidant Capacity (AC) Analysis

The AC analysis was adapted from Bilbao-Sainz et al. [32] with slight modifications. The same methanol extract from the TSP analysis was used for the AC analysis. Sample aliquots of 50 μL were taken from the clear supernatant (equivalent methanol volume as a control) and reacted with 2950 μL of 2,2-diphenyl-1-picrylhydrazyl (DPPH, 103.2 μmol L^{-1} in methanol; absorbance approximately 1.2 at 515 nm) in a covered shaker at room temperature. The samples were allowed to react until steady-state conditions were reached. The AC was calculated with the PharmaSpec UV-1700 spectrophotometer by measuring the decrease in sample absorbance at 515 nm compared with the blank methanol sample. The AC was reported as μg Trolox equivalent from a standard curve developed with Trolox (0–750 μg mL^{-1}) and expressed as mg Trolox g^{-1} FW. Three replicates were performed for each sample.

2.6.7. Total Anthocyanin Content

The total anthocyanin content was determined in duplicate with a PharmaSpec UV-1700 spectrophotometer (Shimadzu Scientific Instruments, Inc., Columbia, MD, USA) following the method reported by Serrano et al. [33]. Results were calculated by Equation (5) and expressed as milligrams 100 g^{-1} FW.

$$\text{Total anthocyanins} = \frac{\left(\frac{\text{ABS}}{\varepsilon \times l} \times \text{MW} \times 1000\right) \times \left(\frac{\text{V}+\text{W}\times\rho}{1000}\right)}{\text{W}} \times 100 \quad (5)$$

where ABS = absorbance of sample; ε = molar absorption coefficient (23,900 L mol^{-1} cm^{-1} for cyanidin-3-glucoside (cyd-3-glu)); l = path length in cm; MW = molecular weight (449.2 g mol^{-1} for cyd-3-glu); V = volume of dilution in mL; W = sample weight in g; ρ = specific weight (0.83 in mL g^{-1}); and 100 = 100 g of FW.

2.7. Statistical Analysis

A completely randomized design was used in the experiments. Statistical analysis was performed with the Statgraphics Centurion XVII software (version 17.1 12, Statgraphics Technologies Inc., The Plains, VA, USA) by a one-way analysis of variance. Significant differences between means were determined by the least significant difference (LSD) test at the 5% significance level ($p < 0.05$).

3. Results and Discussion

3.1. Particle Size and Polydispersity Index (PdI) of Nanoemulsions

After six passes through the microfluidizer, an emulsion was obtained with an average droplet size of 376.89 ± 2.73 nm and a PdI of 0.36 ± 0.04. These results concur with findings reported by other authors [34,35], who indicate that the increased number of passes through a homogenization system produces a reduced particle size and a more homogeneous particle size distribution. Similar results have been found by Artiga-Artigas et al. [36]. They achieved an emulsion with a 261 nm particle size and 0.25 PdI by mixing sodium alginate with an oil-in-water emulsion before the homogenization process (five passes). The smallest droplet size found in that study could be related to the different emulsion compositions because they used Tween 20 as an emulsifier and did not add a plasticizer.

3.2. Microstructural Analysis

Figure 1 shows the cross-sections of fresh sweet cherry (A), the NE coating (B), and NE + $CaCl_2$ coating (C) on the sweet cherry surfaces. The micrograph of the uncoated sweet cherry surface shows a layer of the cuticle membrane over a layer of the regular epidermal cells, similar to the findings reported by Bargel et al. [37]. The layer of epidermal cells can be observed in all three micrographs. Subepidermal cells increased in size below the layer of epidermal cells because they were located farther away from the surface.

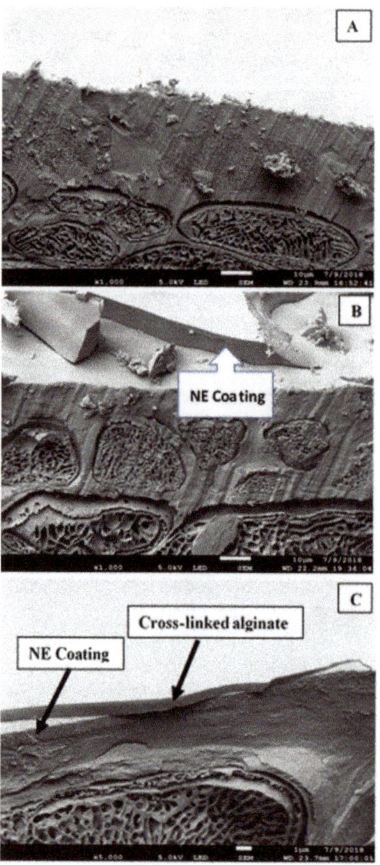

Figure 1. Cross-section micrographs of dried coatings on sweet cherry fruits. (**A**) Control, (**B**) nanoemulsion (NE), and (**C**) NE + $CaCl_2$.

There was a continuous layer of NE coating on the sweet cherry surface (Figure 1B) as a result of the good adhesion of the NE. This adhesion could be attributed to the low surface tension of the coating formation solution because Tween 80 [38] and ethanol [27] were added. Meanwhile, the sweet cherries coated with NE + $CaCl_2$ showed two layers. One layer was the NE on the cuticular membrane and the other was more compact and corresponded to the alginate cross-linked with calcium ions (arrows in Figure 1C). This second layer could reinforce the barrier properties of the cuticular membrane in sweet cherries.

3.3. Barrier Properties

3.3.1. Cracking Index (CI)

Figure 2 shows the CI of coated and uncoated sweet cherries. The NE, which was expected to provide protection against water absorption by cherries and reduce their cracking, had the opposite effect and its application significantly increased the percentage of cracked sweet cherries (71.1%) compared with uncoated sweet cherries (65.6%). This effect could be due to the dissolution of the cuticular waxes by the Tween 80 emulsifier in the NE formulation, resulting in the increased water permeability of the cuticle [39]. This higher cuticle permeability could have induced the water diffusion inside the fruit, causing a localized burst of the cells that led to cracking [6].

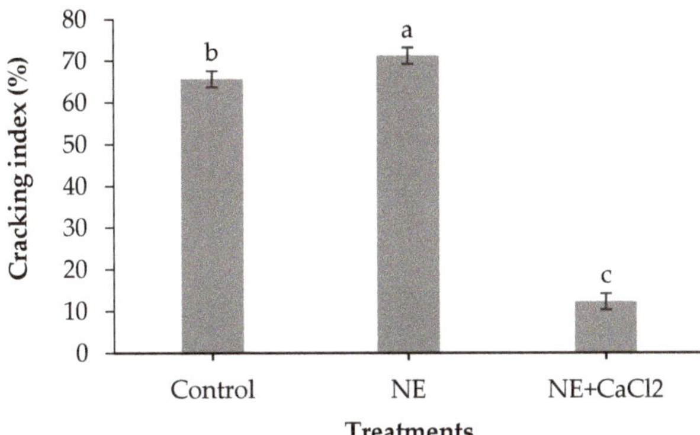

Figure 2. Effect of coating sweet cherries with nanoemulsion (NE) and nanoemulsion plus $CaCl_2$ (NE + $CaCl_2$) on the laboratory-induced cracking of sweet cherries. (a–c) Different letters indicate significant differences between treatments ($p < 0.05$).

Adding calcium ions to the NE coating dramatically increased the cracking tolerance and the percentage of cracked sweet cherries decreased from 65.5% to 12.2% (53.3% reduction). This result could be attributed to three different combined mechanisms. The first could be related to the decrease in osmotic potential on the fruit surface produced by the presence of calcium ions that did not react with the alginate [40]. Other authors mention that incorporating calcium ions on the surface of sweet cherries increased their cracking tolerance because of the decreased osmotic potential [41,42]. The second mechanism could involve strengthening the cuticular wax layer by Ca^{2+}, hardening the cell walls to tolerate greater osmotic pressure before rupturing [43]. The third mechanism could be associated with the formation of a more compact and insoluble cross-linked alginate layer (Figure 1C) that increased the hydrophobicity of the sweet cherry surface [10,21]. This could have decreased the water diffusion from the outside to the inside of the sweet cherries, thus reducing the cracking percentage.

Other researchers have studied the application of hydrophobic coatings in the preharvest stage to reduce rain cracking in sweet cherries. Torres et al. [44] applied RainGard (mixture of fatty acids and vegetable oil) three times on cherry trees and reported 40.5%, 40.0%, and 52.0% reductions in rain cracking at harvest for Bing, Sweetheart, and Van cherries, respectively. They indicated that the coating waterproofed the surface of the cherries and acted as a filler for the micro-cracks in the cuticle. The application of cellulose nanofiber-based hydrophobic coatings (Innofresh) to Sweetheart cherry trees decreased the rain cracking between 31.18% and 44.60%, depending on the level of the surfactant mixture (Tween 80 and Span 80 at a 1:1 ratio) used in the coating [45]. The surfactant mixture was the most critical factor affecting the wettability, hydrophilicity, and elasticity of the coatings [45]. However, in other study an opposite result was found when spraying an anti-transpirant (Vaporgard) on Royal Ann sweet cherry trees 7 d before harvest [46]. They revealed that applying Vaporgard produced more cracking than in the controls by increasing the overall turgor in the trees and causing the cherries to exceed the strength of the cuticle or wall against rupture with minimal water absorption.

3.3.2. Resistance to Water Vapor Transfer (RWVT)

Figure 3 shows the RWVT of coated and uncoated sweet cherries. All the coated fruits were less resistant to water loss than the control (Figure 3). The significant decrease in RWVT in the coated sweet cherries could be due to the emulsifier (Tween 80) that altered some epicuticle sites by changing, partially damaging, or extracting wax from the cuticle. These alterations could cause the dilation of the hydrophilic pores and lead to greater cuticle permeability [39,47]. Crosslinking with $CaCl_2$ did not increase the resistance to water loss despite the presence of an additional insoluble cross-linked alginate layer. This behavior could be associated with the swelling of the NE layer produced by water vapor transfer from the inside to the outside of the fruit. This increased volume of the NE layer can cause mechanical damage in the outer layer of cross-linked alginate, thus reducing its water vapor barrier capacity.

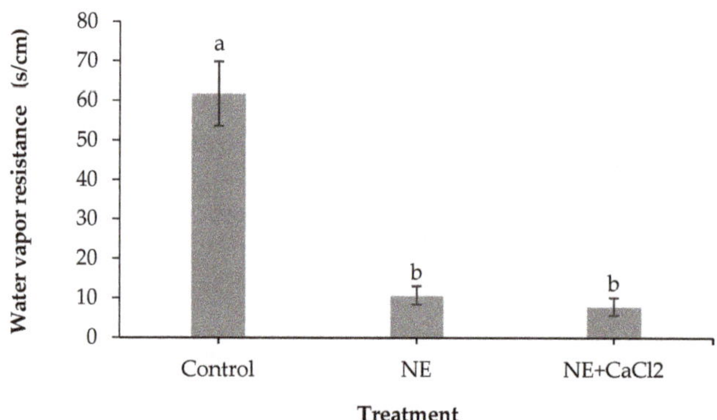

Figure 3. Effect of coating sweet cherries with nanoemulsion (NE) and nanoemulsion plus $CaCl_2$ (NE + $CaCl_2$) on the resistance to water vapor transfer. (a–b) Different letters indicate significant differences between treatments ($p < 0.05$).

3.4. Fruit Quality Parameters

3.4.1. Weight Loss

Fruit weight loss during postharvest is due to the gradient of water vapor pressure between the fruit and the surrounding air [48]. Both the layer of epidermal cells and the cuticle are responsible for controlling this weight loss. Sweet cherries are characterized by rapid senescence and a cuticle with low resistance to water vapor diffusion, which promotes rapid water loss from the fruit and stem [49]. In the present study, the weight loss of coated and uncoated sweet cherries progressively increased with storage time (Figure 4). Contrary to the expected effect, coated sweet cherries experienced greater weight loss (16% and 14% after 28 d at 4 °C for cherries coated with NE and NE + $CaCl_2$, respectively) than uncoated cherries (4% after 28 d at 4 °C).

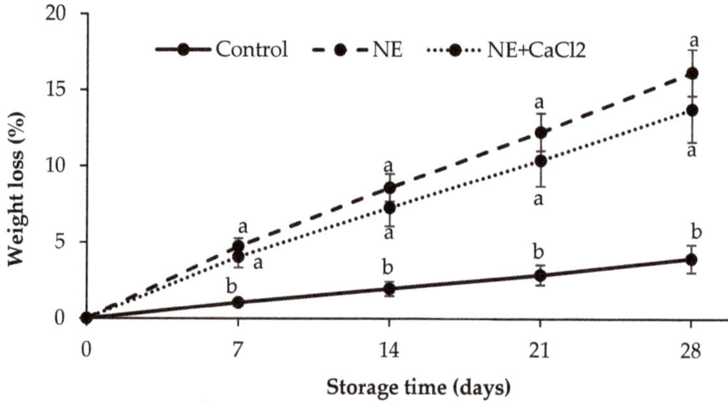

Figure 4. Effect of coting sweet cherries with nanoemulsion (NE) and nanoemulsion plus $CaCl_2$ (NE + $CaCl_2$) on weight loss during cold storage. (a–b) Different letters indicate significant differences between treatments ($p < 0.05$).

Similar results have been reported by Chiabrando and Giacalone [50] when applying alginate coatings at 1%, 3%, and 5%. These authors found that applying these coatings was not effective to reduce the expected weight loss. When using Big Lory sweet cherries, they obtained 8.15%, 7.40%, and 8.25% weight loss, respectively, after 21 d at 2 °C, while uncoated cherries reached 7.35%. In Grace sweet cherries, the control fruits and those coated with 1% alginate achieved a 10% weight loss, while cherries coated with 3% and 5% alginate reached 12%. However, Díaz-Mula et al. [1] obtained weight losses of 5.93%, 4.88%, and 3.71% in Sweetheart cherries when was applying an alginate coating at 1%, 3%, and 5%, respectively, after 16 d at 2 °C. Uncoated cherries had a weight loss of 6.81%. Some previous studies have indicated that gum arabic, almond gum [51], and chitosan [52] coatings can reduce the weight loss of sweet cherries.

In the present study, the higher weight loss of coated sweet cherries compared with uncoated sweet cherries can be attributed to the low RWVT of coated sweet cherries. This lower barrier capacity was caused by the emulsifier present in the NE coating, which increased the permeability of the sweet cherry cuticle, as described in Section 3.3.2. On the other hand, the crosslinking of the NE coating with $CaCl_2$ only slightly reduced the weight loss of the sweet cherries, due to the increase in the water vapor permeability of the alginate layer caused by its swelling.

3.4.2. Color Attributes

Changes in the skin color parameters of uncoated and coated sweet cherries during storage are shown in Figure 5. The hue angle was correlated with the anthocyanin content and the lowest hue angle values corresponded to high anthocyanin contents [53]. Hue angle values slightly decreased during storage in all the samples; the reduction was more pronounced from day 21 onward, especially in coated fruits (Figure 5). The coated fruits had lower hue values than uncoated fruits, and there were no significant differences between NE and NE + $CaCl_2$. Decreased hue values represent the progress of the fruit ripening process, reaching darker red colors in more advanced stages of maturity. This decrease in hue values during storage has also been described for other sweet cherry cultivars [54] and in sweet cherries coated with alginate [1] and Semperfresh (sucrose esters and mono-diglycerides of fatty acids and sodium carboxymethyl cellulose) [14].

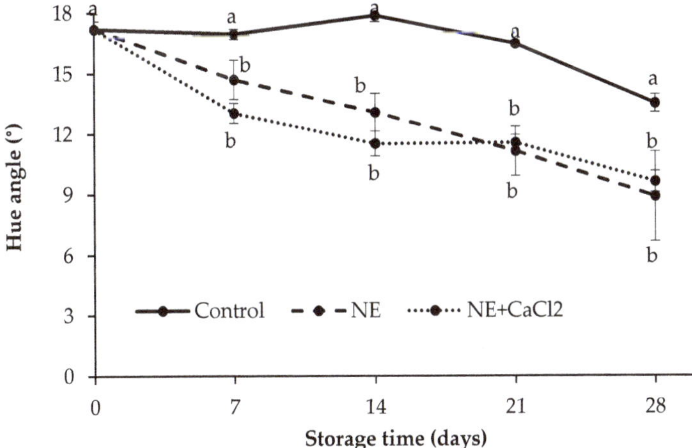

Figure 5. Color evolution (hue angle) of uncoated sweet cherries (control), sweet cherries coated with nanoemulsion (NE) and nanoemulsion plus $CaCl_2$ (NE + $CaCl_2$) during cold storage. (a–b) Different letters indicate significant differences between treatments ($p < 0.05$).

The lower hue values of the coated fruits compared with the uncoated fruits can be related to the greater water loss experienced by the coated fruits during storage and promoted by the emulsifier, as discussed in Section 3.4.1. As a consequence, the anthocyanin content in these fruits increased, thus producing a darker red color. In turn, no significant differences were observed in the hue values in the fruits coated with NE and NE + $CaCl_2$, mainly due to the fact that the sweet cherries with these coatings presented similar weight losses (Figure 4).

3.4.3. Fruit Firmness

Changes in postharvest firmness can be produced by moisture loss and enzymatic changes [55]. All the coated and uncoated fruits showed decreased firmness during storage (Figure 6). The coated sweet cherries exhibited higher firmness values than the uncoated fruits; however, no significant differences were observed between them as of day 21 of storage.

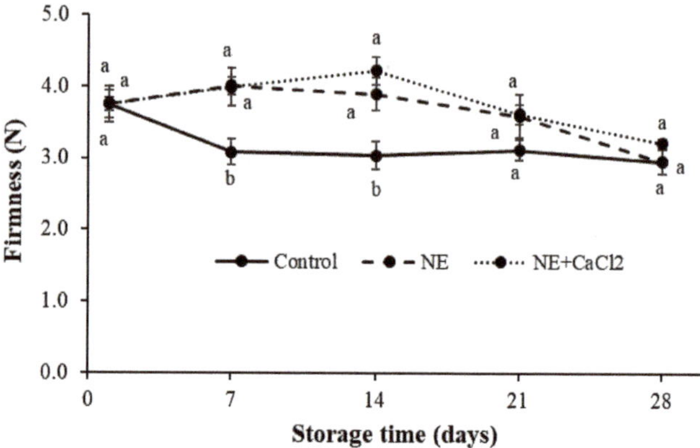

Figure 6. Changes in the firmness of uncoated sweet cherries (control), cherries coated with nanoemulsion (NE), and cherries coated with nanoemulsion and the application of $CaCl_2$ (NE + $CaCl_2$) during cold storage. (a–b) Different letters indicate significant differences between treatments ($p < 0.05$).

Similar results have been described in several studies applying different edible coatings on sweet cherries such as Semperfresh [14], alginate [1,50], almond gum, gum arabic [51], and guar gum with ginseng extract [56]. In these studies, the greater firmness retention of coated sweet cherries has been explained by the delayed enzymatic degradation of the components responsible for fruit structural rigidity caused by a decreased respiratory rate and cold temperature; it is also associated with reduced fruit moisture loss or maintained cellular turgor pressure.

No significant differences in firmness were observed between the fruits coated with NE and NE + $CaCl_2$. Even when a crosslinked alginate layer was formed on the surface of sweet cherries, it was not able to improve its water barrier capacity, obtaining a similar weight loss levels as those coated with NE and producing the same mechanical behavior.

3.4.4. Determination of Titratable Acidity (TA) and Total Soluble Solids (TSS)

The TA and TSS of coated and uncoated sweet cherries are shown in Figure 7. The TA value at harvest was 1.18 $\pm$ 0.1% malic acid equivalent, which decreased during storage for all cherries reaching at the end of the storage period values of 0.97 $\pm$ 0.02%, 1.02 $\pm$ 0.03%, and 1.03 $\pm$ 0.04% for control, NE, and NE + $CaCl_2$-coated sweet cherries, respectively (Figure 7A). The TA value decreased over time (Figure 7A) because organic acids are substrates for the enzymatic reactions of respiration [57]. From day 7 onwards, the uncoated sweet cherries showed a greater reduction in TA than the coated sweet cherries;

however, these differences were not significant. These results indicate that the coatings used were not able to significantly reduce the respiratory rate of fruits because they did not delay the use of organic acids, which are used in the enzymatic reactions of respiration [14]. Similar results have also been described when using 1% alginate coatings [50] and coatings of different types of 1% chitosan [58] in sweet cherries.

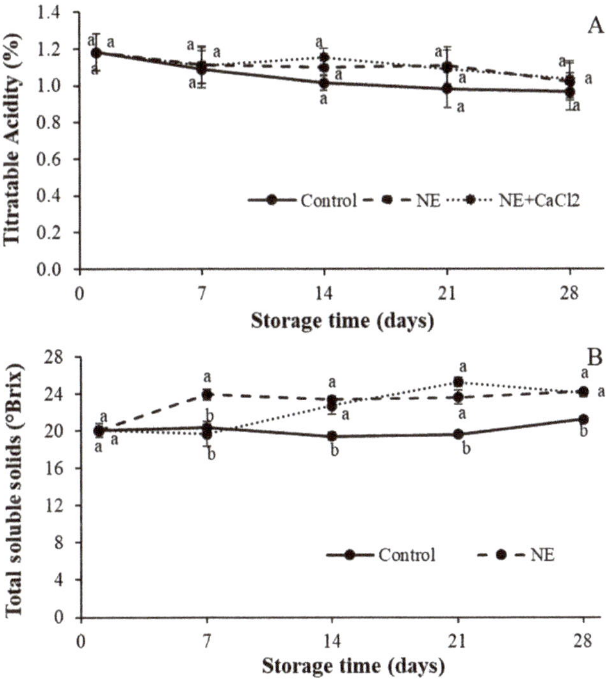

Figure 7. Effects of coating sweet cherries with nanoemulsion (NE) and nanoemulsion plus $CaCl_2$ (NE + $CaCl_2$) on (**A**) the titratable acidity and (**B**) total soluble solids during cold storage. (a–b) Different letters indicate significant differences between treatments ($p < 0.05$).

Additionally, no significant differences were observed between the NE and NE + $CaCl_2$ coatings. These results would also indicate that the formation of a cross-linked layer of alginate fails to improve the gas barrier capacity of the coating and thus reduce the respiratory rate of the coated fruits. Possibly, the nanoemulsified coating and its cross-linked layer undergo a plasticization process as water vapor is lost to the environment, reducing the gas barrier capacity of these coatings.

The initial TSS value was 20.0° Brix, but it increased during the storage period (Figure 7B). The coated sweet cherries exhibited a greater increase in TSS than the uncoated fruits, reaching values of 24.1 and 21.20° Brix, respectively, at the end of storage. The application of edible coatings on sweet cherries usually produces a delayed ripening process and a low increase in TSS compared with uncoated sweet cherries [54,55]. However, the opposite result was obtained in the present work. The higher TSS values for coated sweet cherries can be largely explained by the higher water loss in these fruits. Several authors point out that the loss of water during storage produces the concentration of sugars in coated fruits [55,59]. In the present study, the greater weight loss shown by the coated fruits compared to the uncoated fruits (Figure 4), due to the presence of emulsifier in the NE coatings as described in Sections 3.3.2 and 3.4.1, would be the main cause of these higher TSS values. This is also the reason why no significant differences were detected between the fruits coated with NE and NE + $CaCl_2$, since the fruits with both coatings showed similar levels of weight loss. As previously mentioned, the formation of an additional layer of cross-linked alginate on the sweet cherries was not able to reduce their water loss, because the nanoemulsified coating undergoes a swelling process during storage, reducing its water vapor barrier capacity.

The sourness and sweetness of sweet cherries are important for consumer acceptance. The TSS can be used to measure sweetness and TA to measure sourness [60]. Crisosto et al. [61] indicated that consumer acceptance and the level of satisfaction with sweet cherries increased with higher acidity (TA > 0.80%) and sweetness (TSS > 20.0%). According to our results, coated sweet cherries could have better consumer acceptance and level of satisfaction than the uncoated sweet cherries after 14 d because they had higher TA and TSS values.

3.5. Total Soluble Phenolics (TSP), Anthocyanin Content, and Antioxidant Capacity (AC)

The consumption of fruit and vegetables with high phytochemical contents such as anthocyanins and other polyphenolics, carotenoids, and vitamins C and D have been associated with the prevention of different chronic diseases [62]. Phenolic compounds also contribute to the sensory and organoleptic quality of sweet cherries, such as flavor and astringency [2] and their antioxidant potential [54].

Figure 8 shows that coatings affected the TSP content and AC of sweet cherries but not their anthocyanin content. The initial TSP content in uncoated sweet cherries was 1.62 ± 0.25 mg GAE/g (Figure 8A). The TSP values of uncoated and coated sweet cherries progressively increased over time. Coated sweet cherries had higher TSP values than uncoated sweet cherries. The TSP value after 28 d for uncoated sweet cherries was 2.11 ± 0.10 mg GAE/g, whereas the TSP values for sweet cherries coated with NE and NE + $CaCl_2$ were 2.47 ± 0.08 and 2.56 ± 0.22 mg GAE/g, respectively.

The increase in the total polyphenol content in sweet cherries in the present study was contrary to the expected results. Several studies have reported a decrease in the total polyphenol content in sweet cherries during storage as a result of peroxidase and polyphenol oxidase enzyme activity during the ripening process. They have also mentioned that applying edible coatings on sweet cherries has achieved a higher total polyphenol retention compared with uncoated fruits due to the formation of a protective barrier to gases on their surface [49,50,63].

In our study, the increased total polyphenol content can be explained by the development of two phenomena. First, the formation of a high gas barrier coating might have reduced phenol enzymatic oxidation. Second, the significant water loss in the coated fruits during storage due to the emulsifier in the coatings might have produced a concentration of the soluble polyphenolic compounds.

Anthocyanins are responsible for the red color in sweet cherries; their content increases during postharvest storage because the ripening process progresses and they are used as a quality indicator of cherries [64]. The anthocyanin content for all sweet cherries progressively increased during the storage period, and there were no significant differences between the coated and uncoated fruits (Figure 8B). These results were consistent with the decrease in hue angles observed in sweet cherries during the same period (Figure 5), which indicates a color change in sweet cherries from reddish red to more violet red [53,54]. However, even when the coated fruits had lower hue values than the uncoated fruits, it was not possible to detect these differences in anthocyanin content, which likely occurred because of the increased anthocyanin concentration in these coated fruits caused by significant water loss during storage.

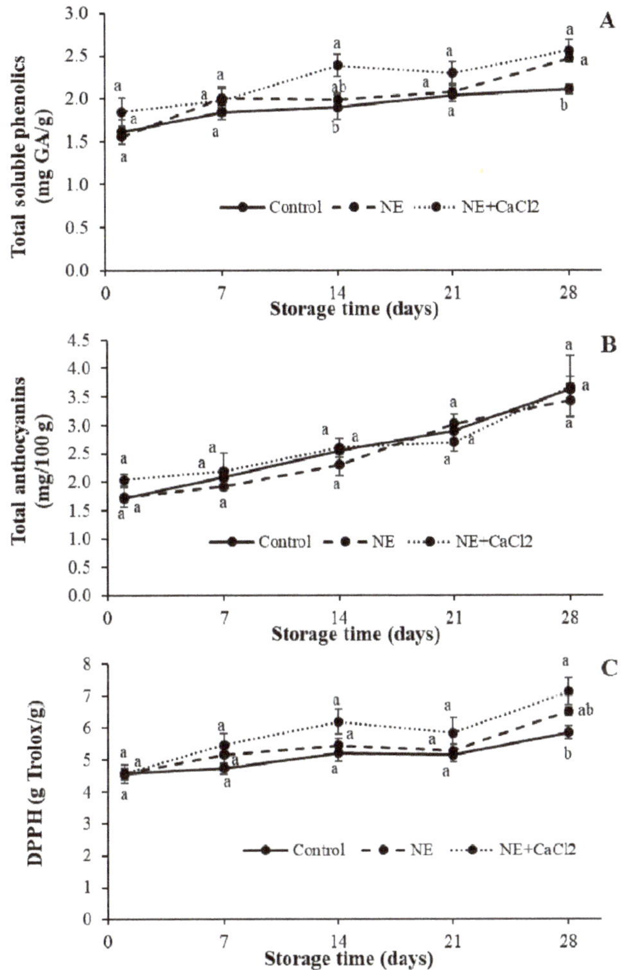

Figure 8. Effects of coating sweet cherries with nanoemulsion (NE) and nanoemulsion plus $CaCl_2$ (NE + $CaCl_2$) on (**A**) total soluble phenolics, (**B**) total anthocyanins, and (**C**) antioxidant capacity (DDPH) during cold storage. (a–b) Different letters indicate significant differences between treatments ($p < 0.05$).

The AC values of stored uncoated and coated sweet cherries progressively increased over time (Figure 8C). In general, sweet cherries coated with NE + $CaCl_2$ exhibited a higher antioxidant activity than uncoated sweet cherries and those coated with NE. At the end of storage, the inhibition of DPPH radicals was 7.13 ± 0.74, 6.50 ± 0.24, and 5.83 ± 0.37 g Trolox/g for NE + $CaCl_2$, NE, and uncoated sweet cherries, respectively. The higher CA of the coated fruits compared with the uncoated fruits could be explained by the increase in the TSP due to the modified internal atmosphere caused by the coatings [63] and the water loss indicated in Section 3.4.1.

4. Conclusions

The results presented in this study showed that the nanoemulsified coatings based on alginate and soybean oil presented different effects on the barrier properties and quality parameters of sweet cherries. The NE + $CaCl_2$ coatings were able to significantly reduce the cracking of sweet cherries, achieving a 53.3% reduction compared to the control fruits, due to the formation of a cross-linked layer on the surface of the coatings caused by the addition of $CaCl_2$ as a cross-linking agent.

The Ne coatings had a limited effect on the delay of the ripening process and the quality parameters of the cherries. The presence of an emulsifier in these coating could have altered the cuticle waxes and caused increased weight loss in sweet cherries coated, reducing their potential effect on the quality parameters of the cherries. This behavior was not improved either with the formation of the cross-linked alginate layer (NE + $CaCl_2$). However, a higher retention of total polyphenols and antioxidant capacity of the sweet cherries coated with NE + $CaCl_2$ was verified. Future studies should focus on optimizing the amount of emulsifier in nanoemulsified coatings to improve the barrier properties and make them more effective in delaying the ripening of sweet cherries.

Author Contributions: Conceptualization, R.V.-C. and C.G.-J.; methodology, R.V.-C., B.-S.C., C.B.-S., and T.W.; software, B.-S.C.; validation, C.G.-J.; formal analysis, C.G.-J. and C.B.-S.; investigation, C.G.-J. and T.W.; resources, T.M.; data curation, B.-S.C.; writing—original draft preparation, C.G.-J. and C.B.-S.; writing—review and editing, R.V.-C., C.B.-S., T.M., and B.-S.C. All authors have read and agreed to the published version of the manuscript.

Funding: This research was funded by Western Regional Research Center (WRRC) of the United States Department of Agriculture (USDA), CONICYT-Chile (Doctoral Scholarship PFCHA/DOCTORADO BECAS CHILE/2016-21161275) and Universidad del Bío-Bío.

Institutional Review Board Statement: Not applicable.

Informed Consent Statement: Not applicable.

Data Availability Statement: Not applicable.

Acknowledgments: The authors thank CONICYT for the Scholarship provided to the Camilo Gutierrez-Jara (PFCHA/DOCTORADO BECAS CHILE/2017—21170948) and the Graduate School of the Universidad del Bío-Bío, Chile for a research grant. The authors thank the Healthy Processed Foods Research Unit (HPFR) in the Western Regional Research Center (WRRC) of the United States Department of Agriculture (USDA), Albany, California, USA, for providing the infrastructure to carry out the research.

Conflicts of Interest: The authors declare no conflict of interest.

References

1. Díaz-Mula, H.M.; Serrano, M.; Valero, D. Alginate Coatings Preserve Fruit Quality and Bioactive Compounds during Storage of Sweet Cherry Fruit. *Food Bioprocess Technol.* **2012**, *5*, 2990–2997. [CrossRef]
2. Ferretti, G.; Bacchetti, T.; Belleggia, A.; Neri, D. Cherry antioxidants: From farm to table. *Molecules* **2010**, *15*, 6993–7005. [CrossRef] [PubMed]
3. Dziedzic, E.; Błaszczyk, J. Evaluation of sweet cherry fruit quality after short-term storage in relation to the rootstock. *Hortic. Environ. Biotechnol.* **2019**, *60*, 925–934. [CrossRef]
4. Simon, G. 2006 Review on rain induced fruit cracking of sweet cherries (*Prunus avium* L.), its causes and the possibilities of prevention. *Int. J. Hortic. Sci.* **2006**, *12*, 27–35. [CrossRef]
5. Correia, S.; Schouten, R.; Silva, A.P.; Gonçalves, B. Sweet cherry fruit cracking mechanisms and prevention strategies: A review. *Sci. Hortic.* **2018**, *240*, 369–377. [CrossRef]
6. Knoche, M.; Winkler, A. Chapter 7: Rain-induced cracking of sweet cherries. In *Cherries: Botany, Production and Uses*; Quero-García, J., Iezzoni, A., Puławska, J., Lang, G., Eds.; CABI: Wallingford, UK, 2017; pp. 140–165.
7. Measham, P.F.; Gracie, A.J.; Wilson, S.J.; Bound, S.A. Vascular flow of water induces side cracking in sweet cherry (*Prunus avium* L.). *Adv. Hortic. Sci.* **2010**, *24*, 243–248.
8. Balbontín, C.; Ayala, H.; Bastías, R.M.; Tapia, G.; Ellena, M.; Torres, C.; Yuri, J.A.; Quero-García, J.; Ríos, J.C.; Silva, H. Cracking in sweet cherries: A comprehensive review from a physiological, molecular, and genomic perspective. *Chil. J. Agric. Res.* **2013**, *73*, 66–72. [CrossRef]
9. Bourtoom, T. Review Article Edible films and coatings: Characteristics and properties. *Int. Food Res. J.* **2008**, *15*, 237–248.
10. Embuscado, M.E.; Huber, K.C. *Edible Films and Coatings for Food Applications*; Springer Science Business Media, LLC.: New York, NY, USA, 2009; pp. 135–168.
11. Romanazzi, G.; Nigro, F.; Ippolito, A. Short hypobaric treatments potentiate the effect of chitosan in reducing storage decay of sweet cherries. *Postharvest Biol. Technol.* **2003**, *29*, 73–80. [CrossRef]

12. Martínez-Romero, D.; Alburquerque, N.; Valverde, J.M.; Guillén, F.; Castillo, S.; Valero, D.; Serrano, M. Postharvest sweet cherry quality and safety maintenance by Aloe vera treatment: A new edible coating. *Postharvest Biol. Technol.* **2006**, *39*, 93–100. [CrossRef]
13. Yaman, Ö.; Bayoındırlı, L. Effects of an edible coating, fungicide and cold storage on microbial spoilage of cherries. *Eur. Food Res. Technol.* **2001**, *213*, 53–55. [CrossRef]
14. Yaman, Ö.; Bayoındırlı, L. Effects of an edible coating and cold storage on shelf-life and quality of cherries. *LWT Food Sci. Technol.* **2002**, *32*, 146–150. [CrossRef]
15. McClements, D.J. Thickening Agents. In *Food Emulsions. Principles Practices and Techniques*; McClements, D.J., Ed.; CRC Press: Boca Raton, FL, USA, 2016; pp. 151–154.
16. Parreidt, T.S.; Müller, K.; Schmid, M. Alginate-based edible films and coatings for food packaging applications. *Foods* **2018**, *7*, 170. [CrossRef]
17. Zapata, P.J.; Guillén, F.; Martínez-Romero, D.; Castillo, S.; Valero, D.; Serrano, M. Use of alginate or zein as edible coatings to delay postharvest ripening process and to maintain tomato (*Solanum lycopersicon* Mill) quality. *J. Sci. Food Agric.* **2008**, *88*, 1287–1293. [CrossRef]
18. Maftoonazad, N.; Ramaswamy, H.S.; Marcotte, M. Shelf-life extension of peaches through sodium alginate and methyl cellulose edible coatings. *Int. J. Food Sci. Technol.* **2008**, *43*, 951–957. [CrossRef]
19. Lin, D.; Zhao, Y. Innovations in the development and application of edible coatings for fresh and minimally processed fruits and vegetables. *Compr. Rev. Food Sci. Food Saf.* **2007**, *6*, 60–75. [CrossRef]
20. Niknejad, H.; Mahmoudzadeh, R. Comparison of different crosslinking methods for preparation of docetaxel-loaded albumin nanoparticles. *Iran. J. Pharm. Res.* **2015**, *14*, 385–394. [CrossRef] [PubMed]
21. Rhim, J.W. Physical and mechanical properties of water resistant sodium alginate films. *LWT Food Sci. Technol.* **2004**, *37*, 323–330. [CrossRef]
22. Jiang, Y.; Zhuang, C.; Zhong, Y.; Zhao, Y.; Deng, Y.; Gao, H.; Chen, H.; Mu, H. Effect of bilayer coating composed of polyvinyl alcohol, chitosan, and sodium alginate on salted duck eggs. *Int. J. Food Prop.* **2018**, *21*, 867–877. [CrossRef]
23. Parreidt, T.S.; Lindner, M.; Rothkopf, I.; Schmid, M.; Müller, K. The development of a uniform alginate-based coating for cantaloupe and strawberries and the characterization of water barrier properties. *Foods* **2019**, *8*, 203. [CrossRef] [PubMed]
24. Syarifuddin, A.; Hasmiyani; Dirpan, A.; Mahendradatta, M. Physical, mechanical, and barrier properties of sodium alginate/gelatin emulsion based-films incorporated with canola oil. *IOP Conf. Ser. Earth Environ. Sci.* **2017**, *101*, 12–19. [CrossRef]
25. Jamróz, E.; Kopel, P.; Tkaczewska, J.; Dordevic, D.; Jancikova, S.; Kulawik, P.; Milosavljevic, V.; Dolezelikova, K.; Smerkova, K.; Svec, P.; et al. Nanocomposite Furcellaran Films—the Influence of Nanofillers on Functional Properties of Furcellaran Films and Effect on Linseed Oil Preservation. *Polymers* **2019**, *11*, 2046. [CrossRef]
26. Morillon, V.; Debeaufort, F.; Blond, G.; Capelle, M.; Voilley, A. Factors affecting the moisture permeability of lipid-based edible films: A review. *Crit. Rev. Food Sci. Nutr.* **2002**, *42*, 67–89. [CrossRef]
27. Kurek, M.; Brachais, C.H.; Ščetar, M.; Voilley, A.; Galić, K.; Couvercelle, J.P.; Debeaufort, F. Carvacrol affects interfacial, structural and transfer properties of chitosan coatings applied onto polyethylene. *Carbohydr. Polym.* **2013**, *97*, 217–225. [CrossRef] [PubMed]
28. Gutiérrez-Jara, C.; Bilbao-Sainz, C.; McHugh, T.; Chiou, B.-S.; Williams, T.; Villalobos-Carvajal, R. Physical, mechanical and transport properties of emulsified films based on alginate with soybean oil: Effects of soybean oil concentration, number of passes and degree of surface crosslinking. *Food Hydrocoll.* **2020**, *109*. [CrossRef]
29. Christensen, V.J. Cracking in Cherries: III. Determination of Cracking Susceptibility. *Acta Agric. Scand.* **1972**, *22*, 128–136. [CrossRef]
30. Avena-Bustillos, R.J.; Cisneros-Zevallos, L.A.; Krochta, J.M.; Saltveit, M.E. Application of casein-lipid edible film emulsions to reduce white blush on minimally processed carrots. *Postharvest Biol. Technol.* **1994**, *4*, 319–329. [CrossRef]
31. Tapia, M.S.; Rojas-Graü, M.A.; Carmona, A.; Rodríguez, F.J.; Soliva-Fortuny, R.; Martin-Belloso, O. Use of alginate- and gellan-based coatings for improving barrier, texture and nutritional properties of fresh-cut papaya. *Food Hydrocoll.* **2008**, *22*, 1493–1503. [CrossRef]
32. Bilbao-Sainz, C.; Sinrod, A.; Powell-Palm, M.J.; Dao, L.; Takeoka, G.; Williams, T.; Wood, D.; Ukpai, G.; Aruda, J.; Bridges, D.F.; et al. Preservation of sweet cherry by isochoric (constant volume) freezing. *Innov. Food Sci. Emerg. Technol.* **2019**, *52*, 108–115. [CrossRef]
33. Serrano, M.; Guillén, F.; Martínez-Romero, D.; Castillo, S.; Valero, D. Chemical constituents and antioxidant activity of sweet cherry at different ripening stages. *J. Agric. Food Chem.* **2005**, *53*, 2741–2745. [CrossRef] [PubMed]
34. Donsí, F.; Wang, Y.; Li, J.I.; Huang, Q. Preparation of curcumin sub-micrometer dispersions by high-pressure homogenization. *J. Agric. Food Chem.* **2010**, *58*, 2848–2853. [CrossRef]
35. Salvia-Trujillo, L.; Rojas-Graü, M.A.; Soliva-Fortuny, R.; Martín-Belloso, O. Effect of processing parameters on physicochemical characteristics of microfluidized lemongrass essential oil-alginate nanoemulsions. *Food Hydrocoll.* **2013**, *30*, 401–407. [CrossRef]
36. Artiga-Artigas, M.; Acevedo-Fani, A.; Martín-Belloso, O. Effect of sodium alginate incorporation procedure on the physicochemical properties of nanoemulsions. *Food Hydrocoll.* **2017**, *70*, 191–200. [CrossRef]
37. Bargel, H.; Spatz, H.-C.; Speck, T.; Neinhuis, C. Two-Dimensional Tension Tests in Plant Biomechanics—Sweet Cherry Fruit Skin as a Model System. *Plant Biol.* **2004**, *6*, 432–439. [CrossRef]
38. Hershko, V.; Nussinovitch, A. Relationships between Hydrocolloid Coating and Mushroom Structure. *J. Agric. Food Chem.* **1998**, *46*, 2988–2997. [CrossRef]

39. Pham, T.T.M.; Singh, Z.; Behboudian, M.H. Different surfactants improve calcium uptake into leaf and fruit of "Washington Navel" sweet orange and reduce albedo breakdown. *J. Plant Nutr.* **2012**, *35*, 889–904. [CrossRef]
40. Wójcik, P.; Akgül, H.; Demirtaş, I.; Sarisu, C.; Aksu, M.; Gubbuk, H. Effect of Preharvest Sprays of Calcium Chloride and Sucrose on Cracking and Quality of "Burlat" Sweet Cherry Fruit. *J. Plant Nutr.* **2013**, *36*, 1453–1465. [CrossRef]
41. Fernandez, R.T.; Flore, J.A.; Ystaas, J. Intermittent application of CaCl2 to control rain cracking of sweet cherry. *Acta Hortic.* **1998**, *468*, 683–689. [CrossRef]
42. Mitre, I.; Mitre, V.; Roman, I.; Pop, A. Campfo-Ca 000—Leaf Fertilizer Effective in Preventing Cracking in the Sweet Cherry. *Bull. Univ. Agric. Sci. Vet. Med. Cluj-Napoca Hortic.* **2008**, *65*, 273–277. [CrossRef]
43. Glenn, G.M.; Poovaiah, B.W. Cuticular properties and postharvest calcium applications influence cracking of sweet cherries. *J. Am. Soc. Hortic. Sci.* **1989**, *114*, 781–788.
44. Torres, C.A.; Yuri, A.; Venegas, A.; Lepe, V. Use of a lipophilic coating pre-harvest to reduce sweet cherry (*Prunus avium* L.) Rain-Cracking. *Acta Hortic.* **2014**, *1020*, 537–544. [CrossRef]
45. Jung, J.; Deng, Z.; Simonsen, J.; Bastías, R.M.; Zhao, Y. Development and preliminary field validation of water-resistant cellulose nanofiber based coatings with high surface adhesion and elasticity for reducing cherry rain-cracking. *Sci. Hortic.* **2016**, *200*, 161–169. [CrossRef]
46. Richardson, D.G. Rain-cracking of "Royal Ann" sweet cherries: Fruit physiological relationships, water temperature, orchard treatments, and cracking index. *Acta Hortic.* **1998**, *468*, 677–682. [CrossRef]
47. Riederer, M.; Schönherr, J. Effects of surfactants on water permeability of isolated plant cuticles and on the composition of their cuticular waxes. *Pestic. Sci.* **1990**, *29*, 85–94. [CrossRef]
48. Laurin, E.; Nunes, M.; Emond, J.P.; Brecht, J.K. Vapor pressure deficit and water loss patterns during simulated air shipment and storage of Beit Alpha cucumbers. *Acta Hortic.* **2005**, *682*, 1697–1704. [CrossRef]
49. Petriccione, M.; De Sanctis, F.; Pasquariello, M.S.; Mastrobuoni, F.; Rega, P.; Scortichini, M.; Mencarelli, F. The Effect of Chitosan Coating on the Quality and Nutraceutical Traits of Sweet Cherry During Postharvest Life. *Food Bioprocess Technol.* **2015**, *8*, 394–408. [CrossRef]
50. Chiabrando, V.; Giacalone, G. Effects of alginate edible coating on quality and antioxidant properties in sweet cherry during postharvest storage. *Ital. J. Food Sci.* **2015**, *27*, 45–52. [CrossRef]
51. Mahfoudhi, N.; Hamdi, S. Use of Almond Gum and Gum Arabic as Novel Edible Coating to Delay Postharvest Ripening and to Maintain Sweet Cherry (*Prunus avium*) Quality during Storage. *J. Food Process. Preserv.* **2015**, *39*, 1499–1508. [CrossRef]
52. Dang, Q.F.; Yan, J.Q.; Li, Y.; Cheng, X.J.; Liu, C.S.; Chen, X.G. Chitosan Acetate as an Active Coating Material and Its Effects on the Storing of *Prunus avium* L. *J. Food Sci.* **2010**, *75*, S125–S131. [CrossRef] [PubMed]
53. Gonçalves, B.; Silva, A.P.; Moutinho-Pereira, J.; Bacelar, E.; Rosa, E.; Meyer, A.S. Effect of ripeness and postharvest storage on the evolution of colour and anthocyanins in cherries (*Prunus avium* L.). *Food Chem.* **2007**, *103*, 976–984. [CrossRef]
54. Serrano, M.; Díaz-Mula, H.M.; Zapata, P.J.; Castillo, S.; Guillén, F.; Martínez-Romero, D.; Valverde, J.M.; Valero, D. Maturity stage at harvest determines the fruit quality and antioxidant potential after storage of sweet cherry cultivars. *J. Agric. Food Chem.* **2009**, *57*, 3240–3246. [CrossRef]
55. Aday, M.S.; Caner, C. Understanding the effects of various edible coatings on the storability of fresh cherry. *Packag. Technol. Sci.* **2010**, *23*, 441–456. [CrossRef]
56. Dong, F.; Wang, X. Guar gum and ginseng extract coatings maintain the quality of sweet cherry. *LWT Food Sci. Technol.* **2018**, *89*, 117–122. [CrossRef]
57. Petriccione, M.; Mastrobuoni, F.; Pasquariello, M.; Zampella, L.; Nobis, E.; Capriolo, G.; Scortichini, M. Effect of Chitosan Coating on the Postharvest Quality and Antioxidant Enzyme System Response of Strawberry Fruit during Cold Storage. *Foods* **2015**, *4*, 501–523. [CrossRef]
58. Tokatlı, K.; Demirdöven, A. Effects of chitosan edible film coatings on the physicochemical and microbiological qualities of sweet cherry (*Prunus avium* L.). *Sci. Hortic.* **2020**, *259*, 108656. [CrossRef]
59. Vieira, J.M.; Flores-López, M.L.; de Rodríguez, D.J.; Sousa, M.C.; Vicente, A.A.; Martins, J.T. Effect of chitosan-Aloe vera coating on postharvest quality of blueberry (*Vaccinium corymbosum*) fruit. *Postharvest Biol. Technol.* **2016**, *116*, 88–97. [CrossRef]
60. Correia, S.; Schouten, R.; Silva, A.P.; Gonçalves, B. Factors affecting quality and health promoting compounds during growth and postharvest life of sweet cherry (*Prunus avium* L.). *Front. Plant Sci.* **2017**, *8*, 2166. [CrossRef] [PubMed]
61. Crisosto, C.H.; Crisosto, G.M.; Metheney, P. Consumer acceptance of "Brooks" and "Bing" cherries is mainly dependent on fruit SSC and visual skin color. *Postharvest Biol. Technol.* **2003**, *28*, 159–167. [CrossRef]
62. Schreiner, M.; Huyskens-Keil, S. Phytochemicals in fruit and vegetables: Health promotion and postharvest elicitors. *CRC. Crit. Rev. Plant Sci.* **2006**, *25*, 267–278. [CrossRef]
63. Zam, W. Effect of Alginate and Chitosan Edible Coating Enriched with Olive Leaves Extract on the Shelf Life of Sweet Cherries (*Prunus avium* L.). *J. Food Qual.* **2019**. [CrossRef]
64. Serradilla, M.J.; Lozano, M.; Bernalte, M.J.; Ayuso, M.C.; López-Corrales, M.; González-Gómez, D. Physicochemical and bioactive properties evolution during ripening of "Ambrunés" sweet cherry cultivar. *LWT Food Sci. Technol.* **2011**, *44*, 199–205. [CrossRef]

 *foods*

MDPI

Article

Effects of Whey Protein Isolate-Based Film Incorporated with Tarragon Essential Oil on the Quality and Shelf-Life of Refrigerated Brook Trout

Maria-Ioana Socaciu [1], Melinda Fogarasi [2], Elemér Lajos Simon [1], Cristina Anamaria Semeniuc [2,*], Sonia Ancuţa Socaci [1], Andersina Simina Podar [1] and Dan Cristian Vodnar [1,*]

1 Department of Food Science, University of Agricultural Sciences and Veterinary Medicine Cluj-Napoca, 3-5 Mănăştur St., 400372 Cluj-Napoca, Romania; maria-ioana.socaciu@usamvcluj.ro (M.-I.S.); elemer.simon@usamvcluj.ro (E.L.S.); sonia.socaci@usamvcluj.ro (S.A.S.); andersina-simina.podar@usamvcluj.ro (A.S.P.)

2 Department of Food Engineering, University of Agricultural Sciences and Veterinary Medicine Cluj-Napoca, 3-5 Mănăştur St., 400372 Cluj-Napoca, Romania; melinda.fogarasi@usamvcluj.ro

* Correspondence: cristina.semeniuc@usamvcluj.ro (C.A.S.); dan.vodnar@usamvcluj.ro (D.C.V.); Tel.: +40-264-596-384 (C.A.S. & D.C.V.)

Abstract: The efficiency of some films prepared from heat-denatured whey protein isolate solutions on the quality and shelf-life of brook trout samples during storage at 4 °C was studied in this research (WPIf-a film based on whey protein isolate and WPIf+2.5%TEO-a film based on whey protein isolate incorporated with 2.5% tarragon essential oil). The control and covered fish samples were periodically assessed (at 3 days) over 15 days of storage for the physicochemical (pH; EC, electrical conductivity; TVB-N, total volatile basic nitrogen; TBARS, thiobarbituric acid reactive substances; color), microbiological (TVC, total viable count; PTC, psychrotrophic count; LAB, lactic acid bacteria; H_2S-producing bacteria), and sensory properties (color discoloration; odor; overall acceptability). The WPIf+2.5%TEO has proven enhanced quality preservation effects compared to WPIf by showing lower values for physicochemical parameters, lower microbial loads, and higher sensory scores in the fish sample. All these effects have led to an extension of the sample's shelf-life. In conclusion, the tarragon essential oil has conferred antioxidant and antimicrobial properties to the film. Thus, the WPIf+2.5%TEO could be a promising material for the packaging of fresh brook trout during refrigerated storage.

Keywords: film; whey protein isolate; tarragon essential oil; brook trout; refrigerated storage; physicochemical quality; microbiological quality; sensory quality; shelf-life

Citation: Socaciu, M.-I.; Fogarasi, M.; Simon, E.L.; Semeniuc, C.A.; Socaci, S.A.; Podar, A.S.; Vodnar, D.C. Effects of Whey Protein Isolate-Based Film Incorporated with Tarragon Essential Oil on the Quality and Shelf-Life of Refrigerated Brook Trout. *Foods* **2021**, *10*, 401. https://doi.org/10.3390/foods10020401

Academic Editors: Matteo Alessandro Del Nobile and Amalia Conte

Received: 6 January 2021
Accepted: 7 February 2021
Published: 11 February 2021

1. Introduction

The brook trout (*Salvelinus fontinalis*) is a species of freshwater fish in the genus *Salvelinus* of the family Salmonidae [1]. Originally found only in north-eastern North America, it was subsequently introduced to western Canada and the United States, as well as South America, New Zealand, Asia, and several regions of Europe [2]. Fish is one of the most commercialized, but also perishable, food products due to high water activity, nutrient availability, nearly neutral-pH, and the presence of autolytic enzymes [3,4]. Under normal refrigerated storage conditions, its shelf-life is limited by the development of enzymatic and chemical reactions [4]. Therefore, keeping fish quality during the supply chain is a challenge for food manufacturers [5].

In recent years there has been an increase in the development of active packaging materials, based on natural polymers, by incorporating essential oils [4,6]. It is well known that essential oils possess antimicrobial and antioxidant properties [7,8]. However, depending on the scent intensity or the level of incorporation, these may impart foreign taste and aroma to food. Thyme and oregano essential oils, the active substances most used in packaging materials, have strong scent intensities. Others, like tarragon essential oil, have a weaker scent.

Previous research has shown that tarragon essential oil (*Artemisia dracunculus* L.) possesses antibacterial, antifungal, and antioxidant properties [9,10]. Russian tarragon and French tarragon are the best known regional "varieties" of *A. dracunculus*. The chemical composition of their essential oils was found to be different; terpinen-4-ol, sabinene, and elemicin are the main components of Russian tarragon and estragole is the predominant compound of French tarragon [11]. The essential oil of tarragon has been successfully used as an active ingredient in the development of a hake protein-based edible film [12], of a chitosan-based coating (for maintaining the postharvest quality of kumquats fruits) [13], and of a chitosan-gelatin coating (for the preservation of pork slices) [14].

Natural polymers, such as protein and polysaccharides, are commonly used to prepare films and coatings for food applications. Several studies have investigated the efficiency of such active packaging materials in retarding fish spoilage during refrigerated storage: a gelatin-alginate film containing oregano essential oil on rainbow trout slices [15], a whey protein concentrate coating incorporated with cinnamon essential oil on beluga sturgeon fillets [16], a coating based on chitosan and gelatin on golden pomfret fillets [17], a whey protein coating incorporated with lactoperoxidase and α-tocopherol on pike-perch fillets [18], and a gelatin-chitosan coating incorporated with clove essential oil on tuna fillets [19]. The antimicrobial effect of chitosan has been widely reported, but the acetic acid used to prepare its solution gives the coating a strong acidic flavor [20,21]. Moreover, chitosan is substantially more expensive than other polymeric matrices such as whey protein.

Films based on whey protein are usually obtained by casting and drying of aqueous WPI (whey protein isolate) and WPC (whey protein concentrate) solutions; these films have shown moderate moisture barrier properties but good oxygen barrier properties [22]. In a previous study, we have developed a WPI-based film incorporated with 2.5% tarragon essential oil [6]. This study aimed to investigate the effectiveness of this film in maintaining the quality, thereby extending the shelf-life of refrigerated brook trout. As far as we know, no research work has considered such a case study on brook trout. The packaging materials incorporated with essential oils previously developed were not investigated for their effect on fish sensory properties. For this purpose, physicochemical, microbiological, and sensory properties of brook trout samples were evaluated during 15 days of storage at 4 °C in the current study.

2. Materials and Methods

2.1. Materials

Live brook trout specimens (*Salvelinus fontinalis*), with an average weight of 650 ± 20 g, were harvested from a local aquaculture farm (Gilău, Cluj County, Romania) in July 2020.

Whey protein isolate (Prolacta 95 LL Instant, Lactalis, France) was purchased from REDIS C.O. S.R. (Bucharest, Romania) and glycerol from Chempur (Piekary Śląskie, Poland). Tarragon essential oil (*Artemisia dracunculus* L.) was purchased from Aroma-Zone (Cabrières, France). n-Hexane for gas chromatography was purchased from Merck KGaA (Darmstadt, Germany). Perchloric acid, antifoam silicone, sodium hydroxide, boric acid, methyl red, methylene blue, and 95% ethanol were purchased from VWR International, LLC (Fontenay-sous-Bois, France). The hydrochloric acid standardized solution was purchased from Alfa Aesar (Kandel, Germany).

Phosphate buffered saline tablets were purchased from VWR International, LLC (Solon, OH, USA). Ethylenediaminetetraacetic acid disodium salt dehydrate was purchased from Sigma Chemical Co. (St. Louis, MO, USA); L(+)-ascorbic acid from Carl Roth GmbH+Co. KG (Karlsruhe, Germany), trichloroacetic acid glacial from VWR International, LLC (Leuven, Belgium), 2-thiobarbituric acid from Thermo Fisher (Kandel) GmbH (Karlsruhe, Germany), and 1,1,3,3-tetramethoxypropane from Tokyo Chemical Industry Co., Ltd. (Tokyo, Japan).

Strainer bags (BA6040/STR, 105 mm × 155 mm) were purchased from Seward Ltd. (Worthing, United Kingdom). Sodium chloride and plate count agar (granulated) were purchased from VWR Chemicals BDH Prolabo (Leuven, Belgium), Lactobacillus MRS agar (MRS agar) from HiMedia Laboratories Pvt. Ltd. (Mumbai, India), and iron agar (Lyngby) from Laboratorios Conda S.A. (Madrid, Spain).

2.2. Preparation and Treatment of Fish Samples

After being immediately slaughtered by percussive stunning, fishes were gutted and washed in potable water. The gutted fishes were then transferred to the laboratory within 20 min of slaughtering, and packed in insulated boxes containing ice. Here, fishes were further headed, filleted, and skinned by hand.

The prepared fillets were cut and then minced using a kitchen meat grinder (Philips Food processor HR7764/13, Amsterdam, Netherlands) equipped with a plate disc knife (4mm diameter holes). The minced fish meat was weighed into 15 ± 0.01 g portions; these were shaped into patties of approximately 43 mm diameter and 6 mm thickness using a customized burger mold.

The fish patties were divided into three batches (90 patties per batch); samples from the first batch were covered (top and bottom) with whey protein isolate-based films incorporated with 2.5% tarragon essential oil (WPIf+2.5%TEO), from the second batch with whey protein isolate-based films (WPIf), and from the third batch were left uncovered (Control). Next, all fish samples were aerobically packed into 100 × 150 mm sterile polyethylene ziplock bags and stored at 4 °C for 15 days. Physicochemical, microbiological and sensory analyses were performed at 3-day intervals to measure the quality of brook trout.

2.3. Preparation of Films

Two types of films (with and without tarragon essential oil) were prepared from heat-denatured whey protein isolate solutions. Film-forming solutions were obtained by dissolving 5% (*w*/*w*) WPI in distilled water, according to the protocol described by Socaciu et al. (2020) [6]. Glycerol was added to filmogenic solutions at a concentration of 5% (*w*/*w*).

Solutions were subsequently heated for 30 min at 90 ± 2 °C while being continuously stirred using a magnetic stirrer with heating (MSH-300, Biosan Ltd., Riga, Latvia). Heated solutions were then cooled at room temperature for 1.5 h and filtered to remove any air incorporated during stirring. For the preparation of active packaging films, the essential oil of tarragon was added to their solution to reach a final concentration of 2.5% (*w*/*w*).

Next, both solutions (with and without tarragon essential oil) were homogenized at 23,000 rpm for 2.5 min using a laboratory dispenser (T 18 digital Ultra-Turrax, IKA-Werke GmbH & Co. KG, Staufen, Germany). The final film-forming solutions were poured (4.8 g) into disposable weighing dishes (43 × 13 mm) and then dried in an oven (Digitheat, J.P. Selecta S.A., Barcelona, Spain) at 37 °C for 42 h. Once formed, the films were peeled off and conditioned for 72 h at 20 ± 2 °C and 50 ± 2% relative humidity in a climatic test cabinet (TK 252; Nüve Sanayi Malzemeleri İmalat ve Ticaret A.Ş., Akyurt/Ankara, Turkey), then used to cover the brook trout samples.

2.4. GC-MS Analysis of Essential Oil

The GC-MS analysis (gas chromatography coupled with mass spectrometry) of the volatile profile of tarragon essential oil was carried out as described in our previous paper, with some modification [23]. In this regard, the tarragon essential oil sample was diluted in n-hexane before the injection into the GC-MS injector.

A Zebron ZB-5 ms capillary column (30 m × 0.25 mm i.d. × 0.25 µm film thickness; Phenomenex, Torrance, CA, USA) was used for the chromatographic separation of volatile constituents. The column oven temperature program was set as follows, from 50 °C (kept at this temperature for 2 min) to 160 °C at 4 °C/min, then raised to 250 °C at 10 °C/min (kept at this temperature for 10 min). The temperature of the ion source, injector, and interface was set at 250 °C. Helium was used as the carrier gas at a flow rate of 1 mL/min.

The split ratio was 1:200. The ion trap mass spectrometer was operated in EI-MS mode; the acquisition mode was set in the range of 40–500 m/z.

The volatile compounds were tentatively identified by comparing their recorded mass spectra with those of standard compounds (β-myrcene, limonene, 1,8-cineol, α-pinene, and caryophyllene) and with those from NIST27 and NIST147 spectra libraries (considering a minimum similarity of 95%) as well as by retention indices obtained from www.pherobase.com (accessed on 29 April 2020) [24] or www.flavornet.org (accessed on 29 April 2020) (for columns with a similar stationary phase to ZB-5 ms) [25]. The results were expressed as the relative percentage of each compound from the total ion chromatograms (TIC) area (100%).

2.5. Physicochemical Analysis of Fish

2.5.1. Determination of Proximate Composition

The fat content, protein content, and moisture of the fish sample were measured using a near-infrared meat analyzer (FoodScan, FOSS, Hillerød, Denmark). Two replicates were run for each fish sample. The results were displayed as g/100 g fish sample.

The ash content was determined by incineration of the fish sample in a muffle furnace (L3/11/B170, Nabertherm GmbH, Bremen, Germany), according to the method described by Nagy et al. (2017) [26]. A 3.0 g portion of the fish sample was weighed in a porcelain melting pot and kept at 600 °C for 6 h in the muffle furnace. The ash content was calculated with the following Equation (1):

$$\text{Ash } (\text{g}/100 \text{ g fish sample}) = \frac{w_a}{w_{fs}} \times 100 \qquad (1)$$

where w_a is the weight of the ash (g) and w_{fs} is the weight of the fish sample (g). Two replicates were run for each fish sample.

2.5.2. Determination of pH and Electrical Conductivity (EC)

The pH and EC of the fish sample were determined according to the method described in the ISO 2917:1999 standard [27]. A 10.0 g portion of the fish sample was homogenized in 100 mL distilled water for 30 sec using a glass rod. The mixture was left undisturbed at room temperature for 30 min and then filtered. The pH and EC of the extract were measured using a digital multi-parameter meter (InoLab® Multi 9310 IDS, WTW, Weilheim, Germany). Two replicates were run for each fish sample. The result for EC was expressed in µS/cm.

2.5.3. Determination of Total Volatile Basic Nitrogen (TVB-N)

The TVB-N concentration of the fish sample was determined by the reference procedure described in the European Commission's Decision 95/149/EC [28]. A 10.0 g portion of the fish sample was homogenized in 90 mL of 6% (w/v) perchloric acid solution at 20,000 rpm for 2 min using a laboratory dispenser (T 18 digital Ultra-Turrax, IKA-Werke GmbH & Co. KG, Staufen, Germany). The mixture was then filtered and the obtained extract kept in the refrigerator until use.

A 50 mL aliquot of the extract was mixed with a few drops of silicone antifoaming agent and 6.5 mL of 20% (w/v) sodium hydroxide solution. The mixture was subjected to steam distillation using a Kjeldahl distillation unit (UDK 140, VELP Scientifica, Milan, Italy) till 100 mL of distillate had been produced. The distillate was collected in a flask containing 100 mL of 3% (w/v) boric acid solution and 3-5 drops of Tashiro mixed indicator (0.2 g methyl red and 0.1 g methylene blue dissolved in 100 mL 95% ethanol). Next, the receiver solution was titrated with a 0.01 N hydrochloric acid standardized solution until the endpoint (pH 5.0 $\pm$ 0.1) was reached. A blank sample was prepared with 50 mL of 6%

(w/v) perchloric acid solution instead of extract and treated identically to the fish sample. The TVB-N concentration was calculated with the following Equation (2):

$$\text{TVB} - \text{N (mg N/100 g fish sample)} = \frac{(V_1 - V_0) \times 0.14 \times 2 \times 100}{w_{fs}} \quad (2)$$

where V_1 is the volume of 0.01 N hydrochloric acid solution consumed for the fish sample at endpoint (mL), V_0 is the volume of 0.01 N hydrochloric acid solution consumed for the blank at endpoint (mL), and w_{fs} is the weight of the fish sample (g). Three replicates were run for each fish sample.

2.5.4. Determination of Thiobarbituric Acid Reactive Substances (TBARS)

The TBARS concentration of the fish sample was determined according to the modified method of Semeniuc et al. (2016a, 2016b) [29,30].

A 2.0 g grams portion of the fish sample was homogenized in 4 mL of PBS [phosphate buffered saline (1X solution, pH 7.4) containing 0.1% (w/v) ethylenediaminetetraacetic acid disodium salt and 0.1% (w/v) ascorbic acid] at 21,500 rpm for 30 sec using a laboratory dispenser (T 18 digital Ultra-Turrax, IKA-Werke GmbH & Co. KG, Staufen, Germany); 2 mL of TCA [30% (w/v) trichloroacetic acid solution] was then added and again homogenized at 17,500 rpm for 30 sec. The mixture was transferred into a 10-mL volumetric flask, brought to volume with PBS and vortexed (Vortex V-1 Plus, Biosan Ltd., Riga, Latvia). The precipitate formed was removed by filtration, using a Whatman no. 1 filter paper, and the extract was collected.

A 5 mL aliquot of the extract was mixed with 5 mL of TBA [0.02 M 2-thiobarbituric acid solution] into a 10-mL test tube with screw cap. The test tube was immersed in a water bath at 90 °C and kept it for 20 min. Subsequently, the test tube was placed into a refrigerator (~4 °C) for 30 min to cool. A blank sample was prepared with 4 mL of PBS and 1 mL of TCA instead of extract and treated identically to the fish sample. The absorbance of each fish sample was read against the blank at 530 nm using a double beam UV-Vis spectrophotometer (UV-1900i, Shimadzu Scientific Instruments, Inc., Columbia, MD, USA). Three replicates were run for each fish sample.

Seven standard solutions of 1,1,3,3-tetra methoxy propane (TMP), a precursor of malondialdehyde, were prepared and subjected to the TBARS assay to construct a calibration curve. The calibration curve was linear over the range of 1.88 to 112.80 nmol MDA/mL. The TBARS concentration was calculated with the following Equation (3):

$$\text{TBARS (mg MDA/kg fish sample)} = \frac{(Abs. - b)}{m} \times 10^{-10} \times \frac{10}{w_{fs}} \times 72.06 \times 10^6 \quad (3)$$

where $Abs.$ is the absorbance value at 530 nm, b is the y-intercept of the linear equation, m is the slope of the regression line, 10 is the volume of the mixture before filtration (mL), w_{fs} is the weight of the fish sample (g), and 72.06 is the molecular weight of malondialdehyde (g/mol).

2.5.5. Measurement of Color

The color of the fish sample was measured using an NH300 portable colorimeter (3NH, Shenzhen, China) based on the CIE $L^*a^*b^*$ color system. The L^*-value represents lightness and ranges from zero (the darkest black) to 100 (the brightest white). The a^*-value represents "redness" or "greenness" and ranges from +60 for absolute red to −60 for absolute green, while b^*-value represents "yellowness" or "blueness" and ranges from +60 for absolute yellow to −60 for absolute blue. Measurements were performed using a D65 illuminant with an opening of 8 mm and a 10° standard observer. The colorimeter was subjected to automatic black and white calibration. Twelve readings were taken on each fish sample.

2.6. Microbiological Analysis of Fish

An amount of 5.0 g fish sample was aseptically weighed into a sterile stomacher bag and homogenized in 45 mL of 0.85% (*w*/*v*) sodium chloride solution for 1 min using a bag mixer (MiniMix® 100 P CC®, Interscience, Saint Nom, France). Seven ten-fold serial dilutions (10^{-2}, 10^{-3}, 10^{-4}, 10^{-5}, 10^{-6}, 10^{-7}, and 10^{-8}) were prepared from this stock solution (10^{-1}).

The total viable count (TVC) and psychrotrophic count (PTC) were determined according to the method described in the ISO 4833-1:2013 standard using plate count agar [31]. Enumeration of TVC and TPC was performed after incubation of inoculated plates at 30 °C for 72 h [31], respectively at 7 °C for 10 days [32].

Lactic acid bacteria (LAB) were determined according to the method described in the ISO 7889 | IDF 117:2003 standard using MRS agar [33]. Enumeration of colonies was performed after anaerobic incubation of inoculated plates at 37 °C for 72 h.

Hydrogen sulfide (H_2S)-producing bacteria (including *Shewanella putrefaciens*) were determined according to the method described by Yu et al. (2017) using iron agar in double-layer [34]. Enumeration of colonies (*Pseudomonas fluorescens* as white colonies and *Shewanella putrefaciens* as black colonies) was performed after incubation of inoculated plates at 30 °C for 4 days.

All counts were expressed as $\log_{10}$ CFU/g and performed in duplicate.

2.7. Sensory Evaluation of Fish

Sensory evaluation was carried out as described by Bahram et al. (2016), with minor modification [16]. The fish samples were evaluated by 5 trained panelists from the laboratory staff (four women and one man, aged 25–41 years). The fish samples were taken out of the bags and films removed from the covered ones. These were then individually placed on small white ceramic plates, coded with 3-digit random numbers, and presented to panelists. They were asked to assess the patties in the following order: (1) the fish patty from control batch, (2) the fish patty from WPIf batch, and (3) the fish patty from WPIf+2.5%TEO batch. The judges were not informed about the batch and storage time of fish patties. The panelists were asked not to eat, drink, or smoke for at least 1 h before the evaluation session; they evaluated the sensory attributes of fish patties at 20 °C (air -conditioning) under white light, in individual cabins.

The sensory evaluation was based on a five-point scale to determine: color discoloration (5, no discoloration; 4, slight discoloration; 3, moderate discoloration; 2, strong discoloration; 1, extreme discoloration), odor (5, extremely desirable; 4, slightly desirable; 3, neither desirable nor unacceptable/off-odor; 2, slightly unacceptable/off-odor; 1, extremely unacceptable/off-odor), and overall acceptability (5, extremely desirable; 4, slightly desirable; 3, neither desirable nor unacceptable; 2, slightly unacceptable; 1, extremely unacceptable). Shelf-life criteria assumed that rejection would occur when the sensory attributes declined below 4.0.

2.8. Statistical Analysis

Data analysis was carried out using Minitab statistical software (version 16.1.0, LEAD Technologies, Inc., Charlotte, NC, USA). The differences in results among different groups were determined using one-way ANOVA. Post-hoc pairwise comparisons were performed with Tukey's test at a 95% confidence level ($p < 0.05$). Correlations among data were calculated using Pearson's correlation coefficient.

3. Results and Discussion

3.1. Volatile Composition of Essential Oil

The volatile compounds detected in tarragon essential oil by GC-MS analysis are listed in Table 1. Nine compounds, representing 100% of the total detected constituents, were identified in the essential oil of tarragon and grouped based on their chemical structure into four classes (monoterpene hydrocarbons, oxygenated monoterpenes, phenylpropanoids, and sesquiterpene hydrocarbons). The most abundant constituents were phenylpropanoids (81.84%), followed by monoterpene hydrocarbons (17.47%), the oxygenated monoterpenes (0.42%), and sesquiterpene hydrocarbons (0.27%). The major components identified in tarragon essential oil were estragole (81.84%), cis-β-ocimene (6.51%), trans-β-ocimene (6.19%), and D-limonene (4.05%).

Table 1. Relative contents (%) of volatile constituents identified in tarragon essential oil.

Crt. No.	Compound	Chemical Class	Retention Time	Relative Content
1	α-Pinene	M.Hc.	7.949	0.54
2	β-Myrcene	M.Hc.	9.920	0.18
3	**D-Limonene**	M.Hc.	11.406	**4.05**
4	1,8-cineole	O.M.	11.534	0.10
5	**trans-β-Ocimene**	M.Hc	11.664	**6.19**
6	**cis-β-Ocimene**	M.Hc	12.065	**6.51**
7	α-Terpineol	O.M.	17.787	0.32
8	**Estragole**	Phe.P.	17.917	**81.84**
9	Caryophyllene	S.Hc.	25.869	0.27
-	TOTAL	-	-	100.00

M.Hc.—monoterpene hydrocarbons; O.M.—oxygenated monoterpenes; Phe.P.—phenylpropanoids; S.Hc.—sesquiterpene hydrocarbons.

3.2. *Physicochemical Properties of Fish*

3.2.1. Proximate Composition

The results of the proximate analysis of brook trout showed a mean value of 19.8 ± 0.078 g/100 g fish sample for protein content, 8.0 ± 0.255 g/100 g fish sample for fat content, 1.4 ± 0.014 g/100 g fish sample for ash content, and 70.2 ± 0.290 g/100 g fish sample for moisture content. The values reported in other studies [35,36] showed some differences, especially in the lipid and protein contents. Variations in the chemical composition of fish muscle among individuals in the same species are related to the state of nutrition, the reproductive cycle of the animal, fish size, as well as other environmental conditions [37].

3.2.2. pH and Electrical Conductivity (EC)

The pH value is an indicator of the freshness of meat, which is fundamental to fish quality [38]. In the flesh of live fish, pH is close to 7.0. After their death, pH can vary from 6.0 to 7.1, depending on the season, species, and other factors [16].

The values of pH in brook trout samples stored at 4 °C for 15 days are shown in Figure 1. Changes in pH showed the same trend during storage in all treatments. The initial value of pH in the fish sample was 6.3, comparable with the value (of 6.5) found by Shen et al. (2015) [39] in rainbow trout fillets but lower than values (of 7.0 and 7.2, respectively) reported by Nistor et al. (2014) [40] and Linhartová et al. (2018) [36] in brook trout fillets. The pH stayed relatively stable up to the third day of storage (6.4 in the control sample, 6.4 in WPIf sample, and 6.3 in WPIf+2.5%TEO sample), decreased from the third day to the sixth day of storage (to 6.1 in the control sample, 6.0 in WPIf sample, and 5.9 in the WPIf+2.5%TEO sample), then increased up to the 15th day of storage (to 6.7 in the control sample, 6.6 in WPIf sample, and 6.5 in WPIf+2.5%TEO sample). The initial decrease in pH could be attributed to the post-mortem breakdown of glycogen, ATP, creatine phosphate [39,41,42], and the dissolution of CO_2 in the fish sample [43]; the later increase in pH was probably due to the production of volatile bases [39,41,44]. A similar decreasing-increasing pH trend during refrigerated storage was also reported by Shen et al. (2015) in rainbow trout fillets [39].

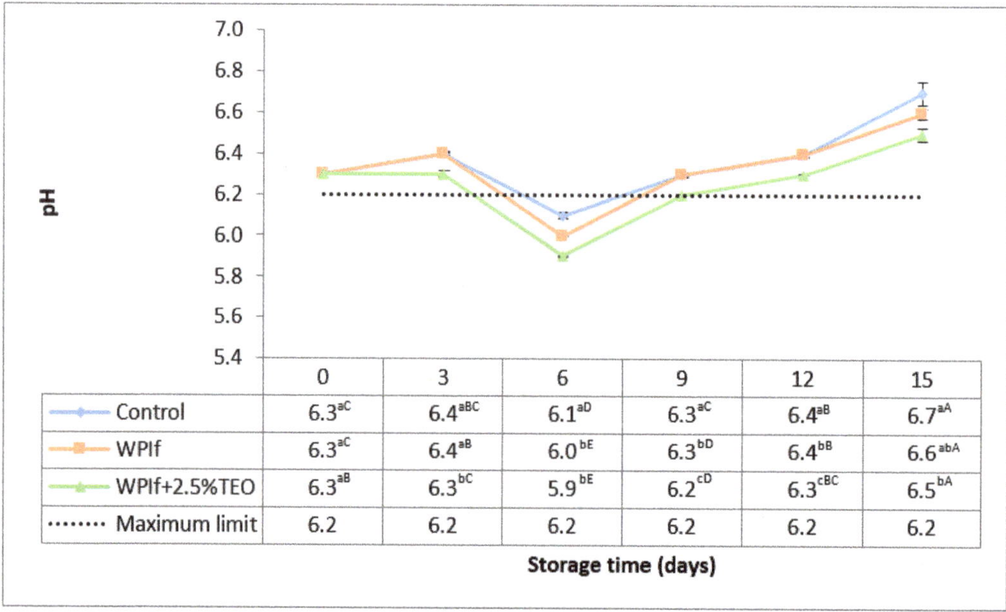

	0	3	6	9	12	15
Control	6.3^{aC}	6.4^{aBC}	6.1^{aD}	6.3^{aC}	6.4^{aB}	6.7^{aA}
WPIf	6.3^{aC}	6.4^{aB}	6.0^{bE}	6.3^{bD}	6.4^{bB}	6.6^{abA}
WPIf+2.5%TEO	6.3^{aB}	6.3^{bC}	5.9^{bE}	6.2^{cD}	6.3^{cBC}	6.5^{bA}
Maximum limit	6.2	6.2	6.2	6.2	6.2	6.2

Figure 1. Changes in pH of brook trout samples during refrigerated storage. Control-uncovered fish samples; WPIf-fish samples covered with WPI-based films; WPIf+2.5%TEO-fish samples covered with whey protein isolate-based films incorporated with 2.5% tarragon essential oil. Values are expressed as mean ± standard deviation of two replicates. Means that do not share a letter (lowercase letters on column and uppercase letters on row) are significantly different.

There were significant differences in pH values between batches from the 3rd day to the 15th day of storage; the lowest pH values were found in WPIf+2.5%TEO samples, followed by WPIf samples, and by control samples. These results show that films, particularly the active film, delayed enzymes activity keeping thus the freshness of fish samples.

The maximum permitted value for pH in fresh fish is 6.2, as set by Romanian standard STAS 5386-86 [45]. The value of pH exceeded the limit level in the 9th day of storage for control and WPIf samples, respectively, in the 12th day of storage for the WPIf+2.5%TEO sample.

EC is an index of the concentration of electrolytes in the muscle tissue of fish; it can impact body fluid balance, survival, and meat quality [17,39]. The autolytic spoilage that post-mortem occurs in fish, mainly caused by enzymes, progressively disrupts the muscle cell membranes [46]. As a consequence, the intracellular fluid leak into the intercellular space and, being an electrolyte solution, increases the EC of the tissue [47].

An opposite behavior to that of pH was noticed for electrical conductivity in fish samples during storage (see Figure 2); hence the strong negative correlation ($r = -0.688$; $p < 0.05$) found between pH and EC values. The initial value of EC in the fish sample was 1417 μS/cm, comparable with that reported by Shen et al. (2015) in rainbow trout fillets (1344 μS/cm) [39]. The EC value decreased in the 3rd day of storage (to 1146 μS/cm in the control sample, to 1094 μS/cm in WPIf sample, and to 1282 μS/cm in WPIf+2.5%TEO sample), then increased from the 3rd day to the 6th day of storage (to 1284 μS/cm in the control sample, to 1420 μS/cm in WPIf sample, and to 1415 μS/cm in WPIf+2.5%TEO sample), and decreased again up to the 15th day of storage (to 1119 μS/cm in the control sample, to 1007 μS/cm in WPIf sample, and to 941 μS/cm in WPIf+2.5%TEO sample). In the final stages of storage, slight drip losses were observed in fish samples. This could be the reason for the reduction of electrolyte concentration in fish samples. Contrary to our findings, other studies have reported an increase in EC of rainbow trout [39] and beluga sturgeon fillets [16] during refrigerated storage. The mobility ratios of positive and negative ions in the fish patties could be the reason for "in the mirror" behaviors of EC and pH during refrigerated storage, negative ions being smaller and more mobile than positive ions [48,49]. Variations of pH and EC up to the 6th day of refrigerated storage are due to pre-rigor, in-rigor, and post-rigor changes in brook trout samples.

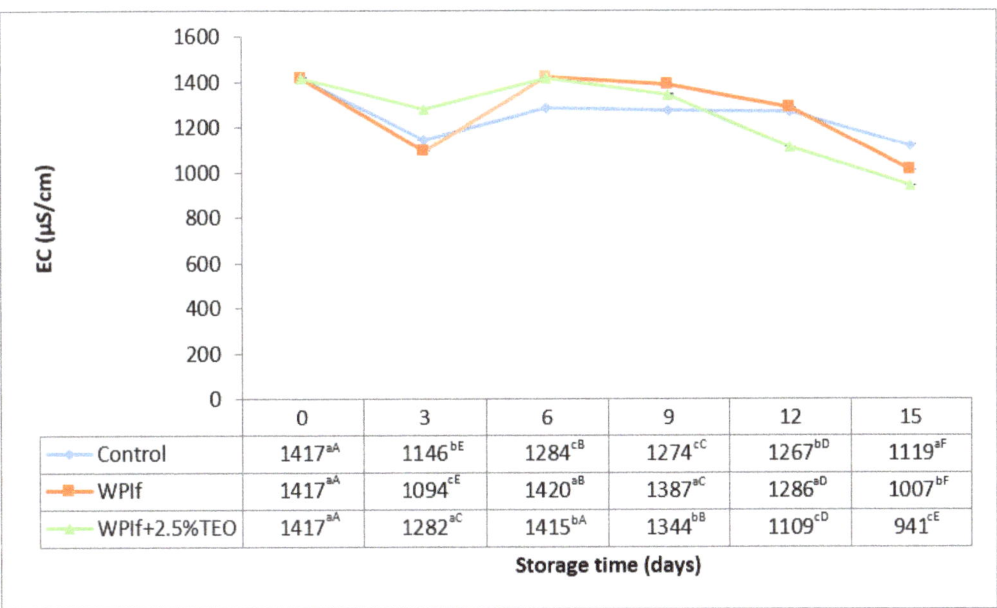

	0	3	6	9	12	15
Control	1417^{aA}	1146^{bE}	1284^{cB}	1274^{cC}	1267^{bD}	1119^{aF}
WPIf	1417^{aA}	1094^{cE}	1420^{aB}	1387^{aC}	1286^{aD}	1007^{bF}
WPIf+2.5%TEO	1417^{aA}	1282^{aC}	1415^{bA}	1344^{bB}	1109^{cD}	941^{cE}

Figure 2. Changes in electrical conductivity (EC) of brook trout samples during refrigerated storage. Control-uncovered fish samples; WPIf-fish samples covered with WPI-based films; WPIf+2.5%TEO-fish samples covered with whey protein isolate-based films incorporated with 2.5% tarragon essential oil. Values are expressed as mean ± standard deviation of two replicates. Means that do not share a letter (lowercase letters on column and uppercase letters on row) are significantly different.

3.2.3. Total Volatile Basic Nitrogen (TVB-N)

The TVB-N quantifies the presence of nitrogenous compounds (ammonia, dimethyl amine, and trimethyl amine) in fish from the sea or from river, revealing the degree of freshness [50]. Its increase during storage is related to the activity of spoilage bacteria and endogenous enzymes [32]. The Commission of the European Union, through the Regulation (EC) No. 1022/2008, [51] has set limit values for TVB-N just in redfish, flatfish, Atlantic salmon, hake, and gadoids; values ≤25 mg N/100 g flesh for *Sebastes* spp., *Helicolenus dactylopterus*, and *Sebastichthys capensis*, ≤30 mg N/100 g flesh for species belonging to the Pleuronectidae family (with the exception of halibut: *Hippoglossus* spp.), and ≤35 mg N/100 g flesh for *Salmo salar*, species belonging to the Merlucciidae family, and species belonging to the Gadidae family. However, the TVB-N value as an indicator of fish freshness has been recently disputed as it was reported below the maximum limit, even when the fish has been rejected by sensory evaluation [52].

The TVB-N values of brook trout samples during refrigerated storage are presented in Figure 3. Initially, the TVB-N value was 2.23 mg N/100 g fish sample; a comparable value (3.59 mg N/100 mg fish sample) was found by Feng et al. (2016) in golden pomfret fillets [17] and much higher values by Kazemi and Rezaei (2015) in rainbow trout slices (10.37 mg N/100 g fish sample) [15], by Bahram et al. (2016) in beluga sturgeon fillets (17.97 mg N/100 mg fish sample) [16], and by Shokri and Ehsani (2017) in pike-perch fillets

(10.99 mg N/100 g fish sample) [18]. To our knowledge, no prior studies have investigated the level of TVB-N in brook trout during storage.

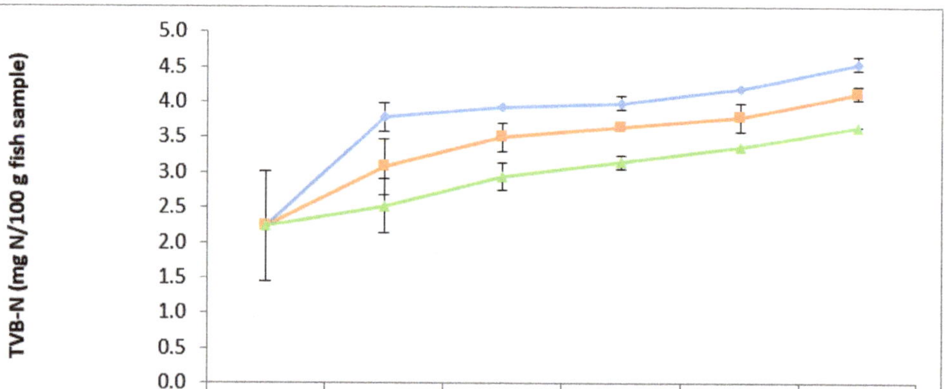

Storage time (days)	0	3	6	9	12	15
Control	2.23^{aB}	3.78^{aA}	3.92^{aA}	3.99^{aA}	4.20^{aA}	4.55^{aA}
WPIf	2.23^{aB}	3.08^{aAB}	3.50^{abAB}	3.64^{bAB}	3.78^{abB}	4.13^{bB}
WPIf+2.5%TEO	2.23^{aA}	2.52^{aA}	2.94^{bA}	3.15^{cA}	3.36^{cA}	3.64^{cA}

Figure 3. Changes in total volatile basic nitrogen (TVB-N) of brook trout samples during refrigerated storage. Control-uncovered fish samples; WPIf-fish samples covered with WPI-based films; WPIf+2.5%TEO-fish samples covered with whey protein isolate-based films incorporated with 2.5% tarragon essential oil. Values are expressed as mean ± standard deviation of three replicates. Means that do not share a letter (lowercase letters on column and uppercase letters on row) are significantly different.

The TVB-N value significantly increased with storage time from 2.23 to 4.55 mg N/100 g fish sample in the control batch, from 2.23 to 4.13 mg N/100 g fish sample in WPIf batch, and from 2.23 to 3.64 mg N/100 g fish sample in WPIf+2.5%TEO batch. Beginning with the sixth day of storage, there were significant differences in TVB-N values between batches; the lowest TVB-N values were found in WPIf+2.5%TEO samples, followed by WPIf samples, and by control samples. These results corroborate the findings from pH measurements and show that fish samples covered with films were more protected toward protein degradation, particularly those covered with active films. Since no limit of acceptability for TVB-N in brook trout has been established by the European Commission [51] or proposed by other researchers, this parameter was not used in establishing the shelf-life of brook trout from the current study.

3.2.4. Thiobarbituric Acid Reactive Substances (TBARS)

The TBARS value is commonly used as an indicator of lipid oxidation, particularly in meat and fish products [53]; thiobarbituric acid reactive substances are formed in the second stage of auto-oxidation when peroxides are oxidized to aldehydes and ketones [54]. Some researchers [18,32,55] have proposed quality criteria for fish: <3 mg MDA/kg for perfect quality material, $3 \leq$ mg MDA/kg < 5 for good quality material, and $5 \leq$ mg MDA/kg < 8 for suitable for human consumption material. Nevertheless, these thresholds have not yet received regulatory approval.

The TBARS values of fish samples during refrigerated storage are shown in Figure 4. The initial value of TBARS in the fish sample was 0.28 mg MDA/kg (perfect quality material); comparable initial levels were reported by Ojagh et al. (2010) in rainbow trout fillets (0.09 mg MDA/kg) [55], by Li et al. (2013) in red drum fillets (0.28 mg MDA/kg) [37], by Jouki et al. (2014) in rainbow trout fillets (0.12 mg MDA/kg) [32], by Ramezani et al. (2015) in silver carp fillets (0.51 mg MDA/kg) [54], by Bahram et al. (2016) in beluga sturgeon fillets (0.02 mg MDA/kg) [16], by Shokri and Ehsani (2017) in pike-perch fillets (0.61 mg MDA/kg) [18], and by Yu et al. (2017) in grass carp fillets (0.20 mg MDA/kg) [34].

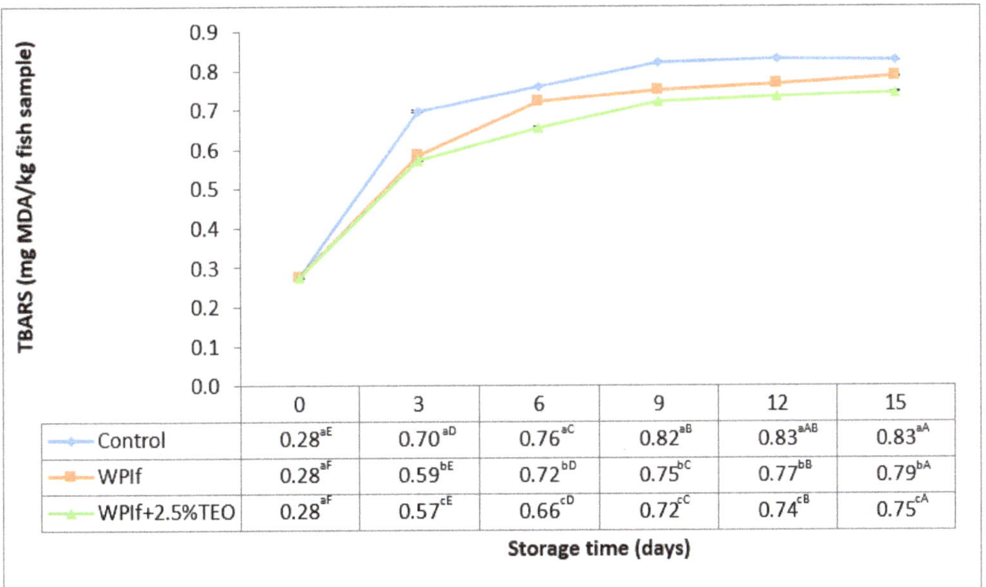

	0	3	6	9	12	15
Control	0.28^{aE}	0.70^{aD}	0.76^{aC}	0.82^{aB}	0.83^{aAB}	0.83^{aA}
WPIf	0.28^{aF}	0.59^{bE}	0.72^{bD}	0.75^{bC}	0.77^{bB}	0.79^{bA}
WPIf+2.5%TEO	0.28^{aF}	0.57^{cE}	0.66^{cD}	0.72^{cC}	0.74^{cB}	0.75^{cA}

Figure 4. Changes in thiobarbituric acid reactive substances (TBARS) of brook trout samples during refrigerated storage. Control-uncovered fish samples; WPIf-fish samples covered with WPI-based films, WPIf+2.5%TEO-fish samples covered with whey protein isolate-based films incorporated with 2.5% tarragon essential oil. Values are expressed as mean ± standard deviation of three replicates. Means that do not share a letter (lowercase letters on column and uppercase letters on row) are significantly different.

The TBARS value significantly increased with storage time from 0.28 to 0.83 mg MDA/kg fish sample in the control batch, from 0.28 to 0.79 mg MDA/kg fish sample in WPIf batch, and from 0.28 to 0.75 mg MDA/kg fish sample in WPIf+2.5%TEO batch. From the 3rd day of storage, there were significant differences in TBARS values between batches; the lowest TBARS values were found in WPIf+2.5%TEO samples, followed by WPIf samples, and by control samples. These results show that fish samples covered with films were less susceptible to lipid oxidation, especially those covered with active films. The TBARS values for all fish samples were below the upper proposed limits throughout the 15-day storage period.

3.2.5. Color

Changes in color attributes (L^*, a^*, and b^*) of brook trout samples during refrigerated storage are shown in Figure 5a–c. The initial values of L^*, a^*, and b^* in the fish sample were 59.33, 4.29, and 8.23, respectively. Shen et al. (2005) have reported initial values of 62.77 for L^*, 7.48 for a^*, and 20.92 for b^* in the brook trout fillet [39], therefore higher.

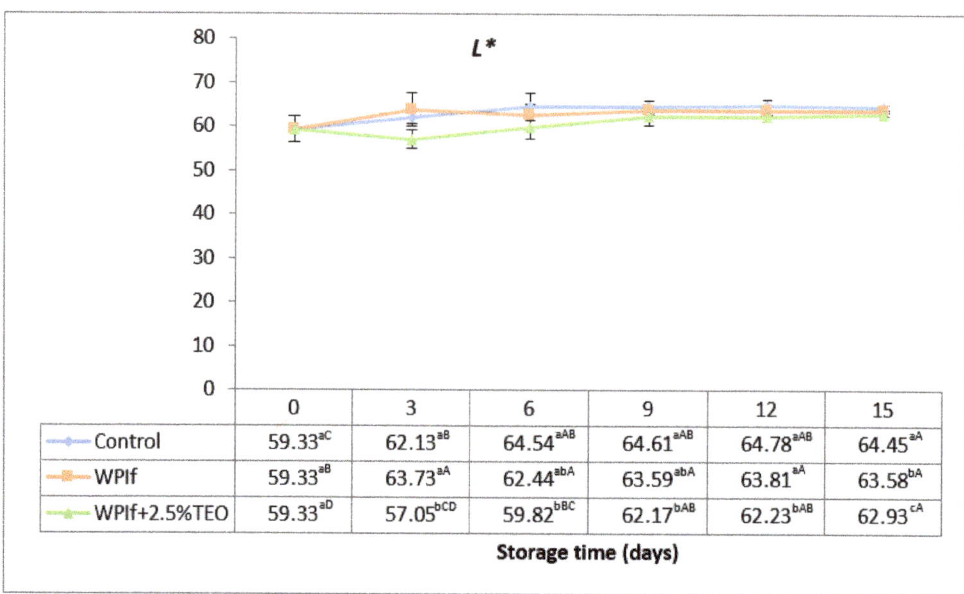

(a)

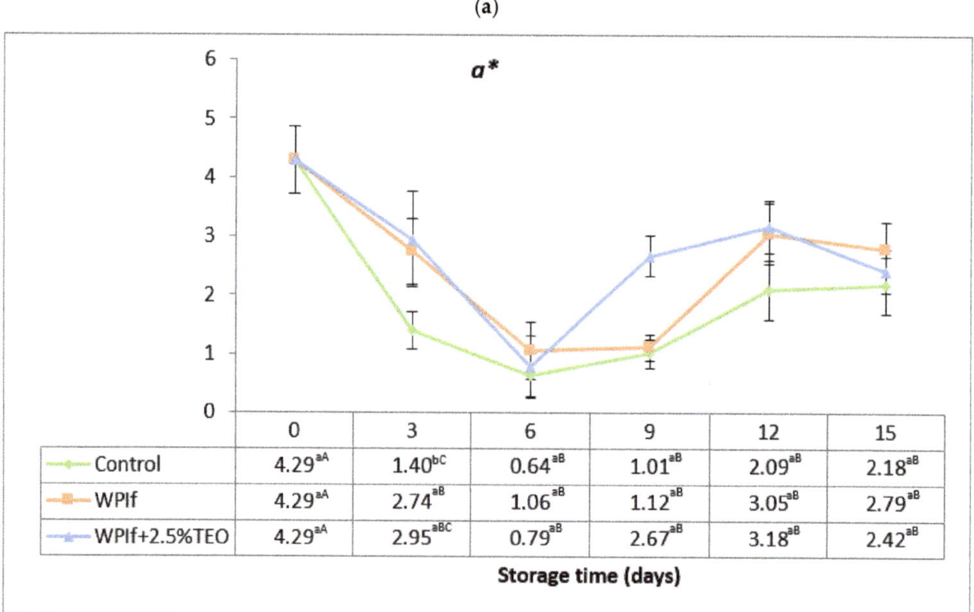

(b)

Figure 5. *Cont.*

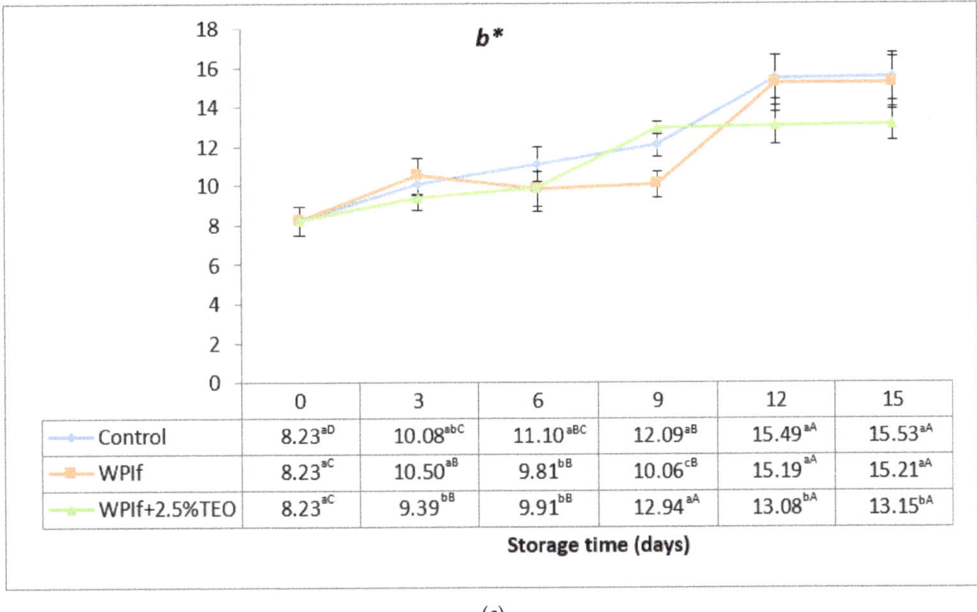

	0	3	6	9	12	15
Control	8.23^{aD}	10.08^{abC}	11.10^{aBC}	12.09^{aB}	15.49^{aA}	15.53^{aA}
WPIf	8.23^{aC}	10.50^{aB}	9.81^{bB}	10.06^{cB}	15.19^{aA}	15.21^{aA}
WPIf+2.5%TEO	8.23^{aC}	9.39^{bB}	9.91^{bB}	12.94^{aA}	13.08^{bA}	13.15^{bA}

(c)

Figure 5. Changes in color attributes of brook trout samples during refrigerated storage. (**a**) *L**; (**b**) *a**; (**c**) *b**. Control-uncovered fish samples; WPIf-fish samples covered with WPI-based films; WPIf+2.5%TEO-fish samples covered with whey protein isolate-based films incorporated with 2.5% tarragon essential oil. Values are expressed as mean ± standard deviation of twelve replicates. Means that do not share a letter (lowercase letters on column and uppercase letters on row) are significantly different.

The *L**-value (brightness) significantly increased with storage time from 59.33 to 64.45 in the control batch, from 59.33 to 63.58 in WPIf batch, and from 59.33 to 62.93 in WPIf+2.5%TEO batch (see Figure 5a). The values of *L** in WPIf+2.5%TEO samples were significantly lower than those in WPIf and control samples, beginning with the third day of storage; there were no significant differences in *L**-values between control and WPIf samples up to the 12th day of storage.

Variations of *a**-values (redness) showed the same trend during storage for all batches (see Figure 5b). The initial value of *a** in the fish sample was 4.29. It decreased up to the sixth day of storage (to 0.64 in the control sample, 1.06 in WPIf sample, and 0.79 in WPIf+2.5%TEO sample), then increased up to the 15th day of storage (to 2.18 in the control sample, 2.79 in WPIf sample, and 2.42 in WPIf+2.5%TEO sample). The reasons for these oscillations are not yet entirely understood. The initial decrease could be caused by the oxidation of red pigments, such as myoglobin and hemoglobin. A possible explanation for the later increased might be the pigments concentrating, as a result of dehydration. The values of *a** in control samples were lower than those in WPIf and WPIf+2.5%TEO samples, but significantly only in the third day of storage; there were no significant differences in *a**-values between WPIf and WPIf+2.5%TEO samples.

The *b**-value (yellowness) significantly increased with storage time, probably due to the oxidation of lipids; from an initial value of 8.23 to 15.53 in the control batch, 15.21 in WPIf batch, and 13.15 in WPIf+2.5%TEO batch (see Figure 5c). There were significant differences in *b**-values between batches from the third day to the ninth day of storage; the values of *b** were generally lower in WPIf+2.5%TEO samples than those in WPIf and control samples.

The strong positive correlation ($r = 0.704$; $p < 0.05$) found between *L**- and *b**-values indicates that the brightness of fish samples increased with their yellowing. Given that

L^*-values increased and a^*-values oscillated during storage, a moderate negative correlation (r = −0.511; $p < 0.05$) was found between L^*- and a^*-values. No significant correlation (r = −0.164; $p \geq 0.05$) was found between a^*- and b^*-values; thus, no relationship between redness and yellowness of the fish samples.

All these results indicate a discoloration of fish samples (increase in brightness) with storage time caused by oxidation of red pigments (decrease in redness) and lipids (increase in yellowness); the least affected were samples covered with whey protein isolate-based films incorporated with 2.5% tarragon essential oil.

3.3. Microbiological Properties of Fish

3.3.1. Total Viable Count (TVC)

The total viable count (TVC) estimates the total number of aerobic mesophilic organisms (such as bacteria, yeasts, and molds) in the fish sample [56]. Changes in TVC of brook trout samples stored at 4 °C for 15 days are shown in Figure 6. The initial TVC in the fish sample was 3.31 $\log_{10}$ CFU/g. Comparable initial counts were reported by Ojagh et al. (2010) [55], Jouki et al. (2014) [32], respectively Volpe et al. (2015) [52] in rainbow trout fillets (3.86 $\log_{10}$ CFU/g, 3.58 $\log_{10}$ CFU/g, and 4.00 $\log_{10}$ CFU/g), by Li et al. (2013) in red drum fillets (3.92 $\log_{10}$ CFU/g) [37], by Kazemi and Rezaei (2015) in rainbow trout slices (2.50 $\log_{10}$ CFU/g) [15], by Bahram et al. (2016) in beluga sturgeon fillets (4.04 $\log_{10}$ CFU/g) [16], and by Yu et al. (2017) in grass carp fillets (4.20 $\log_{10}$ CFU/g) [34]. The TVC stayed relatively stable up to the third day of storage (3.61 $\log_{10}$ CFU/g in the control sample, 3.45 $\log_{10}$ CFU/g in WPIf sample, and 3.35 $\log_{10}$ CFU/g in WPIf+2.5%TEO sample), then significantly increased up to the 15th day of storage (to 8.65 log10 CFU/g in the control sample, 7.68 log10 CFU/g in WPIf sample, and 6.68 $\log_{10}$ CFU/g in WPIf+2.5%TEO sample). Generally, TVC increased with increasing pH in fish samples with decreasing EC; from here there was a moderate positive correlation with the pH (r = 0.518; $p < 0.05$) and moderate negative correlation (r = −0.517; $p < 0.05$) with the EC.

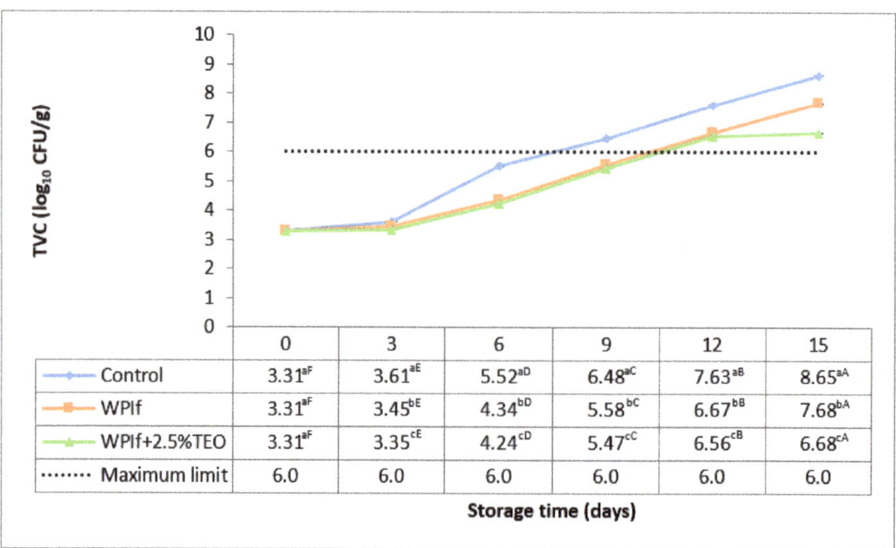

	0	3	6	9	12	15
Control	3.31^{aF}	3.61^{aE}	5.52^{aD}	6.48^{aC}	7.63^{aB}	8.65^{aA}
WPIf	3.31^{aF}	3.45^{bE}	4.34^{bD}	5.58^{bC}	6.67^{bB}	7.68^{bA}
WPIf+2.5%TEO	3.31^{aF}	3.35^{cE}	4.24^{cD}	5.47^{cC}	6.56^{cB}	6.68^{cA}
Maximum limit	6.0	6.0	6.0	6.0	6.0	6.0

Figure 6. Changes in total viable count (TVC) of brook trout samples during refrigerated storage. Control-uncovered fish samples; WPIf-fish samples covered with WPI-based films; WPIf+2.5%TEO-fish samples covered with whey protein isolate-based films incorporated with 2.5% tarragon essential oil. Values are expressed as mean ± standard deviation of two replicates. Means that do not share a letter (lowercase letters on column and uppercase letters on row) are significantly different.

There were significant differences in total viable counts between batches from the third day to the 15th day of storage with the lowest values in WPIf+2.5%TEO samples, followed by WPIf samples, and by control samples. These findings show that films, particularly the one incorporated with tarragon essential oil, inhibited microbial growth.

According to the Food Safety Authority of Ireland [57], the level of TVC in refrigerated fish should be less than 10^6 CFU/g (6.0 $\log_{10}$ CFU/g). The TVC exceeded the limit level in the 9th day of storage for the control sample, respectively in the 12th day of storage for WPIf and WPIf+2.5%TEO samples.

3.3.2. Psychrotrophic Count (PTC)

The spoilage of aerobically stored fish is mainly due to the Gram-negative psychrotrophic non-fermenting rods [58]. These bacteria are capable of growth at 0 °C but with optimum around 25 °C [41]. Therefore, psychrotrophic bacteria were counted in brook trout samples during 15 days of storage at 4 °C (see Figure 7).

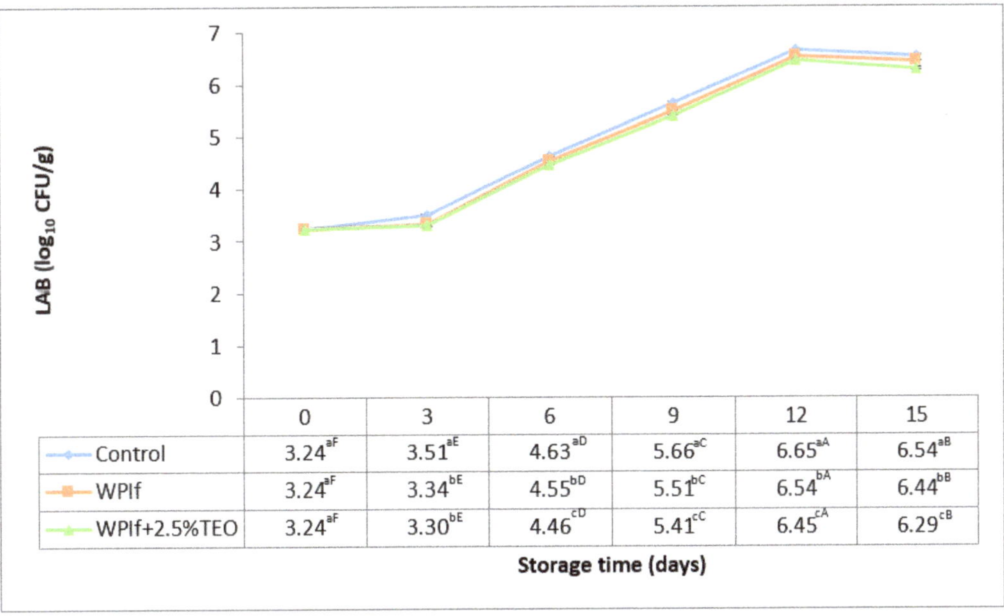

	0	3	6	9	12	15
Control	3.24^{aF}	3.51^{aE}	4.63^{aD}	5.66^{aC}	6.65^{aA}	6.54^{aB}
WPIf	3.24^{aF}	3.34^{bE}	4.55^{bD}	5.51^{bC}	6.54^{bA}	6.44^{bB}
WPIf+2.5%TEO	3.24^{aF}	3.30^{bE}	4.46^{cD}	5.41^{cC}	6.45^{cA}	6.29^{cB}

Figure 7. Changes in psychrotrophic count (PTC) of brook trout samples during refrigerated storage. Control-uncovered fish samples; WPIf-fish samples covered with WPI-based films; WPIf+2.5%TEO-fish samples covered with whey protein isolate-based films incorporated with 2.5% tarragon essential oil. Values are expressed as mean ± standard deviation of two replicates. Means that do not share a letter (lowercase letters on column and uppercase letters on row) are significantly different.

The growth pattern of PTC showed the same behavior in all treatments. The initial value of PTC in the fish sample was 3.24 $\log_{10}$ CFU/g. Other researchers have found initial levels of 2.88 $\log_{10}$ CFU/g and 3.10 $\log_{10}$ CFU/g in rainbow trout fillets [32,55], of 2.50 $\log_{10}$ CFU/g in rainbow trout slices [15], of 3.56 $\log_{10}$ CFU/g in silver carp fillets [54], of 3.89 $\log_{10}$ CFU/g in beluga sturgeon fillets [16], of 3.18 $\log_{10}$ CFU/g in pike-perch fillets [18], respectively of 3.50 $\log_{10}$ CFU/g in grass carp fillets [34]. The TPC significantly increased with storage time from 3.24 to 8.57 $\log_{10}$ CFU/g in the control batch, from 3.24 to 7.64 $\log_{10}$ CFU/g in WPIf batch, and from 3.24 to 7.49 $\log_{10}$ CFU/g in WPIf+2.5%TEO batch. From the third day of storage, significant differences were found between total viable counts of fish samples; the values of WPIf+2.5%TEO samples were lower than those of WPIf and control samples.

3.3.3. Lactic Acid Bacteria (LAB)

Lactic acid bacteria are also associated with the spoilage of fish during refrigerated storage [58]. Initial populations of 2.30 $\log_{10}$ CFU/g, respectively 2.00 $\log_{10}$ CFU/g were found in rainbow trout fillets by Jouki et al. (2014) and by Volpe et al. (2015) [32,52]. In rainbow trout slices, Kazemi and Rezaei (2015) have reported an initial count of 3.00 $\log_{10}$ CFU/g LAB [15]. Changes in LAB of brook trout samples during storage at 4 °C for 15 days are shown in Figure 8. These have shown the same trend during storage, to all treatments. The initial count of LAB in the fish sample was 3.24 $\log_{10}$ CFU/g. It significantly increased up to the 12th day of storage (to 6.65 $\log_{10}$ CFU/g in the control sample, 6.54 $\log_{10}$ CFU/g in WPIf sample, and 6.45 $\log_{10}$ CFU/g in WPIf+2.5%TEO sample), then significantly decreased (to 6.54 $\log_{10}$ CFU/g in the control sample, 6.44 $\log_{10}$ CFU/g in WPIf sample, and 6.29 $\log_{10}$ CFU/g in WPIf+2.5%TEO sample). The later decrease is probably due to the competition of LAB with other microorganisms from the matrix for the remained nutrients [59]. From the 3rd day of storage, there were significant differences between batches regarding LAB counts; the lowest values were found in WPIf+2.5%TEO samples, followed by WPIf samples, and by control samples.

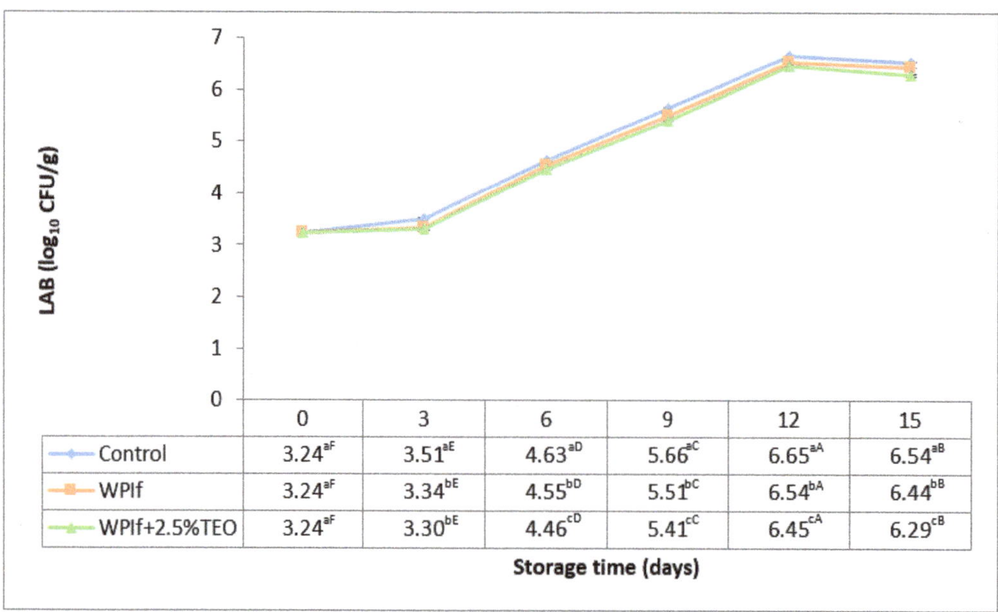

	0	3	6	9	12	15
Control	3.24^{aF}	3.51^{aE}	4.63^{aD}	5.66^{aC}	6.65^{aA}	6.54^{aB}
WPIf	3.24^{aF}	3.34^{bE}	4.55^{bD}	5.51^{bC}	6.54^{bA}	6.44^{bB}
WPIf+2.5%TEO	3.24^{aF}	3.30^{bE}	4.46^{cD}	5.41^{cC}	6.45^{cA}	6.29^{cB}

Figure 8. Changes in lactic acid bacteria (LAB) of brook trout samples during refrigerated storage. Control-uncovered fish samples; WPIf-fish samples covered with WPI-based films; WPIf+2.5%TEO-fish samples covered with whey protein isolate-based films incorporated with 2.5% tarragon essential oil. Values are expressed as mean ± standard deviation of two replicates. Means that do not share a letter (lowercase letters on column and uppercase letters on row) are significantly different.

3.3.4. Hydrogen Sulfide (H_2S)-Producing Bacteria

Some spoilage microorganisms in fish, including *Shewanella putrefaciens*, release hydrogen sulfide (H_2S) upon decomposition of sulfur-containing amino acids [60]. These are so-called H_2S-producing bacteria. Some researchers have reported initial counts of H_2S-producing bacteria by 2.20 $\log_{10}$ CFU/g and 2.00 $\log_{10}$ CFU/g in rainbow trout fillets [32,52], and by 3.20 log_{10} CFU/g in grass carp fillets [34]. Changes in hydrogen sulfide (H_2S)-producing bacteria of brook trout samples during refrigerated storage are shown in Figure 9. All batches revealed the same behavior during storage. The initial count of H_2S-producing bacteria in the fish sample was 2.97 log_{10} CFU/g. It significantly increased with storage time, up to 7.24 log_{10} CFU/g in the control batch, 7.14 log_{10} CFU/g in WPIf batch, and 6.23 log_{10} CFU/g in WPIf+2.5%TEO batch. Starting with the third day of storage, significant differences were observed between counts of H_2S-producing bacteria in fish samples; the values of WPIf+2.5%TEO samples were lower than those of WPIf and control samples.

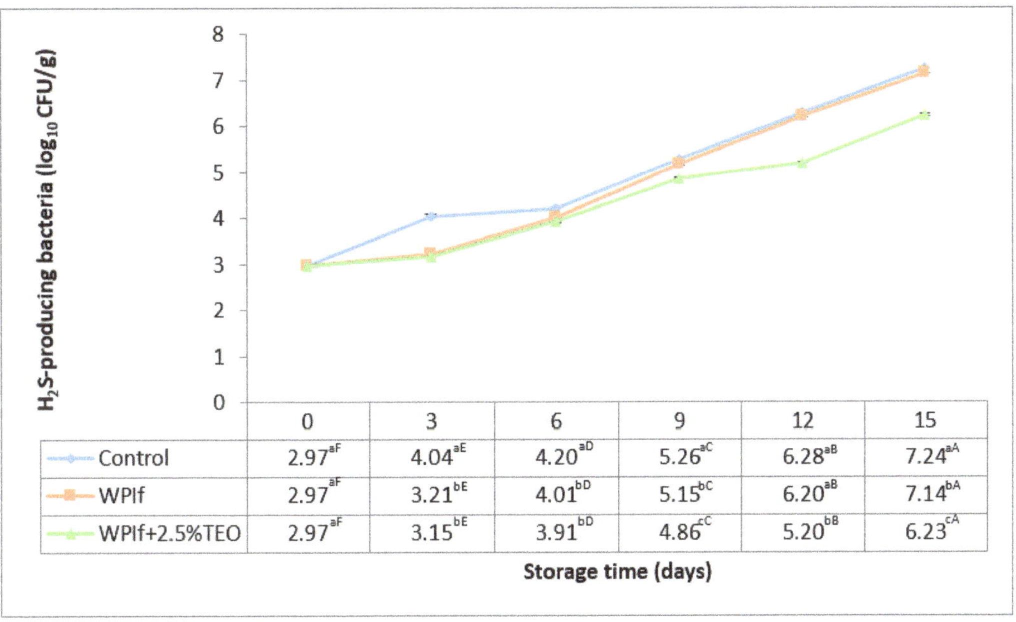

	0	3	6	9	12	15
Control	2.97^{aF}	4.04^{aE}	4.20^{aD}	5.26^{aC}	6.28^{aB}	7.24^{aA}
WPIf	2.97^{aF}	3.21^{bE}	4.01^{bD}	5.15^{bC}	6.20^{aB}	7.14^{bA}
WPIf+2.5%TEO	2.97^{aF}	3.15^{bE}	3.91^{bD}	4.86^{cC}	5.20^{bB}	6.23^{cA}

Figure 9. Changes in hydrogen sulfide (H_2S)-producing bacteria of brook trout samples during refrigerated storage. Control-uncovered fish samples; WPIf-fish samples covered with WPI-based films; WPIf+2.5%TEO-fish samples covered with whey protein isolate-based films incorporated with 2.5% tarragon essential oil. Values are expressed as mean ± standard deviation of two replicates. Means that do not share a letter (lowercase letters on column and uppercase letters on row) are significantly different.

3.4. Sensory Properties of Fish

The color discoloration, odor, and overall acceptability are the sensory attributes chosen to evaluate the quality of fish patties during refrigerated storage as these are significantly changed with fish spoilage. The lower the sensory score of an attribute, the lower the quality of the fish sample. The results of sensory evaluation of fish samples are given in Table 2. The fish samples were considered to be acceptable for human consumption up to a score of 4 [16,54,55]. Variations in scores of sensory attributes (color discoloration, odor, and overall acceptability) showed the same behavior during storage to all treatments; no change up to the sixth day of storage (scores of 5.0 points), then a significant decrease up to the 15th day of storage. From the sixth to the 15th day of storage, the score for color discoloration significantly decreased from 5.0 points to 1.0 point in the control sample, 2.0 points in WPIf sample, and 2.6 points in WPIf+2.5%TEO sample. There were significant differences in scores of color discoloration between all treatments; the lowest scores for color discoloration were found in control samples, followed by WPIf samples, and by WPIf+2.5%TEO samples. Strong negative correlations were found between color discoloration scores and L^*-values (r = −0.605; $p < 0.05$) and between color discoloration scores and b^*-values (r = −0.880; $p < 0.05$). These results strengthen the findings of color measurements.

Table 2. Changes in sensory scores (points) of brook trout samples during refrigerated storage.

Attributes	Treatment	Storage Time (Days)					
		0	3	6	9	12	15
Color discoloration	Control	5.0 ± 0.0 aA	5.0 ± 0.0 aA	5.0 ± 0.0 aA	3.0 ± 0.0 cB	1.6 ± 0.548 bC	1.0 ± 0.0 cD
	WPIf	5.0 ± 0.0 aA	5.0 ± 0.0 aA	5.0 ± 0.0 aA	4.0 ± 0.0 bB	2.6 ± 0.548 aC	2.0 ± 0.0 bD
	WPIf+2.5%TEO	5.0 ± 0.0 aA	5.0 ± 0.0 aA	5.0 ± 0.0 aA	5.0 ± 0.0 aA	3.0 ± 0.0 aB	2.6 ± 0.548 aB
Odor	Control	5.0 ± 0.0 aA	5.0 ± 0.0 aA	5.0 ± 0.0 aA	3.0 ± 0.0 bB	2.6 ± 0.548 aB	1.0 ± 0.0 bC
	WPIf	5.0 ± 0.0 aA	5.0 ± 0.0 aA	5.0 ± 0.0 aA	4.0 ± 0.0 aB	3.0 ± 0.0 aC	2.0 ± 0.0 aD
	WPIf+2.5%TEO	5.0 ± 0.0 aA	5.0 ± 0.0 aA	5.0 ± 0.0 aA	4.0 ± 0.0 aB	3.0 ± 0.0 aC	2.0 ± 0.0 aD
Overall acceptability	Control	5.0 ± 0.0 aA	5.0 ± 0.0 aA	5.0 ± 0.0 aA	3.0 ± 0.0 bB	1.8 ± 0.447 bC	1.0 ± 0.0 bD
	WPIf	5.0 ± 0.0 aA	5.0 ± 0.0 aA	5.0 ± 0.0 aA	4.0 ± 0.0 aB	2.6 ± 0.548 aC	2.0 ± 0.0 aD
	WPIf+2.5%TEO	5.0 ± 0.0 aA	5.0 ± 0.0 aA	5.0 ± 0.0 aA	4.0 ± 0.0 aB	3.0 ± 0.0 aC	2.0 ± 0.0 aD

Control-uncovered fish samples; WPIf-fish samples covered with WPI-based films; WPIf+2.5%TEO-fish samples covered with whey protein isolate-based films incorporated with 2.5% tarragon essential oil. Values are expressed as mean ± standard deviation of five responses. Means that do not share a letter (lowercase letters on column of each attribute and uppercase letters on row) are significantly different.

In terms of odor and overall acceptability, the scores of WPIf+2.5%TEO samples were significantly higher than those of control and WPIf samples; no significant differences were found between scores of the two later batches. Between the sixth and the 15th day of storage, the scores for odor and overall acceptability significantly decreased from 5.0 points to 1.0 point in the control sample, 2.0 points in WPIf sample, and 2.0 points in WPIf+2.5%TEO sample. These results indicate a depreciation of the odor and appearance of fish samples with storage time, to the same extent in samples covered with films and to a greater extent in uncovered samples. To all sensory attributes, the unacceptable score was given on the ninth day of storage for the control sample and in the 12th day of storage for WPIf and WPIf+2.5%TEO samples.

3.5. Shelf-Life of Fish

The shelf-lives resulted from physicochemical, microbiological, and sensory evaluations of brook trout samples are summarized in Table 3. Based on the pH value, the shelf-life of fish sample covered with the active film was extended by three days (from six to nine days), but that of the fish sample covered with the control film was not improved. Based on the total viable count and sensory scores, both the shelf-life of WPIf sample and WPIf+2.5%TEO sample was prolonged by three days. Taking into consideration all parameters, a shelf-life of six days was achieved for the uncovered fish sample, of six–nine days for the fish sample covered with control film, and of nine days for the fish sample covered with active film. These findings demonstrate the research hypothesis of our study.

Table 3. Shelf-lives of brook trout samples at refrigerated storage.

Treatment	pH [a,d]	TVC [b,d]	Color Discoloration [c,d]	Odor [c,d]	Overall Acceptability [c,d]
Control	6	6	6	6	6
WPIf	6	9	9	9	9
WPIf+2.5%TEO	9	9	9	9	9

Control-uncovered fish samples; WPIf-fish samples covered with WPI-based films; WPIf+2.5%TEO-fish samples covered with whey protein isolate-based films incorporated with 2.5% tarragon essential oil. [a] Based on a maximum permitted value of 6.2 for pH. [b] Based on a maximum permitted value of 6.0 $\log_{10}$ CFU/g for TVC. [c] Based on a minimum permitted value of 4 for the acceptance score. [d] Data obtained from Figures 1 and 6 and Table 2, respectively.

4. Conclusions

The WPI-based film incorporated with 2.5% tarragon essential oil has proven to be effective in preserving the quality and, thus, in improving the shelf-life of brook trout sample during storage at 4 °C. This film has shown to possess good antioxidant and antimicrobial properties. The tarragon essential oil from its matrix has caused delays of chemical reactions and microorganisms growth in the fish sample, leading to retention of desirable sensory attributes for a longer period. Due to the low level of incorporation didn't negatively affect the organoleptic properties of the fish sample. The cost of raw materials for the manufacturing of 100 cm^2 WPIf+2.5%TEO reaches 1.4 €. In summary, this active packaging material has good industrial application potential.

Author Contributions: Conceptualization, C.A.S. and D.C.V.; methodology, C.A.S. and D.C.V.; formal analysis, M.-I.S., M.F., E.L.S., S.A.S. and A.S.P.; writing—original draft preparation, M.-I.S.; writing—review and editing, C.A.S.; visualization, D.C.V.; supervision, C.A.S. and D.C.V.; project administration, C.A.S. and D.C.V.; funding acquisition, D.C.V. All authors have read and agreed to the published version of the manuscript.

Funding: This research was funded by MCI-UEFISCDI, grant number PED 386, project PN-III-CERC-CO-PED-2-2019.

Institutional Review Board Statement: Not applicable.

Informed Consent Statement: Not applicable.

Data Availability Statement: Not applicable.

Acknowledgments: This work was supported by a grant of the Romanian Ministry of Education and Research, CNCS-UEFISCDI, project number PN-III-P1-1.1-PD-2019-0475, within PNCDI III.

Conflicts of Interest: The authors declare no conflict of interest. The funders had no role in the design of the study; in the collection, analyses, or interpretation of data; in the writing of the manuscript, or in the decision to publish the results.

References

1. Invasive Species Compendium. Available online: https://www.cabi.org/isc/datasheet/65325 (accessed on 21 November 2020).
2. Québec's 2020-2028 Brook Trout Management Plan. Available online: https://mffp.gouv.qc.ca/wp-content/uploads/quebec_brook_trout_management_plan_2020-2028.pdf (accessed on 21 November 2020).
3. Mazorra-Manzano, M.A.; Ramírez-Suárez, J.C.; Moreno-Hernández, J.M.; Pacheco-Aguilar, R. Seafood proteins. In *Proteins in Food Processing*, 2nd ed.; Yada, R.Y., Ed.; Woodhead Publishing: Duxford, UK, 2018; pp. 445–476.
4. Socaciu, M.I.; Semeniuc, C.A.; Vodnar, D.C. Edible films and coatings for fresh fish packaging: Focus on quality changes and shelf-life extension. *Coatings* **2018**, *8*, 366. [CrossRef]
5. Zarandona, I.; López-Caballero, M.E.; Montero, P.; Guerrero, P.; de la Caba, K.; Gómez-Guillén, M. Horse mackerel (*Trachurus trachurus*) fillets biopreservation by using gallic acid and chitosan coatings. *Food Control* **2021**, *120*, 107511. [CrossRef]
6. Socaciu, M.I.; Fogarasi, M.; Semeniuc, C.A.; Socaci, S.A.; Rotar, M.A.; Mureşan, V.; Pop, O.L.; Vodnar, D.C. Formulation and characterization of antimicrobial edible films based on whey protein isolate and tarragon essential oil. *Polymers* **2020**, *12*, 1748. [CrossRef] [PubMed]
7. Morar, M.I.; Fetea, F.; Rotar, A.M.; Nagy, M.; Semeniuc, C.A. Characterization of essential oils extracted from different aromatic plants by FTIR spectroscopy. *Bull. UASVM Food Sci. Technol.* **2017**, *74*, 37–38. [CrossRef]
8. Semeniuc, C.A.; Pop, C.R.; Rotar, A.M. Antibacterial activity and interactions of plant essential oil combinations against Gram-positive and Gram-negative bacteria. *J. Food Drug. Anal.* **2017**, *25*, 403–408. [CrossRef]
9. Lopes-Lutz, D.; Alviano, D.S.; Alviano, C.S.; Kolodziejczyk, P.P. Screening of chemical composition, antimicrobial and antioxidant activities of Artemisia essential oils. *Phytochemistry* **2008**, *69*, 1732–1738. [CrossRef]
10. Sharafati Chaleshtori, R.; Rokni, N.; Razavilar, V.; Rafieian Kopaei, M. The evaluation of the antibacterial and antioxidant activity of tarragon (*Artemisia dracunculus* L.) essential oil and its chemical composition. *Jundishapur J. Microbiol.* **2013**, *6*, e7877. [CrossRef]
11. Fraternale, D.; Flamini, G.; Ricci, D. Essential oil composition and antigermination activity of *Artemisia dracunculus* (tarragon). *Nat. Prod. Commun.* **2015**, *10*, 1469–1472. [CrossRef] [PubMed]
12. Pires, C.; Ramos, C.; Teixeira, B.; Batista, I.; Nunes, M.L.; Marques, A. Hake proteins edible films incorporated with essential oils: Physical, mechanical, antioxidant and antibacterial properties. *Food Hydrocoll.* **2013**, *30*, 224–231. [CrossRef]
13. Hosseini, S.F.; Amraie, M.; Salehi, M.; Mohseni, M.; Aloui, H. Effect of chitosan-based coatings enriched with savory and/or tarragon essential oils on postharvest maintenance of kumquat (*Fortunella* sp.) fruit. *Food Sci. Nutr.* **2019**, *7*, 155–162. [CrossRef] [PubMed]

14. Zhang, H.; Liang, Y.; Li, X.; Kang, H. Effect of chitosan-gelatin coating containing nano-encapsulated tarragon essential oil on the preservation of pork slices. *Meat Sci.* **2020**, *166*, 108137. [CrossRef]
15. Kazemi, S.M.; Rezaei, M. Antimicrobial effectiveness of gelatin–alginate film containing oregano essential oil for fish preservation. *J. Food Saf.* **2015**, *35*, 482–490. [CrossRef]
16. Bahram, S.; Rezaie, M.; Soltani, M.; Kamali, A.; Abdollahi, M.; Ahmadabad, M.K.; Nemati, M. Effect of whey protein concentrate coating cinamon oil on quality and shelf life of refrigerated beluga sturegeon (*Huso huso*). *J. Food Qual.* **2016**, *39*, 743–749. [CrossRef]
17. Feng, X.; Bansal, N.; Yang, H. Fish gelatin combined with chitosan coating inhibits myofibril degradation of golden pomfret (*Trachinotus blochii*) fillet during cold storage. *Food Chem.* **2016**, *200*, 283–292. [CrossRef] [PubMed]
18. Shokri, S.; Ehsani, A. Efficacy of whey protein coating incorporated with lactoperoxidase and α-tocopherol in shelf life extension of Pike-Perch fillets during refrigeration. *LWT* **2017**, *85*, 225–231. [CrossRef]
19. Kumar, K.S.; Chrisolite, B.; Sugumar, G.; Bindu, J.; Venkateshwarlu, G. Shelf life extension of tuna fillets by gelatin and chitosan based edible coating incorporated with clove oil. *Fish. Technol.* **2018**, *55*, 104–113.
20. Li, J. Characterization and Performance Improvement of Chitosan Films as Affected by Preparation Method, Synthetic Polymers, and Blend Ratios. Ph.D. Dissertation, University of Tennessee, Knoxville, TN, USA, 2008.
21. Azeredo, H.M.C.; de Britto, D.; Assis, O.B.G. Chitosan edible films and coatings—A review. In *Chitosan: Manufacture, Properties, and Usage*; Davis, S.P., Ed.; Nova Science Publishers, Inc.: New York, NY, USA, 2010; pp. 179–194.
22. Ramos, Ó.L.; Reinas, I.; Silva, S.I.; Fernandes, J.C.; Cerqueira, M.A.; Pereira, R.N.; Vicente, A.A.; Poças, M.F.; Pintado, M.E.; Malcata, F.X. Effect of whey protein purity and glycerol content upon physical properties of edible films manufactured therefrom. *Food Hydrocoll.* **2013**, *30*, 110–122. [CrossRef]
23. Semeniuc, C.A.; Socaciu, M.I.; Socaci, S.A.; Mureşan, V.; Fogarasi, M.; Rotar, A.M. Chemometric comparison and classification of some essential oils extracted from plants belonging to Apiaceae and Lamiaceae families based on their chemical composition and biological activities. *Molecules* **2018**, *23*, 2261. [CrossRef]
24. The Pherobase: Database of Pheromones and Semiochemicals. Available online: www.pherobase.com (accessed on 29 April 2020).
25. Flavornet and Human Odor Space. Available online: www.flavornet.org (accessed on 29 April 2020).
26. Nagy, M.; Semeniuc, C.A.; Socaci, S.A.; Pop, C.R.; Rotar, A.M.; Sălăgean, C.D.; Tofană, M. Utilization of brewer's spent grain and mushrooms in fortification of smoked sausages. *Food Sci. Technol. Campinas* **2017**, *37*, 315–320. [CrossRef]
27. International Standard Organization. *ISO 2917:1999: Meat and Meat Products—Measurement of pH—Reference Method*; ISO: Geneva, Switzerland, 1999.
28. European Union. Commission Decision 95/149/EC of 8 March 1995 fixing the total volatile basic nitrogen (TVB-N) limit values for certain categories of fishery products and specifying the analysis methods to be used. *Off. J. Eur. Union* **1995**, *L97*, 84–87.
29. Semeniuc, C.A.; Cardenia, V.; Mandrioli, M.; Muste, S.; Borsari, A.; Rodriguez-Estrada, M.T. Stability of flavoured phytosterol-enriched drinking yogurts during storage as affected by different packaging materials. *J. Sci. Food Agric.* **2016**, *96*, 2782–2787. [CrossRef]
30. Semeniuc, C.A.; Mandrioli, M.; Rodriguez-Estrada, M.T.; Muste, S.; Lercker, G. Thiobarbituric acid reactive substances in flavored phytosterol-enriched drinking yogurts during storage: Formation and matrix interferences. *Eur. Food Res. Technol.* **2016**, *242*, 431–439. [CrossRef]
31. International Standard Organization. *ISO 4833-1:2013: Microbiology of the Food Chain—Horizontal Method for the Enumeration of Microorganisms—Part 1: Colony Count at 30 °C by the Pour Plate Technique*; ISO: Geneva, Switzerland, 2019.
32. Jouki, M.; Yazdi, F.T.; Mortazavi, S.A.; Koocheki, A.; Khazaei, N. Effect of quince seed mucilage edible films incorporated with oregano or thyme essential oil on shelf life extension of refrigerated rainbow trout fillets. *Int. J. Food Microbiol.* **2014**, *174*, 88–97. [CrossRef]
33. International Standard Organization. *ISO 7889:2003 [IDF 117:2003]: Yogurt—Enumeration of Characteristic Microorganisms—Colony-Count Technique at 37 Degrees C.*; ISO: Geneva, Switzerland, 2003.
34. Yu, D.; Jiang, Q.; Xu, Y.; Xia, W. The shelf life extension of refrigerated grass carp (*Ctenopharyngodon idellus*) fillets by chitosan coating combined with glycerol monolaurate. *Int. J. Biol. Macromol.* **2017**, *101*, 448–454. [CrossRef] [PubMed]
35. Coroian, C.O.; Coroian, A.; Răducu, C.M.; Atodiresei, A.C.; Cocan, D.I.; Mireşan, V. Influence of various fat levels on meat quality in rainbow trout (*Oncorhynchus mykiss*) and brook trout (*Salvelinus fontinalis*). *AACL Bioflux* **2015**, *8*, 1064–1071.
36. Linhartová, Z.; Krejsa, J.; Zajíc, T.; Másílko, J.; Sampels, S.; Mráz, J. Proximate and fatty acid composition of 13 important freshwater fish species in central Europe. *Aquacult. Int.* **2018**, *26*, 695–711. [CrossRef]
37. Li, T.; Li, J.; Hu, W.; Li, X. Quality enhancement in refrigerated red drum (*Sciaenops ocellatus*) fillets using chitosan coatings containing natural preservatives. *Food Chem.* **2013**, *138*, 821–826. [CrossRef]
38. pH Meter Line. Meat/Fish Applications. Available online: https://www.wellinq.com/fabrication/meat-and-fish-application/ (accessed on 23 November 2020).
39. Shen, S.; Jiang, Y.; Liu, X.; Luo, Y.; Gao, L. Quality assessment of rainbow trout (*Oncorhynchus mykiss*) fillets during super chilling and chilled storage. *J. Food Sci. Technol.* **2015**, *52*, 5204–5211. [CrossRef]
40. Nistor, C.E.; Pagu, B.I.; Albu, A.; Păsărin, B. Study of meat physical-chemical composition of three trout breeds farmed in salmonid exploitations from Moldova. *Sci. Pap. Anim. Sci. Biotechnol.* **2014**, *47*, 190–195.
41. Huss, H.H. *Quality and Quality Changes in Fresh Fish*; FAO Fisheries Technical Paper No. 348; FAO: Rome, Italy, 1995; pp. 30–53.

42. Zhang, L.; Luo, Y.; Hu, S.; Shen, H. Effects of chitosan coatings enriched with different antioxidants on preservation of grass carp (*Ctenopharyngodon idellus*) during cold storage. *J. Aquat. Food Prod. Technol.* **2012**, *21*, 508–518. [CrossRef]
43. Manju, S.; Jose, L.; Srinivasa Gopal, T.K.; Ravishankar, C.N.; Lalitha, K.V. Effects of sodium acetate dip treatment and vacuum packaging on chemical, microbiological, textural and sensory changes of Pearlspot (*Etroplus suratensis*) during chill storage. *Food Chem.* **2007**, *102*, 27–35. [CrossRef]
44. Goulas, A.E.; Kontominas, M.G. Effect of salting and smoking method on the keeping quality of chub mackerel (*Scomber japonicus*): Biochemical and sensory attributes. *Food Chem.* **2005**, *93*, 511–520. [CrossRef]
45. Romanian Standards Association. *STAS 5386-86: Peşte Proaspăt (Fresh Fish)*; ASRO: Bucharest, Romania, 1986.
46. Oehlenschläger, J. Measurement of freshness quality of fish based on electrical properties. In *Quality of Fish from Catch to Consumer: Labelling, Monitoring and Traceability*; Luten, J.B., Oehlenschläger, J., Ólafsdóttir, G., Eds.; Wageningen Academic Publishers: Wageningen, The Netherlands, 2003; pp. 237–250.
47. Shi, C.; Cui, J.; Luo, Y.; Zhu, S.; Zhou, Z. Post-mortem changes of silver carp (*Hypophthalmichthys molitrix*) stored at 0 °C assessed by electrical conductivity. *Int. J. Food Prop.* **2015**, *18*, 415–425. [CrossRef]
48. Wright, M.D.; Holden, N.K.; Shallcross, D.E.; Henshaw, D.L. Indoor and outdoor atmospheric ion mobility spectra, diurnal variation, and relationship with meteorological parameters. *J. Geophys. Res. Atmos.* **2014**, *119*, 3251–3267. [CrossRef]
49. Chen, X.; Jiang, J. Retrieving the ion mobility ratio and aerosol charge fractions for a neutralizer in real-world applications. *Aerosol Sci. Technol.* **2018**, *52*, 1145–1155. [CrossRef]
50. Determination of the Total Volatile Basic Nitrogen (TVBN) in Fish According to Conway and Byrne Method. Available online: https://www.velp.com/public/file/tvbn-determination-in-fish-udk-139149159169-206277.pdf (accessed on 23 November 2020).
51. European Union. COMMISSION REGULATION (EC) No 1022/2008 of 17 October 2008 amending Regulation (EC) No 2074/2005 as regards the total volatile basic nitrogen (TVB-N) limits. *Off. J. Eur. Union* **2008**, *L277*, 18–20.
52. Volpe, M.G.; Siano, F.; Paolucci, M.; Sacco, A.; Sorrentino, A.; Malinconico, M.; Varricchio, E. Active edible coating effectiveness in shelf-life enhancement of trout (*Oncorhynchusmykiss*) fillets. *LWT* **2015**, *60*, 615–622. [CrossRef]
53. Irwin, J.W.; Hedges, N. Measuring lipid oxidation. In *Understanding and Measuring the Shelf-Life of Food*; Steele, R., Ed.; Woodhead Publishing: Cambridge, UK, 2004; pp. 289–316.
54. Ramezani, Z.; Zarei, M.; Raminnejad, N. Comparing the effectiveness of chitosan and nanochitosan coatings on the quality of refrigerated silver carp fillets. *Food Control* **2015**, *51*, 43–48. [CrossRef]
55. Ojagh, S.M.; Rezaei, M.; Rzavi, S.H.; Hosseini, S.M.H. Effect of chitosan coatings enriched with cinnamon oil on the quality of refrigerated rainbow trout. *Food Chem.* **2010**, *1*, 193–198. [CrossRef]
56. Diez-Gonzalez, F. Total viable counts | Specific techniques. In *Encyclopedia of Food Microbiology*, 2nd ed.; Batt, C.A., Tortorello, M.L., Eds.; Academic Press: Amsterdam, The Netherlands, 2014; pp. 630–635.
57. Food Safety Authority of Ireland. *Guidance Note 3: Guidelines for the Interpretation of Results of Microbiological Testing of Ready-to-Eat Foods Placed on the Market (Revision 4)*; FSAI: Dublin, Ireland, 2020.
58. Gram, L.; Huss, H.H. Microbiological spoilage of fish and fish products. *Int. J. Food Microbiol.* **1996**, *33*, 121–137. [CrossRef]
59. Rotar, M.A.; Semeniuc, C.; Apostu, S.; Suharoschi, R.; Mureşan, C.; Modoran, C.; Laslo, C.; Guş, C.; Culea, M. Researches concerning microbiological evolution of lactic acid bacteria to yoghurt storage during shelf-life. *JAPT* **2007**, *13*, 135–138.
60. Serio, A.; Fusella, G.C.; López, C.C.; Sacchetti, G.; Paparella, A. A survey on bacteria isolated as hydrogen sulfide producers from marine fish. *Food Control* **2014**, *39*, 111–118. [CrossRef]

 *foods* 

Article

Food By-Products to Extend Shelf Life: The Case of Cod Sticks Breaded with Dried Olive Paste

Olimpia Panza, Valentina Lacivita, Carmen Palermo, Amalia Conte * and Matteo Alessandro Del Nobile

Department of Agricultural Sciences, Food and Environment, University of Foggia, via Napoli, 25, 71122 Foggia, Italy; olimpia.panza@unifg.it (O.P.); valentina.lacivita@unifg.it (V.L.); carmen.palermo@unifg.it (C.P.); matteo.delnobile@unifg.it (M.A.D.N.)

* Correspondence: amalia.conte@unifg.it; Tel.: +39-0881-589-240

Received: 17 November 2020; Accepted: 10 December 2020; Published: 19 December 2020

Abstract: Recently, the interest in recovery bioactive compounds from food industrial by-products is growing abundantly. Olive oil by-products are a source of valuable bioactive compounds with antioxidant and antimicrobial properties. One of the most interesting by-products of olive oil obtained by a two-phase decanter is the olive paste, a wet homogeneous pulp free from residuals of the kernel. To valorize the olive paste, ready-to-cook cod sticks breaded with dried olive oil by-products were developed. Shelf-life tests were carried out on breaded cod sticks and during 15 days of storage at 4 °C pH evolution, microbiological aspects, and sensory properties were also monitored. In addition, the chemical quality of both control and active samples was assessed in terms of total phenols, flavonoids, and antioxidant activity. The enrichment with olive paste increased the total phenols, the flavonoids, and the antioxidant activity of the breaded fish samples compared to the control. Furthermore, the bioactive compounds acted as antimicrobial agents, without compromising the sensory parameters. Therefore, the new products recorded a longer shelf life (12 days) than the control fish sample that remained acceptable for nine days.

Keywords: olive oil by-products; breaded fish; fish shelf life; fish quality; sustainable food

---

1. Introduction

Seafood is an important part of a healthy diet, being an excellent source of high biological value proteins, omega-3 polyunsaturated acids, vitamins, and mineral salts [1]. Among the various fishery products, salted cod is appreciated by consumers for its taste and nutritional value [2]. However, due to its high concentration of salt, accounting for about 20%, the cod must be rehydrated for a few days and used quickly because the conditions for bacterial growth are favorable due to the high water content and low salt concentration in the product. The water content can be further reduced by drying and when it falls below 50%, so-called dried salted cod is obtained [3]. While it may be true that the salt content and moisture content of the product can reduce the risk of microbial contamination, on the other hand, the desalination process, through the soaking process, can promote microbial and fungal proliferation. Specifically, the chilled soaking process encourages halotolerant psychrotrophs that have survived salting and drying. Predominant microorganisms from soaked cod at refrigeration temperatures include Enterobacteriaceae, *Pseudomonas fluorescens* and *putida*, *Aeromonas hydrophila*, and *Shewanella putrefaciens* [4].

In addition, its time-consuming preparation contrasts with the trend of the modern consumer, who is looking for ready-to-eat fish products or ready-to-cook products. In this context, the idea of developing new cod fillets could gain consensus among consumers [5,6].

Among the preservation methods reported in the abundant literature dealing with seafood shelf life [7–10], modified atmosphere packaging, chemical preservatives, and non-thermal technologies have been applied to desalted cod [11–13].

However, modern consumers are increasingly looking for sustainable, safe, and healthy products [14]. Therefore, fresh and minimally processed food, more natural, produced with the minimum amounts of additives, microbiologically safe and nutritious but at the same time, sustainable foods are highly preferred [15].

In the context of food sustainability, several pieces of research have been conducted with the aim to valorize industrial by-products. In fact, natural preservatives from fruit and vegetable by-products could be valid agents to guarantee the shelf-life extension, since they are rich in bioactive compounds with antimicrobial and antioxidant properties [16–19]. Among food industrial by-products, there is increasing attention for the application of olive by-products, and in particular semi-solid pomace, being very rich in phenolic compounds with well-known bioactive properties [20–24]. The rising interest in recovering the bioactive compounds of olive-oil industrial by-products for food applications certainly has a dual objective, on the one hand, it allows to adequately manage industrial waste to limit the environmental impact and, on the other hand, it can valorize the beneficial properties and biological activities of polyphenols that are commercially available at very low costs [25,26].

The purpose of our study was to develop breaded ready-to-cook sticks of cod, implementing new and effective combinations between fish and olive-oil by-products. To the aim, the breading was done with and without the olive paste, properly distributed, and then applied on the fish surface. During refrigerated storage, a shelf-life test was carried out, monitoring the quality parameters related to pH evolution, microbiological aspects, and sensory properties to verify if the active breading was able to extend the shelf life of the product. For completeness, the chemical quality of the samples was also assessed, to highlight the difference in terms of phenolic compounds, flavonoids, and antioxidant activity between the control and active products.

2. Materials and Methods

2.1. Raw Materials

Refrigerated salted cod fillets were purchased from a local market (Manfredonia, FG, Italy). The olive paste from Cellina di Nardò cultivar was obtained at the beginning of the year 2020, in January, from a local olive mill (Lecce, BA, Italy) using a Pieralisi Leopard with Multi-Phase Decanter, which combines modern extraction technology without water addition and recovers a certain quantity of by-product paste, named olive paste, made up of wet pulp without any traces of kernel. The olive paste was immediately dried at 35 °C in a dryer (SG600, Namad, Rome, Italy) for 72 h. The dried olive paste was reduced to a fine powder (<500 μm) by a hammer mill (16/BV-Beccaria s.r.l., Cuneo, Italy) and then stored at 4 °C until its utilization in February 2020. All the ingredients to prepare the cod stick breaded fish, such as the breading, the spices, the potato flakes, and the fresh milk were purchased at a local market (Foggia, Italy).

2.2. Breaded Cod Sticks Preparation

Cod fillets were coarsely desalted, soaked, and stored at 4 ± 1 °C for five days, changing the water every day. On the sixth day, cod fillets were drained to eliminate excess water for about half an hour and the skin was removed. Then, the fillets (about 67% *w*/*w* water content and about 1% *w*/*w* NaCl) were cut in sticks of approximately 12 g. Two different mixtures were prepared: (i) the control mixture contained 180 g of breading with fish spices, 180 g of potato flakes; (ii) the active mixture prepared with 180 g of breading with fish spices, 180 g of potato flakes, 90 g of olive paste. The amount of olive paste represents the best optimization with other ingredients to give a final product acceptable after cooking. Control samples (Ctrl) were obtained as follows: after dipping in a solution of water and milk (1:1), samples were breaded in the control mixture, by repeating the passage twice, then were manually compacted, placed above a food tray with a pad and packaged in air using high-barrier film bags (multi-layer film Nylon/Polyethylene) with a thickness of 150 μm, provided by Biochemia (Bari, Italy) and kept under refrigeration (4 ± 1 °C). Active samples (Active) were prepared using the same procedure as for Ctrl samples, using the active mixture. Two sticks were packaged in each bag. A total of 64 sticks were prepared from two different batches: 32 Ctrl samples and 32 Active samples. All samples were stored at 4 ± 1 °C for 15 days.

2.3. Chemicals

Folin–Ciocalteu reagent, gallic acid monohydrate, ethanol, ABTS (2,2-azino-bis (3-ethylbenzothiazoline-6-sulfonic acid) diammonium salt), potassium persulfate ($K_2S_2O_8$), Trolox (6-hydroxy-2,5,7,8-tetramethylchroman-2-carboxylic acid), aluminum chloride ($AlCl_3$), sodium nitrite ($NaNO_2$), sodium hydroxide solution (NaOH), quercetin, were supplied from Sigma-Aldrich (Milan, Italy). Anhydrous sodium carbonate (Na_2CO_3) was supplied from Carlo Erba (Milan, Italy). For the preparation of the phosphate-buffered saline (PBS), the following salts were used: sodium phosphate dibasic heptahydrate ($Na_2HPO_4{\cdot}7H_2O$) and sodium phosphate monobasic monohydrate ($NaH_2PO_4{\cdot}H_2O$). All reagents were of analytical grade.

2.4. Extraction of Bioactive Compounds

For chemical analyses, both raw (R-Ctrl and R-Active) and cooked samples (C-Ctrl and C-Active) (200 °C for 15 min in an electric oven, Europa Forni, Vicenza, Italy), were firstly subjected to drying (35 °C in a ventilated stove, BINDER GmbH, Tuttlingen, Germany), milled to obtain a powder, and then subjected to extraction as reported by Cedola et al. [20].

2.5. Determination of Total Phenols Content, Total Flavonoids, and Antioxidant Activity

The evaluations of total phenols and total flavonoids were carried out on both raw and cooked breaded cod sticks. All the chemical analyses were performed the day after the sample preparation. Total phenol content was determined according to the Folin–Ciocalteu method as reported by Cedola et al. [20]. The colorimetric method allowed to quantify the total phenol content as milligrams of gallic acid equivalents (GAE) per gram of dry weight (dw), according to a calibration curve (3.12–100 mg/L; R^2 = 0.9997).

Total flavonoid content was determined using the aluminum chloride colorimetric method, according to [20]. The measure was carried out at 415 nm with a spectrophotometer (UV1800; Shimadzu Italia s.r.l; Milan, Italy) and the total flavonoids were expressed as milligrams of quercetin equivalent (QE) per gram of dry weight (dw). Quercetin standard solutions were used for constructing the calibration curve (12.5–400 mg/L; R^2 = 0.9955).

The antioxidant activity of breaded cod sticks was assessed using two methods: ABTS (2,2-azino-bis (3-ethylbenzothiazoline-6-sulfonic acid diammonium salt) assay. The test is based on the ability of antioxidants to interact with the radical cation 2,2′-azino-bis (3-ethylbenzothiazoline-6-sulfonic acid) (ABTS·+) inhibiting its absorption at 734 nm [27], and the analysis was carried out according to the methods used by Cedola et al. [20]. A calibration curve was built using Trolox as the standard, at concentrations between 6.25 mg/L and 500 mg/L (R^2 = 0.9977) and the antioxidant activity was expressed as milligrams of Trolox equivalents for gram of dry weight (dw). All tests were carried out in triplicate.

2.6. Microbiological Analyses and pH Determination

Ctrl and Active samples (20 g) were aseptically weighed into a sterile stomacher bag, diluted with peptone water (dilution 1:10), and homogenized for 90 s with a Stomacher LAB Blender 400 (Pbi International, Milan, Italy). Serial dilutions were plated onto specific media in Petri dishes to enumerate *Pseudomonas* spp., hydrogen sulfide-producing bacteria (HSPB), psychrotolerant and heat-labile aerobic bacteria (PHAB), mesophilic and psychrotrophic bacteria, Enterobacteriaceae and lactic acid bacteria (LAB) according to Danza et al. [28]. The conditions used for counting HSPB and PHAB were suggested by the Nordic Committee on Food Analyses [29]. All media and supplements were obtained from Oxoid (Milan, Italy). The microbiological analyses were carried out twice on two different samples and the results are expressed as Log cfu/g. Microbial thresholds were set to 5×10^6 cfu/g for total viable mesophilic and psychrotrophic bacteria, 10^6 cfu/g for *Pseudomonas* spp. and *Shewanella*, 10^7 cfu/g for *Photobacterium* [30]. The fitting of experimental data allowed to quantify the microbiological acceptability limit (MAL), to be intended as the time (day) to reach the specific microbiological threshold. It was calculated according to what was reported by Del Nobile et al. [31].

The measurement of pH was performed in triplicate on the first homogenized dilution of fish samples, using a pH meter (Crison, Barcelona, Spain). Two different samples were used for each measurement. Microbiological analyses and pH were analyzed at the initial time and after 2, 4, 8, 12, and 15 days of refrigerated storage at 4 °C.

2.7. Sensory Analysis

The quantitative descriptive analysis (QDA) was used for a sample comparison, according to the guidelines of the Codex Alimentarius Commission. To the aim, breaded cod sticks were submitted to a panel of five trained judges. The panelists have familiar eating habits with fish and fish products and already had experience in the evaluation of burgers and fillets based on fish. They were retrained for two days (2 h session), to establish the appropriate attributes for sensory evaluation and in order to minimize individual differences and ensure repeatability of results. The panelists were asked to give judgments on odor, color, appearance, texture, and overall quality using a nine-point scale. In the scale, 9 corresponded to excellent, 8 to very good, 7 to good, 6 to reasonable, 5 to not good (acceptable limit), 4 to disliked, 3 to bad, 2 to very bad, and 1 to completely unacceptable [32]. Before the sensory analysis, samples were sliced with a knife without removing the breading crust. Samples were differently coded and presented to each panelist simultaneously in random order. The fitting of the experimental data related to the overall quality allowed to quantify the sensory acceptability limit (SAL), which represents the time (day) necessary to reach the sensory threshold. It was calculated according to Del Nobile et al. [31].

2.8. Statistical Analysis

Fitting of experimental data provided us MAL and SAL parameters. The lowest value among the MAL and SAL parameters gave us the product shelf life. The experimental data were compared by one-way analysis of variance (ANOVA). A Duncan's multiple range test, with the option of homogeneous groups ($p < 0.05$), was carried out to determine significant differences among samples. STATISTICA 7.1 for Windows (StatSoft, Inc, Tulsa, OK, USA) was used.

3. Results and Discussion

3.1. Total Phenols, Total Flavonoids, and Antioxidant Activity of Breaded Cod Sticks

The chemical quality of cod sticks breaded with dry olive paste was assessed in terms of total phenol content (mg GAE/g dw), flavonoids (mg QE/g dw), and antioxidant activity (mg Trolox equivalent/g dw) as shown in Table 1. As can be seen, both raw (R-Ctrl and R-Active) and cooked (C-Ctrl and C-Active) products were analyzed. According to the recorded data, the dry olive paste was able to improve the chemical quality of breaded fish, as total phenols and flavonoids increased in both raw and cooked Active samples. In particular, the total phenol content expressed as Gallic acid equivalents (GAEs) was approximately six times higher in both raw and cooked Active samples (12.63 and 12.46 mg GAE/g dw, respectively) compared to R-Ctrl and C-Ctrl (2.70 and 2.82 mg GAE/g dw, respectively). Moreover, flavonoids are one of the major groups of phenolic compounds present in olive oil by-products [24,33,34]. This trend was also found in our study, in fact, the Active breaded cod sticks showed higher flavonoid content compared to the Ctrl and their concentration varied from 1.69 mg QE/g dw (R-Ctrl) to 13.68 mg QE/g dw (R-Active). Similar results were reported by Cedola et al. [20], who observed that phenols (31.16 mg GAE/g dw) and flavonoids (61.24 mg QE/g dw) recorded in the dry olive paste significantly enriched fish burgers prepared with olive paste in the formulation. As can be seen from Table 1, the flavonoid content in the C-Active samples (10.61 mg QE/g dw) is lower than in R-Active. Probably, the cooking process at 200 °C for 15 min in an electric oven could have destroyed some flavonoids, while keeping them in a higher concentration than the cooked control (1.38 mg QE/g dw). As reported by several authors, the cooking process can affect polyphenols in different ways [35,36]. In some cases, their availability may increase [20,37], in other cases cooking may reduce their content [38]. As a consequence of the enrichment of breading with dry olive paste, an antioxidant capacity was found in the Active breaded cod sticks compared to the Ctrl (20.02 against 5.84 mg Trolox/g dw). Results obtained in our study agree with data reported by Cedola et al. [19], thus confirming that the high concentration of phenols and flavonoids gives fish products a higher antioxidant capacity compared to the control ones. As can be seen, a positive relationship between phenol content and antioxidant activity was also found in our data. This aspect has been confirmed by other authors [22,23,39], who also suggested that the antioxidant capacity of the olive mill waste is related to the content of phenols and to the nature of the phenolic extracts.

Table 1. Total phenols, total flavonoids, and antioxidant activity of raw and cooked sticks.

Samples	Total Phenols (mg GAE/g dw) ± SD	Total Flavonoids (mg QE/g dw) ± SD	Antioxidant Activity (mg Trolox/g dw) ± SD
R-Ctrl	$2.70 \pm 0.15^{a,A}$	$1.69 \pm 0.09^{a,A}$	$5.88 \pm 0.18^{a,A}$
R-Active	$12.63 \pm 0.18^{b,A}$	$13.68 \pm 0.90^{b,A}$	$20.02 \pm 0.43^{b,A}$
C-Ctrl	$2.82 \pm 0.13^{a,A}$	$1.38 \pm 0.13^{a,B}$	$4.40 \pm 0.15^{a,B}$
C-Active	$12.46 \pm 0.26^{b,A}$	$10.61 \pm 0.53^{b,B}$	$12.55 \pm 0.75^{b,B}$

Data in columns with different superscript lowercase letters are significantly different (R-Ctrl and R-Active; C-Ctrl and C-Active), while with different superscript uppercase letters they are different between raw and cooked (R-Ctrl and C-Ctrl; R-Active and C-Active) ($p < 0.05$); GAE: gallic acid equivalent; QE: quercitin equivalent; R-Ctrl: raw breaded Cod sticks without olive paste; C-Ctrl: cooked breaded Cod sticks without olive paste; R-Active: raw breaded Cod sticks with olive paste; C-Active: cooked breaded Cod sticks with olive paste.

3.2. Microbial Quality of Breaded Cod Sticks

In general, the quality of raw fish depends on microbial and sensory quality. The microbial quality decay of the developed breaded cod sticks (Ctrl and Active) was determined by monitoring the viable cell concentration of the total mesophilic and psychrotrophic bacteria (TMB, TPB), *Pseudomonas* spp. (Pse), *Shewanella* (Shew.) and *Photobacterium* (Phot.). The evolution of total mesophilic and psychrotrophic bacteria was reported in Figure 1a,b. As can be seen, the microbial cell concentration in both Ctrl and Active samples gradually increased with time. In particular, the cell concentration of total mesophilic bacteria is very similar between Ctrl and Active samples and both of them did not exceed the microbial limit set at 5×10^6 cfu/g. In contrast, a significant difference was observed for psychrotrophic bacteria, since the cell concentration in the Ctrl samples on the 15th day of monitoring was higher compared to the Active samples. In fact, the Ctrl exceeded the microbial limit (5×10^6 cfu/g) after 9.05 days, while the Active samples maintained the microbial concentration below the limit until the 15th day of storage.

It is well known that the deterioration of fish products could be characterized by the presence of H_2S off-odors and off-flavors, caused by specific microbial groups [5]. For this reason, the microbial concentration of *Pseudomonas* spp., hydrogen-sulfide-producing bacteria (HSPB-*Shewanella*), and psychrotolerant and heat-labile aerobic bacteria (PHAB-*Photobacterium phosphoreum*) were also monitored.

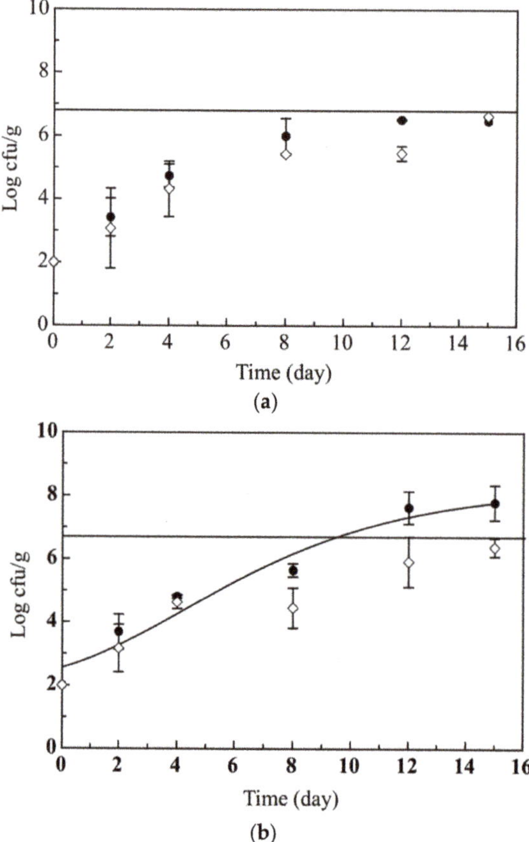

Figure 1. Evolution during 15 days of storage at 4 °C of total mesophilic (**a**) and psychrotrophic (**b**) bacteria in the breaded cod sticks. The curves are the best fitting of the experimental data. (•) Control breaded cod sticks without olive paste (Ctrl) (◊) Active breaded cod sticks with olive paste (Active).

Figure 2 shows the evolution of *Pseudomonas* spp. as a function of storage days. A substantial difference between the Ctrl and the Active samples was also found. Although both samples exceeded the microbial limit set at 10^6 cfu/g, the microbial concentration of Active samples, for almost the entire monitoring time, remained below the control by about one log cycle. As can be seen in Figure 2, *Pseudomonas* spp. growth was slower in the Active samples compared to the control, thus suggesting that the bioactive compounds present in dry olive paste acted as antimicrobial agents [34]. Therefore, the phenolic compounds used to enrich the breading of cod sticks exerted an inhibitory effect against *Pseudomonas* spp. [40] and this could justify the two different trends of data recorded for sticks with and without olive paste.

As regards the behavior of *Shewahella* and *Photobacterium*, the microbial growth was very similar in both Ctrl and Active samples (Figure 3a,b). In both cases, a gradual increase in microbial concentration was observed during the 15 days of storage. Specifically, for *Shewanella* (Figure 3a) the Active samples did not differ substantially from the control, as both of them exceeded the microbiological acceptability limit (10^6 cfu/g) almost on the same day of observation (13.01 and 12.66, respectively). On the other hand, the cell concentration of *Photobacterium* in Ctrl and Active samples was below the microbiological acceptability limit (10^7 cfu/g) for the entire storage time.

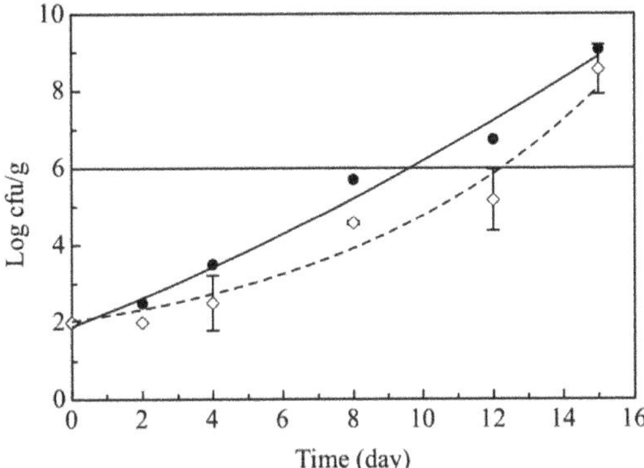

Figure 2. Evolution during 15 days of storage at 4 °C of *Pseudomonas* spp. in the breaded cod sticks. The curves are the best fitting of the experimental data. (•) Control breaded cod sticks without olive paste (Ctrl) (◊) Active breaded cod sticks with olive paste (Active).

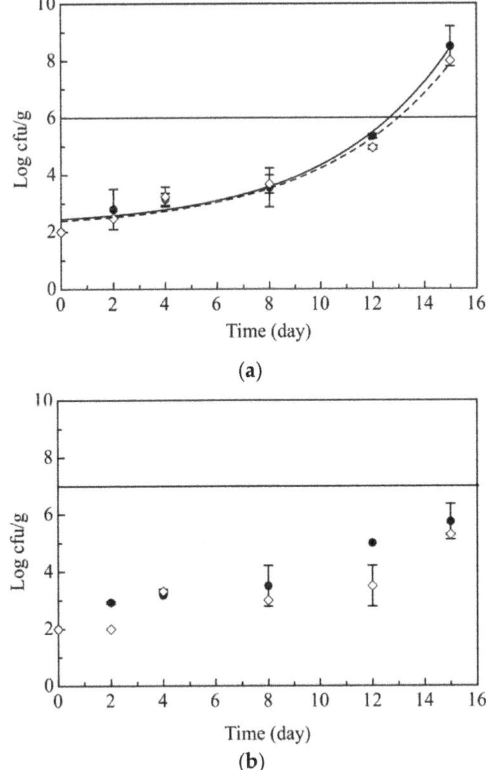

Figure 3. Evolution during 15 days of storage at 4 °C of *Shewanella* (**a**) and *Photobactherium* (**b**) in the breaded cod sticks. The curves are the best fitting of the experimental data. (•) Control breaded cod sticks without olive paste (Ctrl) (◊) Active breaded cod sticks with olive paste (Active).

Based on the results obtained by the microbiological analyses, the fitting of experimental data allowed calculating the microbiological acceptability limit (MAL). All the MAL values were summarized in Table 2 and comparing data recorded for specific spoilage groups, total mesophilic bacteria, and *Photobacterium* may be considered of less importance for the evaluation of the microbial quality of this fish product, with their cell concentrations lower than those required to spoil raw fish [30]. For the other investigated microbial groups, LAB and Enterobacteriaceae, a general inhibition was found in the Active samples. The presence of dry olive paste in the breading exerted an antimicrobial activity, slowing down and keeping the cellular loads lower than those of the control (data not shown). Several studies show that some phenols and flavonoids (hydroxytyrosol, tyrosol, and luteolin) of olive oil by-products have an antimicrobial effect especially against Gram-positive bacteria [22,41]. Our results also showed that phenolic compounds of olive paste affected Gram-negative bacteria such as *Pseudomonas* ssp., Enterobacteriaceae and *Photobacterium*.

Table 2. Shelf life (day) of breaded cod sticks (Ctrl) and Active breaded cod sticks (Active) as the lowest value among microbiological acceptability limit vales, referred to each spoilage group (MAL for total psychrotrophic and mesophilic bacteria—TPB and TMB, *Pseudomonas* spp.—Pse., *Shewanella*—Shew. and *Photobacterium*—Phot.) and sensory acceptability limit (SAL for overall quality).

Samples	Microbiological Acceptability Limit (Day)					Sensory Acceptability Limit (Day)	Shelf Life (Day)
	MAL^{TPB}	MAL^{TMB}	$MAL^{Pse.}$	$MAL^{Shew.}$	$MAL^{Phot.}$	SAL	
Ctrl	9.05 ± 0.87 [a]	>15	9.62 ± 0.55 [a]	12.66 ± 0.57 [a]	>15	13.67 ± 0.38 [a]	9.05 ± 0.87 [a]
Active	>15	>15	12.23 ± 0.72 [b]	13.01 ± 0.68 [a]	>15	14.24 ± 0.45 [a]	12.23 ± 0.72 [b]

Data (±SD $n = 2$) with different superscript letters in each column are significantly different ($p < 0.05$). Ctrl: breaded cod sticks without olive paste; Active: breaded cod sticks with olive paste.

During the 15 days of storage, in the Ctrl the pH remained roughly between values of 7.12 and 6.96, while in the Active samples between 7.0 and 6.40 (data not shown). This slight difference in pH between Ctrl and Active samples could not be attributed to relevant differences in microbial proliferation, but it is most probably ascribed to the olive paste itself which is slightly acidic [42,43].

The antimicrobial effect exerted by olive by-products has also been well evaluated by Kuley et al. [40]. These authors observed the antimicrobial effect of by-products against various food-borne pathogens and fish spoilage bacteria, thus promoting their use as valid food additives. Similar results were also reported by other researchers with other natural extracts. In particular, Martínez et al. [20] observed that the extract obtained from olives, pomegranate, and rosemary reduced microbial growth in fish patties. Danza et al. [28] observed that bio-citrus (500 ppm) in fish burgers exerted an antimicrobial effect against *Pseudomonas* spp., HSPB, and PHAB, keeping their growth lower than the control up to 13 days of storage. On the other hand, Del Nobile et al. [31] evaluated the combined effects of MAP and three natural extracts (thymol, lemon, and grapefruit seed) on fresh blue fish burgers. They observed that the combined effect of MAP and natural extracts maintained the microbial quality of products, exerting an antimicrobial effect against the main spoilage microorganisms of fish.

3.3. Sensory Quality of Breaded Cod Sticks

The expert panel judged the general appearance, the color, the odor, the texture, and the overall quality of breaded cod sticks. Scores recorded from Ctrl and Active samples for the four specific sensory attributes are reported in Table 3, as mean with relative standard deviation. As can be inferred from data in the table, for eight days of storage, both types of cod sticks, with and without olive paste, were perceived fully acceptable. After this storage time, some differences start to appear between Ctrl and Active fish, above all in terms of the odor parameter, generally considered the most critical sensory attribute for fish products [13]. As a fact, Ctrl products after 12 days were considered no more acceptable fue to unpleasant odor, whereas cod sticks with active breading remained acceptable for one more day. The other sensory attributes maintained a better trend in the Active samples than in the Ctrl fish.

Table 3. Sensory quality of breaded cod sticks in terms of appearance, odor, color, and texture.

Sensory Attributes	Samples	Storage Time (Day)					
		0	2	4	8	12	15
Appearance	Ctrl	8.0 ± 0.0 ^{a}	8.0 ± 0.0 ^{a}	8.0 ± 0.0 ^{a}	7.50 ± 0.0 ^{a}	6.17 ± 0.29 ^{a}	4.33 ± 0.29 ^{a}
	Active	8.0 ± 0.0 ^{a}	8.0 ± 0.0 ^{a}	8.0 ± 0.0 ^{a}	7.50 ± 0.0 ^{a}	7.0 ± 0.0 ^{b}	6.17 ± 0.29 ^{b}
Color	Ctrl	8.0 ± 0.0 ^{a}	8.0 ± 0.0 ^{a}	8.0 ± 0.0 ^{a}	7.33 ± 0.29 ^{a}	6.33 ± 0.29 ^{a}	4.83 ± 0.29 ^{a}
	Active	8.0 ± 0.0 ^{a}	8.0 ± 0.0 ^{a}	8.0 ± 0.0 ^{a}	7.33 ± 0.29 ^{a}	7.0 ± 0.0 ^{b}	5.17 ± 0.29 ^{a}
Odor	Ctrl	8.17 ± 0.29 ^{a}	7.50 ± 0.0 ^{a}	7.33 ± 0.29 ^{a}	7.0 ± 0.50 ^{a}	5.33 ± 0.29 ^{a}	3.67 ± 0.29 ^{a}
	Active	8.50 ± 0.0 ^{a}	8.33 ± 0.29 ^{b}	7.83 ± 0.29 ^{a}	7.0 ± 0.50 ^{a}	6.17 ± 0.29 ^{b}	4.50 ± 0.0 ^{b}
Texture	Ctrl	8.50 ± 0.0 ^{a}	8.33 ± 0.29 ^{a}	8.0 ± 0.0 ^{a}	7.83 ± 0.29 ^{a}	6.33 ± 0.29 ^{a}	5.0 ± 0.0 ^{a}
	Active	8.50 ± 0.0 ^{a}	8.33 ± 0.29 ^{a}	8.0 ± 0.0 ^{a}	8.00 ± 0.0 ^{a}	6.83 ± 0.29 ^{a}	6.0 ± 0.50 ^{b}

^{a,b} Data (±SD n = 2) with different superscript letters in each column for each attribute are significantly different ($p < 0.05$). Ctrl = breaded cod sticks without olive paste; Active = breaded cod sticks with olive paste.

In Figure 4 data with fitting curves of global sensory quality were reported. As can be seen, both Ctrl and Active samples were positively judged by the panel, in fact for most of the monitoring days the samples obtained a score between 8 and 6, corresponding to a good quality product. As expected, a gradual decrease was observed, more emphasized after the first week of storage. It is worth noting that the Ctrl samples maintained the Overall Quality score slightly lower than the Active one, and so the samples without any olive paste became unacceptable before the Active fish products. Specifically, the Ctrl exceeded the limit (score = 5) after 13.67 days of storage, while the Active samples exceeded the limit after 14.24 days (SAL values reported in Table 2). These results show that adding the olive paste to breading does not worsen the quality of cod sticks, on the contrary, it contributes to retaining sensory quality. Another striking future of sensory data recorded for cod sticks is that the panelists did not consider negatively the slightly darker color of the breading. On the contrary, Cedola et al. [20] reported that the presence of this olive by-product in fish burgers initially compromised the overall quality, due to worsening of color and texture. Our finding suggests that the application of the olive paste to the breading of cod sticks is very appropriate.

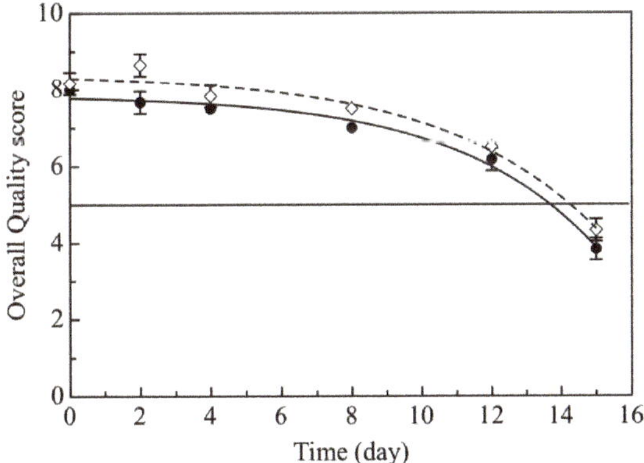

Figure 4. Overall Quality evolution during 15 days of storage of breaded cod sticks. The curves are the best fitting of the experimental data. (•) Control breaded cod sticks without olive paste (Ctrl) (◊) Active breaded cod sticks with olive paste (Active).

3.4. Breaded Cod Sticks Shelf Life

To determine final product shelf life, both fitting values of microbiological (MAL) and sensory quality (SAL) were taken into account. To this aim, data collected in Table 2 can be considered. In particular, the shelf life of breaded cod sticks was reported in the last column of this table as the lowest value between the MAL referred to all spoilage bacteria and the SAL. As a fact, the factor that influenced the breaded cod sticks' shelf life more significantly was the microbial proliferation. In particular, psychrotrophic bacteria, *Pseudomonas* spp. and *Shewanella* affected the shelf life of the Ctrl samples, while the growth of *Pseudomonas* spp. and *Shewanella* influenced the shelf life of the Active cod sticks. These results highlighted that the shelf life of Active samples (12.23 days) was longer than the Ctrl, for which a shelf life of 9.05 days was recorded. Similar results have been reported by Martínez et al. [19], who also applied bioactive compounds from food matrixes of vegetable origin on fish. In particular, these authors observed that the extract obtained from olives, pomegranate, and rosemary acted as a preservative in fish patties, extending the shelf life for 11 days. Danza et al. [28] also noted that dipping whole fish fillets in bio-citrus extract prior to mincing and forming fish burgers proved to be the best preserving strategy for its shelf life.

Based on our findings, breading enriched with olive paste is a useful approach to valorize an industrial by-product, enhance the nutritional properties of cod sticks, and extend their shelf life because the bioactive compounds entailed better stability of the product from both the microbiological and sensory point of view [39,40]. It is also worth considering that the current study was carried out in refrigerated conditions and that proper packaging of cod samples under MAP [12,13] could promote synergic effects with active breading, thus giving further more significant shelf-life prolongation.

4. Conclusions

In this study, dried olive paste, as a valuable by-product of the modern olive oil production process, was applied as breading to cod sticks to prolong their shelf life. To this aim, the microbiological, sensory, and chemical qualities of the breaded cod sticks were assessed during two weeks of refrigerated storage. The results show that the olive paste used in the breading increased the phenol and flavonoid content of cod sticks and, as a consequence, its antioxidant activity. In addition, the fortification maintained the microbiological quality of the active samples better than the control by more than three days and the sensory acceptability of the product was not compromised. In fact, an extension of the shelf life of active samples was observed. While the control cod sticks remained acceptable for about nine days, the active fish products were acceptable up to more than 12 days. Therefore, this was a concrete example of sustainable reuse of olive oil by-products to develop a breaded cod stick with healthful properties.

Author Contributions: Conceptualization, M.A.D.N. and A.C.; methodology, M.A.D.N., A.C. and C.P.; formal analysis, O.P. and V.L.; data curation, M.A.D.N., A.C. and C.P.; writing—original draft preparation, O.P. and V.L.; writing—review and editing, M.A.D.N.; A.C. and C.P.; supervision M.A.D. and A.C. All authors have read and agreed to the published version of the manuscript.

Funding: This research received no external funding.

Conflicts of Interest: The authors declare no conflict of interest.

References

1. Hosomi, R.; Yoshida, M.; Fukunaga, K. Seafood Consumption and Components for Health. *GJHS* **2012**, *4*, 72. [CrossRef] [PubMed]
2. Heredia, A.; Andrés, A.; Betoret, N.; Fito, P. Application of the SAFES (systematic approach of food engineering systems) methodology to salting, drying and desalting of cod. *J. Food Eng.* **2007**, *83*, 267–276. [CrossRef]
3. Bjørkevoll, I.; Olsen, L.R.; Skjerdal, O.T. Origin and spoilage potential of the microbiota dominating genus *Psychrobacter* in sterile rehydrated salt-cured and dried salt-cured cod (Gadus morhua). *Int. J. Food Microbiol.* **2003**, *84*, 175–187. [CrossRef]
4. Rodrigues, M.J.; Ho, P.; Lopez-Caballero, M.E.; Vaz-Pires, P.; Nunes, M.L. Characterization and identification of microflora from soaked cod and respective salted raw materials. *Food Microbiol.* **2003**, *20*, 471–481. [CrossRef]
5. Geeroms, N.; Verbeke, W.; Van Kenhove, P. Consumers' health-related motive orientations and ready meal consumption behaviour. *Appetite* **2008**, *51*, 704–712. [CrossRef]

6. Smaldone, G.; Marrone, R.; Zottola, T.; Vollano, L.; Grossi, G.; Cortesi, M.L. Formulation and shelf-life of fish burgers served to preschool children. *Ital. J. Food Saf.* **2017**, *6*, 6373. [CrossRef]
7. Olatunde, O.O.; Benjakul, S. Natural Preservatives for Extending the Shelf-Life of Seafood: A Revisit: Natural preservatives for seafood. *Compr. Rev. Food Sci. Food Saf.* **2018**, *17*, 1595–1612. [CrossRef]
8. Alves, V.L.C.D.; Rico, B.P.M.; Cruz, R.M.S.; Vicente, A.A.; Khmelinskii, I.; Vieira, M.C. Preparation and characterization of a chitosan film with grape seed extract-carvacrol microcapsules and its effect on the shelf-life of refrigerated Salmon (Salmo salar). *LWT* **2018**, *89*, 525–534. [CrossRef]
9. Perez-Won, M.; Lemus-Mondaca, R.; Herrera-Lavados, C.; Reyes, J.E.; Roco, T.; Palma-Acevedo, A.; Tabilo-Munizaga, G.; Aubourg, S.P. Combined Treatments of High Hydrostatic Pressure and CO_2 in Coho Salmon (Oncorhynchus kisutch): Effects on Enzyme Inactivation, Physicochemical Properties, and Microbial Shelf Life. *Foods* **2020**, *9*, 273. [CrossRef]
10. Senturk, T.; Alpas, H. Effect of High Hydrostatic Pressure Treatment (HHPT) on Quality and Shelf Life of Atlantic Mackerel (*Scomber scombrus*). *Food Bioprocess Technol.* **2013**, *6*, 2306–2318. [CrossRef]
11. Rode, T.M.; Rotabakk, B.T. Extending shelf life of desalted cod by high pressure processing. *Innov. Food Sci. Emerg. Technol.* **2020**, 102476. [CrossRef]
12. Magnússon, H.; Sveinsdóttir, K.; Lauzon, E.H.; Thorkelsdóttir, A.; Martinsdóttir, E. Keeping Quality of Desalted Cod Fillets in Consumer Packs. *J. Food Sci.* **2006**, *71*, 69–76. [CrossRef]
13. Wang, T.; Sveinsdóttir, K.; Magnússon, H.; Martinsdóttir, E. Combined Application of Modified Atmosphere Packaging and Superchilled Storage to Extend the Shelf Life of Fresh Cod (Gadus morhua) Loins. *J. Food Sci.* **2007**, *73*, S11–S19. [CrossRef]
14. Demartini, E.; Gaviglio, A.; La Sala, P.; Fiore, M. Impact of information and Food Technology Neophobia in consumers' acceptance of shelf-life extension in packaged fresh fish fillets. *Sustain. Prod. Consump.* **2019**, *17*, 116–125. [CrossRef]
15. Seto, K.; Fiorella, K.J. From Sea to Plate: The Role of Fish in a Sustainable Diet. *Front. Mar. Sci.* **2017**, *4*, 4. [CrossRef]
16. Alongi, M.; Melchior, S.; Anese, M. Reducing the glycemic index of short dough biscuits by using apple pomace as a functional ingredient. *LWT* **2019**, *100*, 300–305. [CrossRef]
17. Licciardello, F.; Kharchoufi, S.; Muratore, G.; Restuccia, C. Effect of edible coating combined with pomegranate peel extract on the quality maintenance of white shrimps (*Parapenaeus longirostris*) during refrigerated storage. *Food Packag. Shelf Life* **2018**, *17*, 114–119. [CrossRef]
18. Spinelli, S.; Conte, A.; Del Nobile, M.A. Microencapsulation of extracted bioactive compounds from brewer's spent grain to enrich fish-burgers. *Food Bioprod. Process.* **2016**, *100*, 450–456. [CrossRef]
19. Martínez, L.; Castillo, J.; Ros, G.; Nieto, G. Antioxidant and Antimicrobial Activity of Rosemary, Pomegranate and Olive Extracts in Fish Patties. *Antioxidants* **2019**, *8*, 86. [CrossRef]
20. Cedola, A.; Cardinali, A.; Del Nobile, M.A.; Conte, A. Fish burger enriched by olive oil industrial by-product. *Food Sci. Nutr.* **2017**, *5*, 837–844. [CrossRef]
21. Tufariello, M.; Durante, M.; Veneziani, G.; Taticchi, A.; Servili, M.; Bleve, G.; Mita, G. Patè Olive Cake: Possible Exploitation of a By-Product for Food Applications. *Front. Nutr.* **2019**, *6*, 3. [CrossRef] [PubMed]
22. Leouifoudi, I.; Harnafi, H.; Zyad, A. Olive Mill Waste Extracts: Polyphenols Content, Antioxidant, and Antimicrobial Activities. *Adv. Pharmacol. Sci.* **2015**, *2015*, 1–11. [CrossRef] [PubMed]
23. Ramos, P.; Santos, S.A.O.; Guerra, Â.R.; Guerreiro, O.; Felício, L.; Jerónimo, E.; Silvestre, A.J.D.; Neto, C.P.; Duarte, M. Valorization of olive mill residues: Antioxidant and breast cancer antiproliferative activities of hydroxytyrosol-rich extracts derived from olive oil by-products. *Ind. Crops Prod.* **2013**, *46*, 359–368. [CrossRef]
24. Herrero, M.; Temirzoda, T.N.; Segura-Carretero, A.; Quirantes, R.; Plaza, M.; Ibañez, E. New possibilities for the valorization of olive oil by-products. *J. Chromatogr. A* **2011**, *1218*, 7511–7520. [CrossRef]
25. Yener, M.E. Supercritical Fluid Processing for the Recovery of Bioactive Compounds from Food Industry By-Products. In *High Pressure Fluid Technology for Green Food Processing*; Fornari, T., Stateva, R.P., Eds.; Springer International Publishing: Cham, Switzerland, 2015; pp. 305–355. ISBN 978-3-319-10610-6.

26. Roselló-Soto, E.; Koubaa, M.; Moubarik, A.; Lopes, R.P.; Saraiva, J.A.; Boussetta, N.; Grimi, N.; Barba, F.J. Emerging opportunities for the effective valorization of wastes and by-products generated during olive oil production process: Non-conventional methods for the recovery of high-added value compounds. *Trends Food Sci. Technol.* **2015**, *45*, 296–310. [CrossRef]
27. Re, R.; Pellegrini, N.; Proteggente, A.; Pannala, A.; Yang, M.; Rice-Evans, C. Antioxidant activity applying an improved ABTS radical cation decolorization assay. *Free Radic. Biol. Med.* **1999**, *26*, 1231–1237. [CrossRef]
28. Danza, A.; Conte, A.; Del Nobile, M.A. Technological options to control quality of fish burgers. *J. Food Sci. Technol.* **2017**, *54*, 1802–1808. [CrossRef]
29. NMKL Method No. 184. *Aerobic Count and Specific Spoilage Organisms in Fish and Fish Products*; Nordic Committee on Food Analysis: Espoo, Finland, 2006; pp. 1–6.
30. Corbo, M.R.; Speranza, B.; Filippone, A.; Conte, A.; Sinigaglia, M.; Del Nobile, M.A. Natural compounds to preserve fresh fish burgers. *Int. J. Food Sci. Technol.* **2009**, *44*, 2021–2027. [CrossRef]
31. Del Nobile, M.A.; Corbo, M.R.; Speranza, B.; Sinigaglia, M.; Conte, A.; Caroprese, M. Combined effect of MAP and active compounds on fresh blue fish burger. *Int. J. Food Microbiol.* **2009**, *135*, 281–287. [CrossRef]
32. Angiolillo, L.; Conte, A.; Del Nobile, M.A. A new method to bio-preserve sea bass fillets. *Int. J. Food Microbiol.* **2018**, *271*, 60–66. [CrossRef]
33. Aliakbarian, B.; Casazza, A.A.; Perego, P. Valorization of olive oil solid waste using high pressure–High temperature reactor. *Food Chem.* **2011**, *128*, 704–710. [CrossRef]
34. Moudache, M.; Colon, M.; Nerín, C.; Zaidi, F. Phenolic content and antioxidant activity of olive by-products and antioxidant film containing olive leaf extract. *Food Chem.* **2016**, *212*, 521–527. [CrossRef] [PubMed]
35. Murador, D.; Braga, A.R.; Da Cunha, D.; De Rosso, V. Alterations in phenolic compound levels and antioxidant activity in response to cooking technique effects: A meta-analytic investigation. *Crit. Rev. Food Sci. Nutr.* **2018**, *58*, 169–177. [CrossRef] [PubMed]
36. Dolinsky, M.; Agostinho, C.; Ribeiro, D.; Rocha, G.D.S.; Barroso, S.G.; Ferreira, D.; Polinati, R.; Ciarelli, G.; Fialho, E. Effect of different cooking methods on the polyphenol concentration and antioxidant capacity of selected vegetables. *J. Culin. Sci. Technol.* **2016**, *14*, 1–12. [CrossRef]
37. Ng, Z.-X.; Chai, J.-W.; Kuppusamy, U.R. Customized cooking method improves total antioxidant activity in selected vegetables. *Int. J. Food Sci. Nutr.* **2011**, *62*, 158–163. [CrossRef]
38. Lodolini, E.M.; Cabrera-Bañegil, M.; Fernández, A.; Delgado-Adámez, J.; Ramírez, R.; Martín-Vertedor, D. Monitoring of acrylamide and phenolic compounds in table olive after high hydrostatic pressure and cooking treatments. *Food Chem.* **2019**, *286*, 250–259. [CrossRef]
39. Suárez, M.; Romero, M.-P.; Ramo, T.; Macià, A.; Motilva, M.-J. Methods for Preparing Phenolic Extracts from Olive Cake for Potential Application as Food Antioxidants. *J. Agric. Food Chem.* **2009**, *57*, 1463–1472. [CrossRef]
40. Kuley, E.; Durmus, M.; Balikci, E.; Ucar, Y.; Regenstein, J.M.; Özoğul, F. Fish spoilage bacterial growth and their biogenic amine accumulation: Inhibitory effects of olive by-products. *Int. J. Food Prop.* **2017**, *20*, 1029–1043. [CrossRef]
41. Yangui, T.; Sayadi, S.; Gargoubi, A.; Dhouib, A. Fungicidal effect of hydroxytyrosol-rich preparations from olive mill wastewater against Verticillium dahliae. *Crop Prot.* **2010**, *29*, 1208–1213. [CrossRef]
42. Romero-García, J.M.; Niño, L.; Martínez-Patiño, C.; Álvarez, C.; Castro, E.; Negro, M.J. Biorefinery based on olive biomass. State of the art and future trends. *Bioresour. Technol.* **2014**, *159*, 421–432. [CrossRef]
43. Haouache, N.; Bouchaleta, A. Olive mill wastewater: Effect on the survival and reproduction of the ecological indicator "Gammarus gauthieri". *J. Mat. Environ. Sci.* **2016**, *7*, 2288–2294.

Publisher's Note: MDPI stays neutral with regard to jurisdictional claims in published maps and institutional affiliations.

MDPI

Review

State-of-Art on the Recycling of By-Products from Fruits and Vegetables of Mediterranean Countries to Prolong Food Shelf Life

Sara Nardella, Amalia Conte * and Matteo Alessandro Del Nobile

Department of Agricultural Sciences, Food and Environment, University of Foggia, Via Napoli 25, 71121 Foggia, Italy; sara_nardella.547821@unifg.it (S.N.); matteo.delnobile@unifg.it (M.A.D.N.)
* Correspondence: amalia.conte@unifg.it; Tel.: +39-0881-589242

Abstract: Annually, 1.3 billion tons of food are wasted and this plays a major role in increasing pollution. Food waste increases domestic greenhouse gas emissions mainly due to the gas emissions associated with its production. Fruit and vegetable industrial by-products occur in the form of leaves, peel, seeds, pulp, as well as a mixture of them and represent the most abundant food waste. The disposal of agricultural by-products costs a large amount of money under certain governmental regulations. However, fruit and vegetable by-products are rich in valuable bioactive compounds, thus justifying their use as food fortifier, active food packaging or as food ingredients to preserve food quality over time. The present review collects the most recent utilization carried out at lab-scale on Mediterranean fruit and vegetable by-products as valid components to prolong food shelf life, providing a detailed picture of the state-of-art of literature on the topic. Bibliographic research was conducted by applying many keywords and filters in the last 10 years. Several scientific findings demonstrate that by-products, and in particular their extracts, are effectively capable of prolonging the shelf life of dairy food, fresh-cut produce, meat and fish-based products, oil, wine, paste and bakery products. All of the studies provide clear advances in terms of food sustainability, highlight the potential of by-products as a source of bioactive compounds, and promote a culture in which foods are intended to receive a second useful life. The same final considerations were also included regarding the current situation, which still limits by-products diffusion. In addition, a conclusion on a future perspective for by-products recycling was provided. The most important efforts have to be conducted by research since only a multidisciplinary approach for an advantageous investigation could be an efficient method to promote the scale up of by products and encourage their adoption at the industrial level.

Keywords: fruit and vegetable by-products; food shelf life; sustainable food; by-products recycling

Citation: Nardella, S.; Conte, A.; Del Nobile, M.A. State-of-Art on the Recycling of By-Products from Fruits and Vegetables of Mediterranean Countries to Prolong Food Shelf Life. *Foods* **2022**, *11*, 665. https://doi.org/10.3390/foods11050665

Academic Editor: Luís Manuel Lopes Rodrigues da Silva

Received: 13 January 2022
Accepted: 16 February 2022
Published: 24 February 2022

1. Introduction

In later years, the FAO stated that "food loss" comprises a decrease in the quantity or quality of food in the production chain [1]. It typically occurs during the animal and plant production, storage, processing, and distribution stages. In addition, it may be due to an incorrect agricultural practice as well as some phenomena, such as bruising or wilting or inadequate storage [2]. Globally, researchers estimate that, in food supply chains the percentages of food loss in production, postharvest, and consumption stages are 24, 24, and 35%, respectively [3]. "Food waste" is a component of the wider problem of food loss [4] and refers to the decrease in the quantity or quality of food resulting from decisions and actions by retailers, food service providers, and consumers [1]. This is regarding edible food, which is wasted in the second portion of the food supply chain until final consumption. In the UK, for example, the value of wasted products was estimated between USD 1500 to nearly 3000 per ton, depending on whether the waste is generated during processing or final consumption [5]. Food waste causes a large depletion of available land

and other environmental concerns, including the unnecessary gas emissions, which are measured by carbon footprint and water wastage [6,7]. Venkat [8] observed that grains, vegetables, and fruits generate 56% of the US waste, but register relatively low emission footprints.

The current linear model of the economy is based on the concept of the constant supply of products with a short useful life, forcing an increased production to satisfy the consumer's constant needs. It increases the indiscriminate exploitation of limited natural resources that would yield a significant environmental and economic crisis. In 2009, the FAO organized a forum on "How to Feed the World in 2050". Based on the prospects of that time period, the world population will increase up to 9.1 billion in 2050, accompanied by an increase of urbanization. This will lead to a rise in annual food production. However, it will only be possible with the implementation of the right investment and agricultural policies [9]. This is only one of many concrete examples, which demonstrate the real entity of the current growth trends. Apart from reducing losses and wastes at all levels of the food chain, it is possible to recycle rather than throw them [10]. Therefore, both economic and environmental impacts may be limited.

Recycling of Fruit and Vegetable By-Products

Within the framework of food loss, food industries annually produce tons of by-products during food processing. The most abundant part is represented by fruit and vegetable by-products, which can occur during the pre- and post-harvesting process, preparation, and processing of fruits and vegetables. These industrial by-products are very different from one another due to the difference in industrial processes. For example, grape and olive pomace are derived from wine and oil production, while other fruit by-products are derived from the juice, jelly, and jam industry, for example, from apple, kiwifruit, citrus, passion fruit, mango, etc., as well as from the processing of potato, tomato or carrot.

By-products have phytochemical compounds with recognized antioxidant and antimicrobial properties [11–15]. Generally, food by-products are used as animal feed or for the production of biomaterials, biofuels, biogas, platform chemicals, and bio-fertilizers [16,17]. However, over time, thanks to the great potential of their active compounds, they have been utilized in several industrial fields (cosmetic, pharmaceutical, and food) [18].

In regards to the food industry, the goal was to re-introduce the by-products to the production line as raw materials, to obtain new functional products with health benefits or to enhance food preservation or develop active packaging, as presented in Figure 1 [19].

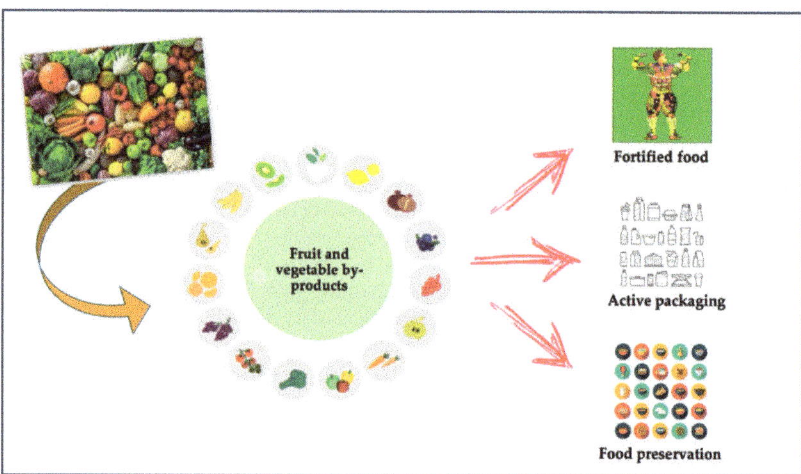

Figure 1. New directions in managing fruit and vegetable by-products.

The above picture represents the potential applications reported in the scientific literature in the field. In fact, many research studies report the usage of these by-products (i) as ingredients in different types of foods to increase the nutritional value, (ii) as natural preservatives to maintain food quality, (iii) as a source of bioactive agents to develop films and coatings with antimicrobial and/or antioxidant properties.

Specifically, with regards to the by-products for food preservation, it is interesting to observe that literature data are very abundant. Some of these studies are related to the active packaging systems, which are aimed at prolonging food shelf life. Additionally, other studies are focused on by-products, which are directly applied to food in order to preserve its quality during storage. Here, we provide some examples. Aloui et al. [20] developed a film incorporated with an extract of tomato pomace, while Kanmani and Rhim [21] added grapefruit seed extract into an agar film. Torres-León et al. [22] and Kanatt and Chawla [23] tested the effect of mango by-product-based films on peaches and chicken meat, respectively. Other authors investigated the usage of winemaking by-products on fruits and vegetables [24,25] and on fishery products [26,27]. Madzuki et al. [28] and Gallego et al. [29] tested tomato by-products added with film and coating on the deterioration of calamansi and pork meat, respectively. Moreover, there are various studies on different types of olive milling by-products [30,31]. In regards to by-products that are added to food in order to prolong the shelf life, the numerous applications found highlight that their effects are strictly dependent on the type of by-products and on the food characteristics. Therefore, the present review aims at collecting all of the information available in the literature during the last 10 years, which deal with specific applications of fruit and vegetable by-products to prolong food shelf life. To better organize the work, the studies were divided by the type of food, thus including dairy food, fresh-cut produce, vegetable-based processed food, meat, fish, and cereal-based products. In each paragraph, the reader can find details regarding the typology, concentration, and technique to apply by-products to food. In addition, the main results regarding the effects of by-products on food quality were highlighted. This provides a real map of the most effective by-products and food sectors, where by-products could find the concrete possibility of recycling. The final considerations regarding the current situation and future trends are also reported in the conclusions.

2. Materials and Methods

In this review, electronic literature searches were conducted using the online library "opac.unifg.it", which collects files from several databases including PubMed, Google Scholar, and Science Direct databases. Numerous search terms have been used, including keywords as "fruit by-products and food shelf life", "vegetable by-products and food shelf life", "antioxidant activity of food by-products", and "antimicrobial activity of food by-products". More specific search terms were also used, as "pomegranate by-products in food products", "grape pomace and food shelf life" or "fruit by-products and meat shelf life". The selected time interval was 2010–2022, since the treated subject is quite recent and the aim of the current review is to collect the latest lab-scale trends. The research process uncovered numerous articles, and among them, 86 articles were used for the current overview. Literature data focused on the potential re-utilization of fruit and vegetable by-products, and how these resources can be used as ingredients in food products to prolong food shelf life. The studies cited in this overview are grouped by the type of food, in which by-products are added (dairy food, horticultural food, vegetable food, meat, fish food, and cereal food).

3. By-Products and Dairy Food Applications

Dairy products represent a large and differentiated group of food products. The quality of raw material is the first aspect that must be ensured when referring to dairy products. Furthermore, each type of product is associated with specific defects. However, it is possible to identify some factors that are common to a range of goods. In fact, high fat products are prone to oxidation. In addition, dairy products can be subjected to enzymatic, chemical degradation or spoilage by bacteria, molds, and yeasts that grow at a refrigerated temperature [32]. At present, there are many innovative treatments and packaging technologies, which can limit the degradation of dairy products and extend their shelf life. In this regard, fruit and vegetable by-products are a valid option. Only two studies regarding the actual use of fruit and vegetable by-products to prolong the dairy product shelf life are reported in the literature, one study with olive by-products and the other with grape by-products. The main research results are presented in Table 1.

Olive and Grape By-Products

Olive oil extraction produces a variety of solid and liquid by-products, which can be grouped in olive pomace, olive mill wastewaters, olive leaves, and olive stones, and seeds. In addition to the economic burden for producers, olive oil by-products represent a severe environmental problem for their safe disposal. Simultaneously, these are rich in bioactive molecules as phenolic compounds with antioxidant and antibacterial activity [33], which can be used in the food industry as a source of natural preservatives. Rolia et al. [34] used olive phenolic extract, obtained by liquid–liquid extraction with ethyl acetate, to limit the growth of spoilage bacteria in "Fior di latte" cheese. Phenolic extract was added to the brine of packaged cheese, in two different concentrations, 250 and 500 μg/mL. The maximum microbial load tolerated for *P. fluorescens* was reached at nearly 13 and 15 days of storage for 250 and 500 μg/mL loaded samples, respectively, compared to about 11 days, which was found for the untreated cheese. For Enterobacteriaceae, the threshold limit was reached at nearly 14 and 16 days of storage for 250 and 500 μg/mL added samples, respectively, whereas it was reached at about 10 days for the traditionally packaged cheese. Therefore, it can be asserted that this olive by-product extract can be advantageously used to extend the shelf life of "Fior di latte" from 2 to 4 days, by limiting the main spoilage microorganisms.

Grape processing by-products are generated from the winemaking process. It typically occurs as pomace consisting of a mixture of residual seeds and skins, which are rich in polyphenols, dietary fibers, citric acid, ascorbic acid, tocopherols, limonoid, and other trace compounds. These mixtures also have a strong antimicrobial activity against both Gram-positive and Gram-negative bacteria, antiseptic, germicidal, fungicidal, and anti-viral properties [35]. For this reason, wine grape pomace can represent an excellent alternative to synthetic compounds in prolonging food shelf life. Tseng and Yanyun [36] examined the feasibility of using wine grape pomace as a source of antioxidant dietary fibers and polyphenols in yogurt for the improvement of its nutritional value and enhancement of its storability. In this study, dried whole grape pomace, pomace liquid extract (LE), and freeze-dried liquid extract (FDE) were investigated. Peroxide value, as an indicator of oxidation, increased during storage and after 3 weeks. In addition, 3% grape pomace added with yogurt showed the lowest oxidation value. The results are interesting, even though it is still necessary to further examine the mechanisms and methods of retention of total phenolic content and radical scavenging activity in the product.

Table 1. Applications of by-products to dairy food.

By-Products	Food	Note	Reference
Olive phenolic extract	Fior di latte cheese	Enhancement of microbial quality and sensory stability.	[34]
Grape pomace extract	Yogurt	Lipid oxidation control, however, the radical scavenging activity decreased during storage.	[36]

4. By-Products Applied to Fresh-Cut Produce

Minimally processed foods represent a great source of vitamins, minerals, dietary fiber, and other constituents. Their peculiarity is that as living entities, their metabolic processes, such as respiration, transpiration, and biochemical transformations continue after the harvest. Therefore, fresh produce quality decreases throughout time, until it becomes unacceptable for consumption. However, these processes can be slowed down by manipulating the process and storage conditions. Due to the advent of increasingly innovative technologies, it is possible to extend their shelf life over and beyond the harvest season [37]. This section is regarding the actual use of fruit and vegetable by-products, which are used as a dipping solution or food ingredient, with the aim of extending the fresh-cut food shelf life. All of the case-studies are presented in Table 2 and reported in detail below.

Tomoto Papaya, Grapefruit, and Pomegranate By-Products

Tomato by-products are a good source of chlorophylls and phenolic compounds with antioxidant and antimicrobial activities [38]. Martínez-Hernández et al. [39] investigated the effects of lycopene microsphere based dipping solutions on the evolution of the physicochemical, microbial, and bioactive quality of fresh-cut apples during refrigerated storage. Lycopene was obtained by thermal extraction from tomato by-products. The Browning Index showed that the treated samples of lycopene microspheres maintained a lower value until the 9th day of storage. In particular, this is shown in the sample dipped in the solution with the highest concentration of lycopene microspheres when compared to the water dipped control sample and the ascorbic acid dipped sample. Microbial loads of the control sample showed mesophilic, yeast, and mold increments after 9 days, similar to fresh-cut apples dipped in low lycopene microspheres solutions. Whereas, the sample dipped in the highest concentration of lycopene microspheres, showed no microbial increment during the storage period.

Papaya peel is also used to inhibit the food browning process. Papaya is a tropical fruit, in which its processing by-products mainly consist of peels and seeds. They are rich in bioactive compounds [40]. Faiq and Theerakulkait [41] evaluated the effects of dipping in the papaya peel crude extract and distilled water-based solution on the browning inhibition in potato, banana, and apple slices. Papaya peel extract was obtained by stirring a mixture of papaya peel with distilled water and then centrifuging. The authors reported that during storage, the extract had an higher percentage of browning inhibition in potato slices when compared to banana and apple slices.

In regards to grapefruit seed extract, Kim et al. [42] conducted a study on its antimicrobial efficacy against *E. coli* O157:H7, *S. typhimurium*, and *L. monocytogenes* inoculated in fresh-cut lettuce, alone and in combination with malic acid. Fresh-cut lettuce was stored at 5 °C, and its microbial quality was monitored over a period of 14 days. Regarding *E. coli* O157:H7, all of the treated samples showed more than 4.4 log CFU/g reductions. Concerning *S. typhimurium*, all of the treatments had more than 4.1 log reductions. Moreover, *L. monocytogenes* was subjected to a significant reduction of its viable cell concentration in all of the treated samples. Furthermore, it was observed that the combined treatment was more effective for the reduction of all the tested foodborne pathogens.

Pomegranate by-products mainly consist of seeds and peels, that originate from the industrial production of juices and jams. They are rich in active compounds, in particular, anthocyanins and hydrolysable tannins, with high antioxidant and antimicrobial activities [43]. Belgacem et al. [44] tested pomegranate peel extract as a natural antimicrobial to reduce the growth of *L. monocytogenes* inoculated on fresh-cut melons, apples, and pears via dipping. Fresh fruits were peeled, cut, and inoculated with the pathogen. Then, the inoculated samples were dipped in solutions at several extract concentrations. Regarding the fresh-cut apples, the sample without the extract showed an increase of the population of *L. monocytogenes* during 7 days of cold storage. On the contrary, the extract efficiently reduced the population of the pathogen. Moreover, the viable cell concentration reduction was reported as higher for samples with the higher concentration of added extract. In regards to fresh-cut pear, except for the sample at the lowest extract concentration, all of the other treated samples showed a significant reduction of *L. monocytogenes* when compared to the control sample. As expected, the sample at the highest concentration was the most effective. Moreover, in fresh-cut melon, after 7 days of storage, the sample at the highest concentration showed the most significant reduction of the tested pathogen. Elsherbiny et al. [45] also investigated the efficacy of a methanol extract of pomegranate peel applied as curative and preventive treatments to control *Fusarium* dry rot on potato tubers, using a dipping in five concentrations of extract solution or distilled water (control). The authors reported that, in curative application, the extract caused a significant reduction of dry rot development of potato tubers when compared to the control. The best reduction was obtained for the sample with the highest extract concentration. Moreover, it significantly reduced the diameter of dry rot lesions in potato tubers in the preventive application when compared to the control. Very recently, other authors tested pomegranate peel as it is, without producing any extract. In this case, Lacivita et al. [46] evaluated the effects of pomegranate peel powder at two concentrations, on the quality decay of a mixture of fresh-cut fruit salad (nectarine and pineapple) in fructose syrup, stored at 4 °C for 4 weeks. Microbial analysis showed that for the first 8 days, all of the samples maintained approximately the initial mesophilic and psychrotrophic bacteria concentration, whereas at the end of the storage period, the control sample had a significantly higher bacterial proliferation. The control salad was microbiologically acceptable for about 24 days, whereas both of the treated samples were below the microbial threshold until the end of storage. Therefore, the lowest concentration of pomegranate by-product is enough to ensure a longer microbial stability of fresh-cut fruit salad when compared to the control. In particular, at the end of the storage period, the treated salad had a significantly lower yeast concentration when compared to the control, especially the sample with the highest concentration of peel. It was also observed that the peel was very effective on lactic acid bacteria, which were lower than the control at the end of the storage. The overall sensory quality of the control sample lasted for 3 weeks, whereas the active samples lasted longer. In conclusion, it can be stated that the pomegranate peel capacity can be considered as a broad-spectrum since it had the capability of exerting a good antimicrobial and antifungal activity. These by-products can represent a great alternative to synthetic compounds, which are generally recognized by the scientific community as effective sanitizers of fresh-cut fruits [44].

Table 2. Applications of by-products to fresh-cut produce.

By-Products	Food	Note	Reference
Tomato lycopene	Fresh-cut apples	Reduced enzymatic browning and microbial growth.	[39]
Papaya peel extract	Potato, banana, and apple slices	Browning inhibition in potato slices when compared to banana and apples.	[41]
Grapefruit seed extract	Fresh-cut lettuce	Significant microbial reduction of food pathogens.	[42]
Pomegranate peel extract	Fresh-cut apples	It was very effective in reducing *L. monocytogenes* population.	[44]
	Potato tubers	It reduced the diameter of dry rot lesions in potato tubers.	[45]
Peel powder	Fresh-cut salad	Microbial control and sensory quality preservation.	[46]

5. By-Products Applied to Vegetable-Based Processed Products

Various types of vegetable-based processed foods, such as wine, olive-based pâté, hazelnut paste, and different refined oils, are presented in this section. Each of these foods has a peculiar mechanism of deterioration and specific shelf life since chemical, microbiological, and physical changes can occur during storage. In recent times, many studies have been conducted on the feasibility of preserving these different food products with natural compounds derived from by-products. Table 3 briefly presents the case-studies that are detailed in the subsequent two paragraphs.

5.1. Olive By-Products

Sulfur dioxide (SO_2) is commonly added to commercial wines at the time of bottling, in order to confer microbiological and oxidative stabilities [47]. Ruiz-Moreno et al. [48]

investigated the antioxidant and antimicrobial activities of hydroxytyrosol-enriched extract obtained from olive mill waste, as a potential alternative to SO_2 in winemaking. The authors concluded that the hydroxytyrosol-enriched extract itself is not sufficient for the replacement of SO_2 in wines, but a combination of SO_2 and hydroxytyrosol-enriched extract could be a valid solution. In fact, the analyses showed that *Saccharomyces cerevisiae* delayed its growth only with SO_2. *Hanseniaspora uvarum* and *Pediococcus damnosus* were completely inhibited by SO_2, whereas the extract only delayed their growth. The extract resulted in inefficiency against *Lactobacillus plantarum*.

Bouaziz et al. [49] conducted a study on different types of olive leaf extracts (enzymatic hydrolysate leaf extract (EHE), acetylated hydrolysate extract (AHE), and pure oleuropein), in refined olive oil (ROo) and refined olive-pomace oil (RPo) during storage, then compared them to extracts with α-tocopherol. Both of the oils that were added with EHE showed the lowest oxidation value. The rise in oxidation of the control was similar to the AHE added with oil. Finally, the oxidative resistance improved with the addition of olive leaf phenolic extracts, especially for the EHE loaded sample. Furthermore, after 6 months of storage, the oxidative resistance was lower for the two refined oils added with oleuropein, whereas it fell to zero in the control sample, as well the extracts enriched with α-tocopherol and AHE. Therefore, it can be stated that enrichment with olive leaf extract reduced oil rancidity.

Other authors [50] evaluated the effect of olive leaf extract on non-thermally stabilized olive-based pâté during storage, as compared to the butylated hydroxytoluene (BHT). In regards to the microbial analysis, the absence of pathogens (*Clostridium* and *Listeria* spp.) and contaminants (*Pseudomonas* spp. and Enterobacteriaceae) was observed. As the samples of olive-based pâté were not subjected to thermal stabilization, cultivable bacteria, yeasts, and molds were detected during sample production and storage. However, their growth was affected by the addition of the extract and refrigeration storage. The main microbial groups registered a significant reduction of the viable cell concentration in samples added with the extract, especially with 1.0 g/kg. Therefore, the authors reported that the samples were suitable for consumption, from a microbiological point of view, during refrigeration storage for 120 days under modified packaging conditions, compared to the 90 days of the untreated or BHT-added samples.

5.2. Potato, Grape, and Pomegranate By-Products

Potato by-products usually occur, among other residues, as peels, which are derived from many industrial potato processes. Potato peel is a rich source of bioactive compounds due to its high content in phenolic compounds with recognized health-promoting properties [51,52]. Samotyja [53] evaluated the antioxidant properties of two different cultivars (Jazzy and Gala) of potato peel extracts on rapeseed and sunflower oil. In regards to rapeseed oil, potato peel extracts of both cultivars inhibited the formation of hydroperoxides, showing higher efficiency when compared to BHT and butylated hydroxyanisole (BHA). Among the tested extracts, 80% ethanol extract was the most efficient. The extracts also delayed the formation of conjugated diene hydroperoxides in oil. In regards to sunflower oil, ethanolic extracts showed a moderate activity against hydroperoxides formation. Moreover, by increasing the extract concentrations, the activity against hydroperoxides formation increased. All of the investigated extracts were capable of delay conjugated diene hydroperoxide formation. Furthermore, all of the extracts significantly reduced hexanal formation when compared to the control. Therefore, potato peel extracts, are not only capable of delaying primary oxidative changes, but also have a positive effect on retarding oil rancidity.

Table 3. Applications of by-products to vegetable-based processed food.

By-Products	Food	Note	Reference
Olive (Hydroxytyrosol-enriched extract)	Wine	The extract alone is not sufficient to replace SO_2, but it was effective in delaying microbial growth.	[48]
Olive leaf extract	Refined olive oils	Oil rancidity reduction.	[49]
Olive leaf extract	Olive-based pâté	The extract enhanced oil oxidative stability.	[50]
Potato peel extract	Rapeseed and sunflower oil	Delay of primary oxidative changes in oil and positive effects on retarding oxidative rancidity.	[53]
Grape marc extract	Hazelnut paste	Capable of inhibiting lipid oxidation.	[54]
Pomegranate peel extract	Pomegranate seed oil	Capable of inhibiting lipid oxidation, above all when combined with other antioxidant compounds.	[55]

In regards to grape by-products, Spigno et al. [54] investigated the feasibility of using a grape marc phenolic freeze-dried extract to protect commercial hazelnut paste against oxidation, crude, and encapsulation into different nano-emulsion-based delivery systems. The authors asserted that phenolic grape marc can inhibit hazelnut paste oxidation and consequently improve its shelf life, even though the antioxidant effect was limited since the paste split in two phases after 5 weeks and the extract was separated. The extract did not solubilize completely in the dark fluid hazelnut paste added with emulsifiers. The encapsulation process reduced the antioxidant activity of almost all of the extracts. All of the tested formulations were no longer active after 83 days. The oil in water nano-emulsion resulted in the best encapsulation solution due to its efficiency.

Pomegranate peel extract exerted a significant role in the preservation of seed oil. This is the result of the case study by Drinić et al. [55]. The authors investigated the effect of pomegranate peel extract, alone and combined with BHT, on the antioxidant stability of pomegranate seed oil. At the end of storage, oil with the combination of antioxidants (0.05% pomegranate peel extract and 0.01% BHT) had a significant lower oxidation when compared to the pomegranate peel extract added with oil. Thiobarbituric acid reactive substances (TBARS) showed that at the end of storage, oil added with pomegranate peel extract showed the highest percentage of oxidation inhibition, followed by the combination of antioxidants and BHT.

6. By-Products and Meat Applications

The shelf life of meat products is influenced by many factors, which are complex and interconnected. Microbial growth is one of them, and the types of microorganisms that can be found in meat depend, among other things, on animal species, personnel sanitation, type of packaging, storage time, and temperature [56]. Lipid oxidation is the major cause of chemical alteration during storage. It causes irreversible changes in taste, flavor, color, and texture of the products, resulting in a shelf life reduction. Moreover, protein oxidation causes the loss of sensory properties, essential amino acids, and protein digestibility. Modified atmosphere packaging and synthetic antioxidant and antimicrobial substances are widely used in the food industry to improve the meat product shelf life. Nevertheless, synthetic compounds are known as potentially toxic. Therefore, natural extracts can represent a good alternative to protect these foods.

6.1. Olive Mill Wastewater, Pomace, and Seed Extract

As previously reported, tomato pomace has shown good bioactive properties. Olive pomace is an important natural source of phenolic compounds. Pomegranate seeds are known to have powerful antioxidant compounds. Finally, grape pomace is very rich in antioxidant and antibacterial agents. All of these by-products were adopted, mainly in the form of extracts, with slight effects on the quality of meat-based food. In general, all of them, and in particular grape by-products, are capable of preserving meat products against lipid oxidation, protein oxidation, and microbial growth (Table 4). Nevertheless, in some cases, it is better to combine them with other preserving methods to increase their effects.

In the study by Andrés et al. [57], the authors evaluated the shelf life of lamb meat patties added with aqueous extracts from tomato, olive, pomegranate, and grape by-products stored in retail sale conditions. The authors reported that free thiols consistently decreased after 7 days of storage, even though the extract-loaded samples showed the highest values at the end of the observation period. Counts of mesophilic and psychrotrophic bacteria in lamb patties with extracts were significantly lower than the control and sodium ascorbate added samples, even though the antimicrobial effect of the extracts was less evident for Enterobacteriaceae and lactic acid bacteria. In a second study, Andres et al. [58] evaluated the in vitro antioxidant potential of tomato pomace extracts and then the effect on the shelf life of lamb meat packaged under modified packaging conditions (MAP). After 7 days of storage, the TBARS values of meat treated with the extract obtained using ethyl acetate and with the extract obtained using ethanol significantly increased, with no significant differences from the control. Free thiols significantly decreased during storage. Moreover, microbiological analysis showed that the aerobic count in lamb meat remained under the microbiological limits in fresh meat, for both mesophilic and psychrophilic bacteria. The final values of lactic acid bacteria were also substantially lower than the imposed limit.

Better results were found by Selani et al. [59], who investigated the effects of Isabel and Niágara grape seed and peel extracts on lipid oxidation of raw and cooked chicken meat vacuum-packed and stored at −18 °C for 9 months. The authors reported that both natural extracts were similar to commercial antioxidants in preventing lipid oxidation in raw and cooked chicken meat and their effects were more evident in cooked samples. According to Selani et al. [59], other authors also assessed the effects of grape seed extracts. Kulkarni et al. [60] compared three levels of grape seed extract as a commonly used antioxidant, in pre-cooked, frozen, and stored beef and pork sausage. Based on sensory characteristics, instrumental color, and TBARS values, the authors concluded that the sample added with 100 and 300 ppm of extract was generally as good as propyl gallate in maintaining product quality during the 4 months of storage.

Lorenzo et al. [61] evaluated the effect of grape seed and chestnut extracts and BHT on physico-chemical and microbiological changes, as well as lipid oxidation during the ripening process of dry-cured sausages. To this aim, the extracts were mixed with meat during the initial phases of processing. The authors asserted that the grape seed extract was the most effective antioxidant of dry-fermented sausages. In fact, they reported that the TBARS values decreased when compared to the control. However, compared to the control and chestnut extract-treated sausages, grape seed extract improved lipid stability. Total viable count and lactic acid bacteria increased during the first 19 days of ripening and remained stable until the end of the curing process. The highest lactic acid bacterial counts were observed in sausages of the chestnut extract group and control batch. At the end of the curing process, mold and yeast counts were higher in control and grape seed extract samples than in BHT and chestnut extract-treated sausages.

Grape pomace extract was found less effective than seed and peel extracts by Garrido et al. [62]. These authors evaluated the effects of two types of grape pomace extracts on the physico-chemical characteristics of pork burgers and their preservative capacity. They reported that total viable count, psychrophilic bacteria, and total coliform count were not affected by the extract addition. In fact, the samples were microbiologically unacceptable after 6 days. Nevertheless, compared to the extract obtained using methanolic extraction, the extracts obtained using the high–low instantaneous pressure and high–low instantaneous pressure + methanolic extraction showed a stronger antioxidant effect when added to meat.

Table 4. Applications of by-products to meat food.

By-Products	Food	Note	References
Tomato, Olive, Pomegranate and Grape Extract	Lamb patties	Capable of controlling the growth of mesophilic and psychrotrophic bacteria.	[57]
Tomato pomace extract	Lamb meat	Capable of controlling both microbial growth and TBARS values.	[58]
GRAPE Seed and peel extract Seed extract Seed extract and chestnut extract Pomace extract Wine pomaces	Chicken Pork sausage Dry-cured sausages Pork burgers Beef patties	Capable of delaying lipid oxidation without affecting color. Capable of maintaining product quality during the 4 months of storage. Capable of maintaining lipid stability and microbial quality. Antioxidant effects on meat quality. Effective against protein oxidation.	[59] [60] [61] [62] [63]
Olive mill wastewater Leaf extracts Hydroxytyrosol extracts	Italian dry fermented sausages Karadi lamb patties Halal minced beef Chilled poultry meat Frozen hamburger Minced beef Low-fat frankfurter chicken sausages	Reduction of undesired fungal growth on the surface, oxidation control. Effective against oxidation and microbial growth. Effective against oxidation and microbial growth. Effective against oxidation and microbial growth. Effective against oxidation. Significant shelf life prolongation. Capable of preventing lipid and protein oxidation of sausages, especially when olive oil was used rather than pork fat.	[64] [65] [66] [67] [68] [69] [70]

Garcia-Lomillo et al. [63] evaluated the effect of milled red and white grape skin wine pomaces on beef patties, which are stored for 15 days in high oxygen modified atmosphere against protein oxidation, in comparison to the protective effect of sulfites. The analysis on beef proteins showed that patties added with sulfites had the lowest accumulation of protein radicals. The addition of white grape skin wine pomace to the beef patties resulted in higher radical intensity, whereas the addition of red pomace caused no effect compared to the sample added with water (control). Moreover, while the radical intensity of control and white pomace added with patties increased during storage, those added with sulfites or red pomace were constant. The authors concluded that red pomace effectively protected against protein oxidation.

In regards to olive mill wastewater, Chaves-López et al. [64] evaluated its effect against the undesired household fungi that may grow on the surface of Italian dry fermented sausages during ripening, considering its impact on the microbiological and physico-chemical characteristics. The authors reported that all of the microbial groups grew during the observation time, except for Enterobacteriaceae, which decreased. The treated samples showed an extra reduction of fungal growth, which was proportional to the extract concentration. In the enriched samples, only *Penicillium nalgiovense* and *Penicillium chrysogenum* were isolated. Micrococcaceae, yeasts, and molds decreased in the treated batches. However, compared to the control, their proteolysis index was slightly higher. In the presence of olive mill wastewater polyphenols, the volatile compounds derived from microbial esterification and lipid oxidation decreased. Furthermore, TBARS of fortified samples showed reduced values when compared to the control batch. Therefore, the authors found that the surface treatment of fermented sausages with 2.5% by-products addition was effective against some undesired fungal species.

6.2. Olive Leaf Extracts

Olive leaf extract was also greatly used at lab-scale to prolong the shelf life of several meat-based foods, which are considered a valid alternative to synthetic additives. The extracts are capable of maintaining both chemical and microbiological safety. Therefore, they can enhance meat quality in the same way or even better than synthetic compounds. In many cases, it was reported that their positive effect was more evident by increasing their concentrations (Table 4).

Specifically, Baker [65] conducted a study to establish the optimum concentration of olive leaf extract in minimizing the oxidative and microbiological deterioration of Karadi lamb patties. The author reported that compared to the untreated sample, the treated samples with 1, 2, 3% of extract had significantly lower TBARS values. The bacterial counts increased during storage. However, for the treated patties, microbial proliferations were significantly lower when compared to the control sample, until the end of the tested period. The authors found that 1% of treated patties also showed the best overall acceptability. In the same context, Djenane et al. [66] used Algerian wild olive leaf extracts as an enrichment of Halal minced beef at two different levels (1 and 5%, *v*/*w*) and evaluated their effects on microbiological safety during 6 days of retail. The authors reported that the addition of 5% extract showed the strongest antimicrobial activity. In addition, at the end of the storage period, the microbial count found in the treated samples was still not near the critical microbiological threshold. The TBARS values of all the tested samples also decreased by increasing the extract concentration. Other authors recently [67] assessed the effects of olive leaf extracts on physico-chemical properties and microbial quality of chilled poultry meat. To this aim, meat was dipped for 15 min in the extract at concentrations of 0.25%, 0.5%, and 1%. Herein, it can be asserted that olive leaf extract has the capability of maintaining the chemical and microbiological quality of chilled poultry meat. In fact, compared to the control (meat without the extract), the total volatile basic nitrogen (TVBN mg/100 g) in poultry meat after the treatment was significantly reduced, especially in 1% of the treated sample. Moreover, this last sample reached the unfit limit for TVN after 15 days over 6 days for the control. The TBARS values of all the samples increased during storage.

However, compared to the control samples, the treated product had noticeably lower TBARS values. The extract also decreased the total aerobic plate count. Compared to the lower concentrations, 1% was more effective. Moreover, the extract, especially 1%, positively delayed the growth of psychrophilic, Enterobacteriaceae, Staphylococcal, as well as the mold and yeast counts. Elama et al. [68] investigated the effects of oleuropein from olive leaf extract on lipid peroxidation in frozen bovine hamburger, compared with sodium erythorbate. Oleuropein was mixed at different concentrations with meat during the hamburger preparation. The authors reported that the amount of oxidation products increased for both control and treated samples (with 0.25, 0.5, and 0.75%) during 6 months of storage. However, at the end of storage, the amount of oxidation products was found as lower in the treated samples than in the control. The sample which showed the best results was the one with 0.5% oleuropein.

The effects of destoned olive cake on the physico-chemical, microbiological, and sensory quality of beef patties during cold storage were also assessed by other authors [69]. Samples were prepared by mixing minced beef with olive by-products at several concentrations. Compared to the untreated sample, the DPPH radical scavenging activity values after 14 days of storage of fortified patties at a concentration above 2% were found as higher. The TBARS values of all the samples increased during the storage period. However, compared to the untreated samples, the TBARS values observed for the treated beef patties were lower. Moreover, it was found that the by-products effect is concentration dependent. Furthermore, the authors observed that all of the fortified samples had a significantly lower total plate count than the control sample. Therefore, the authors reported that the incorporation of destoned olive cake in beef patties could prolong their shelf life up to 14 days. To conclude the state-of-art of olive by-products, the case study by Nieto et al. [70] can be also cited. These authors evaluated the effects of different hydroxytyrosol extracts on chemical and sensory properties of low-fat frankfurter chicken sausages during 21 days of storage at 4 °C in a modified atmosphere. The chicken sausage formulation includes olive oil as a fat substitute, walnut as a macronutrient, and hydroxytyrosol extract as an antioxidant. The extracts were added to the homogenized meat, then mixed. The authors reported that the TBARS values of all the samples increased during the storage period. However, compared to the control sample, the oxidation was significantly lower in extracts added with sausages. Compared to the control with pork fat, control with walnut, and control with walnut and olive oil, the extracts significantly reduced the thiol concentration since the start of the storage period. Therefore, to conclude, the addition of walnuts and extracts was useful for the prevention of lipid and protein oxidation of sausages, especially when olive oil was used rather than pork fat.

7. By-Products Applied to Fish-Based Products

Fish food represents a highly healthy food, as they are an abundant source of high biological value proteins, long chain polyunsaturated fatty acids, and other nutritional components. In fact, color and lipid oxidation are the main factors for quality deterioration in fishery products during storage. The pH values in fish, which do not decrease as in meat products, cause enzymatic systems to be highly active, thus increasing their vulnerability to bacteria [71]. The shelf life of fish products is usually extended with refrigeration, freezing or thermal inactivation. Synthetic additives are frequently used to enhance their shelf life, although nowadays, food industries are searching for natural alternatives and have found them in by-products (Table 5).

Table 5. Applications of by-products to fish food.

By-Products	Food	Note	Reference
Citrus and mint peel extract	Indian mackerel	Considerable effects on fish chemical and microbial quality during storage.	[72]
Pomegranate rind, green tea, and grape seed extract	Minced mackerel	Oxidative stability of fish. Pomegranate extract was the most effective in protecting its quality.	[73]
Grape and pomegranate extract	Fish	Effective against oxidation.	[74]
Grape pomace	Silver carp fillets	Effective against oxidation.	[75]
Citrus peel extract	Ice to store pandora	Effective against oxidation.	[76]
Cabbage and banana peel and leaves	Fish-based products	Partial effects on lipid oxidation.	[77]
Clove flower buds, sage leaves, kiwifruit Peel extracts	Tuna fish fingers	The mixture of antioxidants is effective against oxidation and microbial growth.	[78]
Olive mill wastewater	Rose shrimp	Capable of delaying lipid oxidation, microbial growth, and volatile nitrogen compound formation.	[79]
Olive dry olive paste powder	Breaded cod sticks	Effective in reducing microbial quality decay.	[80]
Pomegranate peel powder	Breaded cod sticks	Very effective in reducing microbial quality decay.	[81]

7.1. Peel, Seed, and Leaf Extracts

Viji et al. [72] investigated and compared the effects of citrus peel extract and mint extracts to prolong the shelf life of washed, beheaded, and eviscerated Indian mackerel during chilled storage. The authors reported that TVBN increased significantly in all of the samples during storage. The rate of lipid hydrolysis was substantially lower in treated fish, whereas the level of free fatty acids and peroxide values in the sample remained slightly lower than the control sample. The TBARS values of all the samples increased gradually during storage. However, at the end of the observation period, the TBARS value of the sample with mint extract was the lowest one. Moreover, the extracts were capable of slowing down bacterial growth. Therefore, the treatments extended the shelf life of refrigerated mackerel by 2 and 5 days with citrus and mint extracts, respectively.

Özen and Soyer [73] evaluated the efficiency of green tea extract, grape seed extract, and pomegranate rind extract in limiting lipid and protein oxidation in minced mackerel, during 6 months of frozen storage. Peroxide values increased for minced mackerel added with the extracts and the control until the 4th month of storage, then declined rapidly. Among the extracts, pomegranate was the most effective. TBARS of the BHT treated sample showed only a slight increase, followed by the sample with pomegranate extract. However, at the end of frozen storage, compared to the control, all of the samples loaded with antioxidants showed significantly lower TBARS values. Compared to the control, the carbonyl content of all the antioxidant-loaded samples was significantly lower at the end of the 6th month. Furthermore, the changes in total protein solubility showed a progressive decrease throughout the storage period and compared to the control, the antioxidant treated samples showed a significantly higher total protein solubility. The authors concluded that these natural antioxidants improved the oxidative stability of fish and pomegranate extract was the most effective in protecting its quality. Prior to the Özen and Soyer study [73], Özen et al. [74] investigated the effects of natural extracts from by-products on lipid oxidation of fish during 3 months of frozen storage. The authors used grape and pomegranate seed extracts added to minced chub mackerel muscle. In addition, they reported that peroxide values increased for all of the tested samples during the storage period, especially for the pomegranate loaded sample, whereas no significant increase was observed for the sample with grape seed extract. Moreover, compared to the antioxidant treated samples, the TBARS values of the control fish were significantly higher at the end of the tested period.

Hasani and Alizadeh [75] evaluated the effects of red grape pomace extract, added at two different percentages (2 and 4%), on quality changes in silver carp fillets during refrigerated storage. The TBARS values of all the samples increased during storage. However, compared to the control, the TBARS values of grape pomace extract-loaded sample were significantly lower, more evidently with 4% of extract.

Yerlikaya et al. [76] produced an ice with different citrus peel extracts and evaluated their effects on the shelf life and quality of common pandora during refrigerated storage. The authors reported that the TVBN concentrations in the fish stored in ice increased during storage in all of the samples. At the end of the storage, the grapefruit flavedo treatment had the highest TVBN value, whereas the bitter orange flavedo treatment showed the lowest value. Moreover, the TBARS values of samples treated with citrus extracts remained low. The count of total psycrophilic bacteria of all the samples significantly increased during storage, but the bitter orange extract-loaded samples showed the lowest value at the end of the storage.

No significant results were found by Ali et al. [77], who evaluated the supplementation of extracts from cabbage leaves and banana peels in fish-based products to prevent the formation of potential oxidized free fatty acid and peroxide compounds. Natural extracts were loaded at different concentrations (0.5, 1, and 1.5%) to fish meat balls and then stored both at 4 and −18 °C for 9 days and 2 months, respectively. All of the treated samples showed an increase of peroxide value and free fatty acid values, more evidently when stored at 4 °C. Contrary to Ali et al. [77], Abdel-Wahab et al. [78] investigated with success the antioxidant and antimicrobial properties of clove flower buds, sage leaves, kiwifruit peels aqueous extracts, as well as their mixture, on tuna fish fingers. The extracts mixed at concentrations of 0.1, 0.25, and 0.5% were mixed during processing. The authors observed that the TBARS values significantly increased in all of the samples during 30 days of storage. However, compared to the control sample, the treated fish reached significantly lower TBARS values. Moreover, compared to the 0.25 and 0.5% mixture of the added samples, the carbonyl content increased significantly in the other samples. TVBN values increased in all of the samples during storage. However, compared to the control, the extract mixture of the treated samples showed lower final values. Furthermore, the total bacterial count increased significantly in all of the samples during storage, less than two orders of magnitude in the mixtures of the treated samples. The control reached the threshold limit at day 3, and the BHT-treated fish at day 6. On the contrary, the 0.5% extract mixture of the treated sample reached the threshold limit at day 24, thus demonstrating the great effects of the extracts on product shelf life.

Miraglia et al. [79] asserted that the extract derived from olive mill wastewater was capable of delaying lipid oxidation, microbial growth, and TVBN on pink shrimp stored at 2 °C. The most remarkable reduction of microbial growth rate in the treated sample was observed for psychrotrophic bacteria.

7.2. Peel Powders

To date, a small number of examples are reported in the literature on the use of peel without any preliminary extraction, before direct application to fish products (Table 5). In this context, two articles can be cited by Panza et al. [80,81]. In the first article, the authors [80] developed a ready-to-cook cod stick breaded with different concentrations of dry olive paste powder and monitored the quality parameters during 15 days of refrigerated storage. The authors reported that compared to the raw and cooked breaded cod sticks without olive paste, the raw and cooked active samples showed an higher antioxidant activity. The microbial cell concentration increased during the storage, in both active and control fish. However, in the control sample, it was significantly higher than in the active samples. The control samples remained slightly lower than the active sample in the overall quality score. When the authors compared both microbiological and sensory limits to determine the product shelf life, they found that the shelf life of active samples was about 12 days and it was longer by 3 days than the control fish. At a later date, Panza et al. [81] evaluated the effects of pomegranate peel powder on the same breaded cod sticks stored at refrigerated conditions. During the 17 days of storage, total mesophilic bacteria of the investigated breaded cod sticks, similar to the psychrotrophic count, gradually increased and reached the unacceptable limit after about 6 days of storage in the sample without the addition of the active powder. At the end of the storage period, the sticks breaded with pomegranate peel powder showed the lowest total bacterial count. The analysis on *Pseudomonas* spp. growth showed that the treated samples never reached the microbial acceptability limit during the storage period, whereas the control sample was already unacceptable after 9 days. Compared to the control, *Shewanella putrefaciens, Photobacterium phosphoreum,* and Enterobacteriaceae growth were limited in all of the treated samples. Therefore, the use of pomegranate peel powder also promoted a significant improvement of microbial stability of cod sticks.

8. By-Products and Cereal Food Applications

Cereals are a huge group of food products, among which bakery one represents the main portion. Its distinguished feature is that the recipes contain a significant proportion of wheat flour. Causes of the deterioration of bakery products are microbial spoilage or physical changes, which of course, also cause changes in their sensory properties, thus limiting the shelf life. Among the physical changes, loss or absorption of moisture from the atmosphere, which also causes crumb firming through starch retrogradation, are the most likely to occur during their storage. For this reason, a key function of the packaging of baked products is to control moisture transfer to and from the product [82]. Synthetic additives and modified atmosphere packaging are also widely used to prolong their shelf life. Recent trends include the use of natural compounds obtained from industrial by-products, which are safer and can add value to food products. The three recent case studies are presented in Table 6.

Table 6. Applications of by-products to cereal-based food.

By-Products	Food	Note	Reference
Pomegranate extract	Cookies	Reduction of microbial load and inhibition of lipid oxidation.	[83]
Tomato pomace	Muffins	Control of microbial spoilage and good antioxidant activity.	[84]
Olive leaf extract	Baked snacks	Capable of inhibiting lipid oxidation of oil in baked snacks.	[85]

Peel, Pomace, and Leaf Extracts

Ismail et al. [83] evaluated the potential of pomegranate extracts and its residues in cookies as a natural preservative and promising food fortifier. The authors reported that higher free radical scavenging properties were observed for extract-loaded cookies when compared to the ones added with residues. Some fortified cookies showed a significant reduction of the microbiological load during the tested period. In addition, samples with extract-supplemented cookies showed a better inhibition rate to lipid oxidation.

Mehta et al. [84] produced bread and muffins by adding tomato pomace to investigate its effect on their nutritional properties and storage stability. The authors reported that the antioxidant activity of tomato pomace incorporated bread and muffin was higher when compared to their controls. In fact, tomato pomace added with bread showed a shelf life of 5 and 4 days at 10 and 25 °C, respectively, which was longer when compared to the control.

Difonzo et al. [85] evaluated the effects of olive leaf extract mixed in baked snacks, during both accelerated oxidation conditions and storage. Two different quality levels of oil were used for baked snacks preparation, EVO1 and EVO2. Results of the oxidative stability evaluation showed that the induction period of both oil-added baked snacks was higher when compared to the control. Snacks with EVO1 had a higher activity of hydrophilic fraction than the samples containing EVO2. Moreover, the authors reported that EVO1 showed a higher quality level than EVO2, as the amount of volatile compounds derived from oxidation was strongly lower in the EVO1 added snack.

9. Conclusions

The awareness of human, economic, and environmental impacts caused by food loss and waste encouraged researchers to investigate the feasibility of using industrial by-products to prolong food shelf life. Many compounds present in fruit and vegetable by-products have proven effective in prolonging food shelf life. In particular, the studies collected in this review demonstrated that by-products can be used for inhibiting oxidation processes, microbial growth, as well as physic degradation of food, without compromising sensory properties. Some critical considerations can be highlighted regarding the current situation and future trends.

Current situation. At present, it is still difficult to find these kinds of innovative products in the market. This represents a huge lack in the modern world, which is continuously aiming at sustainability. Several causes may lead to this issue. Considering this issue, presumably one of the main reasons is the high-risk investment associated with the recycling of by-products. The industrial implementation for recovery processes is complex and requires a careful evaluation. Currently, it seems as a significantly expensive investment, especially for small- and medium-sized enterprises, as the consumer is not inclined to pay the additional cost. Generally, people tend to assume an incoherent behavior on the matter. In fact, they demand an increased production at the lowest possible cost, and consider it even better when disposable. At the same time, they are increasingly interested in environmental and health matters. This is the reason that nowadays people demand natural preserving additives as a replacement of synthetic ones. However, they commonly tend to identify food by-products with something that is not safe or suitable for human consumption, which is at the end of its life cycle and cannot be reintegrated in the food chain. To date, most of the consumers would be discouraged from buying food with by-products.

Future trends. In a novel bio-economy perspective, the promotion of pathways that encourage the recovery of compounds, which are still presenting an added value that otherwise would be lost, is a priority. It is fundamental to dispel the misconceptions, which lead to identifying industrial by-products as trash. As in many other fields, information is the basis of concrete progress. In this perspective, a very important role is played by a holistic research approach, which is capable of identifying the advantages of by-products and their real efficacy. Research needs to simultaneously focus the attention on main interdisciplinary factors that could make industrial food by-products an effective entry point to mitigate the greater food waste problem. The recycling of by-products needs to be approved by the current legislation, and the prospect as new ingredients will depend on new safety and regulatory assessments. In addition, social, environmental, cultural, and psychological influences on consumers' food choices need to be considered. In particular, consumers should be increasingly aware of the great potential of by-products as a valid support in extending the food shelf life. In this regard, consumers will most likely be willing to pay for the additional cost.

Author Contributions: Conceptualization, A.C. and M.A.D.N.; data curation, S.N.; writing—original draft preparation, S.N.; writing—review and editing, A.C. and M.A.D.N.; supervision, A.C. All authors have read and agreed to the published version of the manuscript.

Funding: This research received no external funding.

Data Availability Statement: Data are available on request.

Conflicts of Interest: The authors declare no conflict of interest.

References

1. FAO. *Global Food Losses and Waste. Extent, Causes and Prevention*; FAO: Rome, Italy, 2011.
2. Lipinski, B.; Hanson, C.; Lomax, J.; Kitinoja, L.; Waite, R.; Searchinger, T. *Reducing Food Loss and Waste*; World Resource Institute: Washington, DC, USA, 2013; pp. 1–40.
3. Xue, L.; Liu, G.; Parfitt, J.; Liu, X.; Van Herpen, E.; Stenmarck, Å.; O'Connor, C.; Ostergren, K.; Cheng, S. Missing food, missing data? A critical review of global food losses and food waste data. *Environ. Sci. Technol.* **2017**, *51*, 6618–6633. [CrossRef] [PubMed]
4. Vieri, S.; Calabrò, G. Food Waste: An expression of the evolution of current agricultural development systems. In *Food Waste: Practices, Management and Challenges*; Riley, G.L., Ed.; Nova Science Publishers: Hauppauge, NY, USA, 2016; pp. 1–22.
5. FAO. *Food Wastage Footprint. Impacts on Natural Resources*; FAO: Rome, Italy, 2013.
6. WRAP. *Estimates of Waste in the Food and Drink Supply Chain*; WRAP: Branbury, UK, 2013.
7. Chapagain, A.; James, K. Accounting for the impact of food waste on water resources and climate change. In *Food Industry Wastes: Assessment and Recuperation of Commodities*; Kosseva, M., Webb, C., Eds.; Academic Press: Cambridge, MA, USA, 2013; pp. 217–236.
8. Venkat, K. The Climate Change and Economic Impacts of Food Waste in the United States. *Int. J. Food Syst. Dynam.* **2011**, *2*, 431–446.
9. FAO. *How to Feed the World in 2050*; FAO: Rome, Italy, 2009.
10. Yeo, J.; Oh, J.I.; Cheung, H.H.L.; Lee, P.K.H.; Kyoungjin, A. Smart food waste recycling bin (S-FRB) to turn food waste into green energy resources. *J. Environ. Manag.* **2019**, *234*, 290–296. [CrossRef] [PubMed]
11. Helkar, P.B.; Sahoo, A.; Patil, N. Review: Food industry by-products used as functional food ingredients. *Int. J. Waste Resour.* **2016**, *6*, 248.
12. Galali, Y.; Omar, Z.A.; Sajadi, S.M. Biologically active components in by-products of food processing. *Food Sci. Nutr.* **2020**, *8*, 3004–3022. [CrossRef]
13. Mwaurah, P.W.; Kumar, S.; Kumar, N.; Panghal, A.; Attkan, A.K.; Singh, V.K.; Garg, M.K. Physicochemical characteristics, bioactive compounds and industrial applications of mango kernel and its products: A review. *Compr. Rev. Food Sci. Food Saf.* **2020**, *19*, 2421–2446. [CrossRef]
14. Villacís-Chiriboga, J.; Elst, K.; Van Camp, J.; Vera, E.; Ruales, J. Valorization of byproducts from tropical fruits: Extraction methodologies, applications, environmental, and economic assessment: A review (Part 1: General overview of the byproducts, traditional biorefinery practices, and possible applications). *Compr. Rev. Food Sci. Food Saf.* **2020**, *19*, 405–447. [CrossRef]
15. Bordiga, M.; Travaglia, F.; Locatelli, M. Valorisation of grape pomace: An approach that is increasingly reaching its maturity—A review. *Int. J. Food Sci. Technol.* **2019**, *54*, 933–942. [CrossRef]
16. Banerjee, J.; Singh, R.; Vijayaraghavan, R.; MacFarlane, D.; Patti, A.F.; Arora, A. Bioactives from fruit processing wastes: Green approaches to valuable chemicals. *Food Chem.* **2017**, *225*, 10–22. [CrossRef]
17. Dahiya, S.; Kumar, A.N.; Sravan, J.S.; Chatterjee, S.; Sarkar, O.; Mohan, S.V. Food waste biorefinery: Sustainable strategy for circular bioeconomy. *Bioresour. Technol.* **2018**, *248*, 2–12. [CrossRef]
18. Osorio, L.L.D.R.; Flórez-López, E.; Grande-Tovar, C.D. The potential of selected agri-food loss and waste to contribute to a circular economy: Applications in the food, cosmetic and pharmaceutical industries. *Molecules* **2021**, *26*, 515. [CrossRef] [PubMed]
19. Majerska, J.; Michalska, A.; Figiela, A. A review of new directions in managing fruit and vegetable processing by-products. *Trends Food Sci. Technol.* **2019**, *88*, 207–219. [CrossRef]
20. Aloui, H.; Baraket, K.; Sendon, R.; Silva, A.S.; Khwaldia, K. Development and characterization of novel composite glycerol-plasticized films based on sodium caseinate and lipid fraction of tomato pomace by-product. *Int. J. Biol. Macromol.* **2019**, *139*, 128–139. [CrossRef] [PubMed]
21. Kanmani, P.; Rhim, J.W. Antimicrobial and physical-mechanical properties of agar-based films incorporated with grapefruit seed extract. *Carbohydr. Polym.* **2014**, *102*, 708–716. [CrossRef]
22. Torres-León, C.; Vicente, A.A.; Flores-López, M.L.; Rojas, R.; Serna-Cock, L.; Alvarez-Pérez, O.B.; Aguilar, C.N. Edible films and coatings based on mango (var. Ataulfo) by-products to improve gas transfer rate of peach. *LWT-Food Sci. Technol.* **2018**, *97*, 624–631. [CrossRef]
23. Kanatt, S.R.; Chawla, S.P. Shelf life extension of chicken packed in active film developed with mango peel extract. *J. Food Saf.* **2018**, *38*, 12385. [CrossRef]
24. Won, J.S.; Lee, S.J.; Park, H.H.; Song, K.B.; Min, S.C. Edible coating using a chitosan-based colloid incorporating grapefruit seed extract for cherry tomato safety and preservation. *J. Food Sci.* **2018**, *83*, 138–146. [CrossRef]
25. Hajer, A.; Khwaldia, K.K.; Sánchez-González, L.; Muneret, L.; Jeandel, C.; Hamdi, M.; Desobry, S. Alginate coatings containing grapefruit essential oil or grapefruit seed extract for grapes preservation. *Int. J. Food Sci. Technol.* **2014**, *49*, 952–959.
26. Kim, J.H.; Hong, W.S.; Oh, S.W. Effect of layer-by-layer antimicrobial edible coating of alginate and chitosan with grapefruit seed extract for shelf-life extension of shrimp (*Litopenaeus vannamei*) stored at 4 °C. *Int. J. Biol. Macromol.* **2018**, *120*, 1468–1473. [CrossRef]
27. Song, H.Y.; Shin, Y.J.; Song, K.B. Preparation of a barley bran protein–gelatin composite film containing grapefruit seed extract and its application in salmon packaging. *J. Food Eng.* **2012**, *113*, 541–547. [CrossRef]
28. Madzuki, I.N.; Tan, J.M.; Mohamad Shalan, N.A.A.; Mohd, I.N.S. Tomato peel-cutin based film mitigates the deterioration of calamansi (*Citrofortunella microcarpa*). *IOP Conf. Ser. Earth Environ. Sci.* **2021**, *765*, 012013. [CrossRef]

29. Gallego, M.; Arnal, M.; Talens, P.; Toldrá, F.; Mora, L. Effect of gelatin coating enriched with antioxidant tomato by-products on the quality of pork meat. *Polymers* **2020**, *12*, 1032. [CrossRef] [PubMed]
30. Khalifa, I.; Barakat, H.; El-Mansy, H.A.; Soliman, S.A. Improving the shelf-life stability of apple and strawberry fruits applying chitosan-incorporated olive oil processing residues coating. *Food Packag. Shelf Life* **2016**, *9*, 10–19. [CrossRef]
31. Khalifa, I.; Barakat, H.; El-Mansy, H.A.; Soliman, S.A. Enhancing the keeping quality of fresh strawberry using chitosan-incorporated olive processing wastes. *Food Biosci.* **2016**, *13*, 69–75. [CrossRef]
32. Muir, D.D. The stability and shelf life of milk and milk products. In *Food and Beverage Stability and Shelf Life*; Kilcast, D., Subramaniam, P., Eds.; Woodhead Publishing: New Delhi, India, 2011; pp. 755–778.
33. Nunes, M.A.; Pimentel, F.B.; Costa, A.S.G.; Alves, R.C.; Oliveira, M.B.P.P. Olive by-products for functional and food applications: Challenging opportunities to face environmental constraints. *Inn. Food Sci. Emerg. Technol.* **2016**, *35*, 139–148. [CrossRef]
34. Roila, R.; Valiani, A.; Ranucci, D.; Ortenzi, R.; Servili, M.; Veneziani, G.; Branciari, R. Antimicrobial efficacy of a polyphenolic extract from olive oil by-product against "Fior di latte" cheese spoilage bacteria. *Int. J. Food Microbiol.* **2019**, *295*, 49–53. [CrossRef]
35. Spigno, G.; Marinoni, L.; Garrido, G.D. *State of the Art in Grape Processing By-Products*; Galanakis, C.M., Ed.; Academic Press: Cambridge, MA, USA, 2017; pp. 1–27.
36. Tseng, A.; Zhao, Y. Wine grape pomace as antioxidant dietary fibre for enhancing nutritional value and improving storability of yogurt and salad dressing. *Food Chem.* **2013**, *138*, 356–365. [CrossRef]
37. Sousa-Gallagher, M.J.; Tank, A.; Sousa, R. Emerging technologies to extend the shelf life and stability of fruits and vegetables. In *The Stability and Shelf Life of Food*; Subramaniam, P., Ed.; Woodhead Publishing: New Delhi, India, 2016; pp. 399–430.
38. Anibarro-Ortega, M.; Pinela, J.; Ciric, A.; Martins, V.; Rocha, F.D.; Sokovic, M.; Barata, A.M.; Carvalho, A.M.; Barros, L.; Ferreira, I. Valorisation of table tomato crop by-products: Phenolic profiles and in vitro antioxidant and antimicrobial activities. *Food Bioprod. Proc.* **2020**, *124*, 307–319. [CrossRef]
39. Martínez-Hernández, G.B.; Castillejo, N.; Artés-Hernández, F. Effect of fresh–cut apples fortification with lycopene microspheres, revalorized from tomato by-products, during shelf life. *Postharvest. Biol. Technol.* **2019**, *156*, 110925. [CrossRef]
40. Ang, Y.K.; Sia, W.C.M.; Khoo, H.E.; Yim, H.S. Antioxidant potential of *Carica papaya* peel and seed. *Focus Mod. Food Ind.* **2012**, *1*, 11–16.
41. Faiq, M.; Theerakulkait, C. Effect of papaya peel extract on browning inhibition in vegetable and fruit slices. *Ital. J. Food Sci.* **2017**, *30*, 57–61.
42. Kim, J.; Kwon, K.; Oh, S. Effects of malic acid or/and grapefruit seed extract for the inactivation of common food pathogens on fresh-cut lettuce. *Food Sci. Biotechnol.* **2016**, *25*, 1801–1804. [CrossRef]
43. Andradea, M.A.; Lima, V.; Silva, A.S.; Vilarinho, F.; Castilho, M.C.; Khwaldia, K.; Ramos, F. Pomegranate and grape by-products and their active compounds: Are they a valuable source for food applications? *Trends Food Sci. Technol.* **2019**, *86*, 68–84. [CrossRef]
44. Belgacem, I.; Schena, L.; Teixidó, N.; Romeo, F.V.; Ballistreri, G.; Abadias, M. Effectiveness of a pomegranate peel extract (PGE) in reducing *Listeria monocytogenes* in vitro and on fresh-cut pear, apple and melon. *Eur. Food Res. Technol.* **2020**, *246*, 1765–1772. [CrossRef]
45. Elsherbiny, E.A.; Aminb, B.H.; Baka, Z.A. Efficiency of pomegranate (*Punica granatum* L.) peels extract as a high potential natural tool towards Fusarium dry rot on potato tubers. *Postharvest. Biol. Technol.* **2016**, *111*, 256–263. [CrossRef]
46. Lacivita, V.; Incoronato, A.L.; Conte, A.; Del Nobile, M.A. Pomegranate peel powder as a food preservative in fruit salad: A sustainable approach. *Foods* **2021**, *10*, 1359. [CrossRef] [PubMed]
47. Waterhouse, A.L. Chemical and physical deterioration of wine. In *Chemical Deterioration and Physical Instability of Food and Beverages*; Skibsted, L.H., Risbo, J., Andersen, M.L., Eds.; Woodhead Publishing: New Delhi, India, 2010; pp. 466–482.
48. Ruiz-Moreno, M.J.; Raposo, R.; Moreno-Rojas, J.M.; Zafrilla, P.; Cayuela, J.M.; Mulero, J.; Puertas, B.; Guerrero, R.F.; Pineiro, Z.; Giron, F.; et al. Efficacy of olive oil mill extract in replacing sulfur dioxide in wine model. *LWT-Food Sci. Technol.* **2015**, *61*, 117–123. [CrossRef]
49. Bouaziz, M.; Feki, I.; Ayadi, M.; Jemai, H.; Sayadi, S. Stability of refined olive oil and olive-pomace oil added by phenolic compounds from olive leaves. *Eur. J. Lipid Sci. Technol.* **2010**, *112*, 894–905. [CrossRef]
50. Difonzo, G.; Squeo, G.; Calasso, M.; Pasqualone, A.; Caponio, F. Physico-chemical, microbiological and sensory evaluation of ready-to-use vegetable pâté added with olive leaf extract. *Foods* **2019**, *8*, 138. [CrossRef]
51. Albishi, T.; John, J.A.; Al-Khalifa, A.S.; Shahidi, F. Phenolic content and antioxidant activities of selected potato varieties and their processing by-products. *J. Funct. Foods* **2013**, *5*, 590–600. [CrossRef]
52. Friedman, M.; Huang, V.; Quiambao, Q.; Noritake, S.; Liu, J.; Kwon, O.; Chintalapati, S.; Young, J.; Levin, C.E.; Tam, C.; et al. Potato peels and their bioactive glycoalkaloids and phenolic compounds inhibit the growth of pathogenic trichomonads. *J. Agric. Food Chem.* **2018**, *66*, 7942–7947. [CrossRef] [PubMed]
53. Samotyja, U. Potato peel as a sustainable resource of natural antioxidants for the food industry. *Potato Res.* **2019**, *62*, 435–451. [CrossRef]
54. Spigno, G.; Donsì, F.; Amendola, D.; Sessa, M.; Ferrari, G.; De Faveri, D.M. Nanoencapsulation systems to improve solubility and antioxidant efficiency of a grape marc extract into hazelnut paste. *J. Food Eng.* **2013**, *114*, 207–214. [CrossRef]
55. Drinić, Z.; Mudrić, J.; Zdunić, G.; Bigović, D.; Menković, N.; Šavikin, K. Effect of pomegranate peel extract on the oxidative stability of pomegranate seed oil. *Food Chem.* **2020**, *333*, 127501. [CrossRef] [PubMed]

56. O'Sullivan, M.G. The stability and shelf life of meat and poultry. In *Food and Beverage Stability and Shelf Life*; Kilcast, D., Subramaniam, P., Eds.; Woodhead Publishing: New Delhi, India, 2011; pp. 793–816.
57. Andrés, A.I.; Petrón, M.J.; Adámez, J.D.; López, M.; Timón, M.L. Food by-products as potential antioxidant and antimicrobial additives in chill stored raw lamb patties. *Meat Sci.* **2017**, *129*, 62–70. [CrossRef]
58. Andres, A.I.; Perton, M.J.; Delgado-Adamez, J.; Lopez, M.; Timon, M. Effect of tomato pomace extracts on the shelf-life of modified atmosphere-packaged lamb meat. *J. Food Process. Preserv.* **2017**, *41*, 13018. [CrossRef]
59. Selani, M.M.; Contreras-Castillo, C.J.; Shirahigue, L.D.; Gallo, C.R.; Plata-Oviedo, M.; Montes-Villanueva, N.D. Wine industry residues extracts as natural antioxidants in raw and cooked chicken meat during frozen storage. *Meat Sci.* **2011**, *88*, 397–403. [CrossRef]
60. Kulkarni, S.; DeSantos, F.A.; Kattamuri, S.; Rossi, S.J.; Brewer, M.S. Effect of grape seed extract on oxidative, color and sensory stability of a pre-cooked, frozen, re-heated beef sausage model system. *Meat Sci.* **2011**, *88*, 139–144. [CrossRef]
61. Lorenzo, J.M.; González-Rodríguez, R.M.; Sánchez, M.; Amado, I.R.; Franco, D. Effects of natural (grape seed and chestnut extract) and synthetic antioxidants (buthylatedhydroxytoluene, BHT) on the physical, chemical, microbiological and sensory characteristics of dry cured sausage "chorizo". *Food Res. Int.* **2013**, *54*, 611–620. [CrossRef]
62. Garrido, M.D.; Auqui, M.; Martí, N.; Linares, M.B. Effect of two different red grape pomace extracts obtained under different extraction systems on meat quality of pork burgers. *LWT-Food Sci. Technol.* **2011**, *44*, 2238–2243. [CrossRef]
63. Garcia-Lomillo, J.; González-SanJosé, M.L.; Skibsted, L.H.; Jongberg, S. Effect of skin wine pomace and sulfite on protein oxidation in beef patties during high oxygen atmosphere storage. *Food Bioprocess Technol.* **2016**, *9*, 532–542. [CrossRef]
64. Chaves-López, C.; Serio, A.; Mazzarrino, G.; Martuscelli, M.; Scarpone, E.; Paparella, A. Control of household microflora in fermented sausages using phenolic fractions from olive mill wastewaters. *Int. J. Food Microbiol.* **2015**, *207*, 49–56. [CrossRef] [PubMed]
65. Baker, I.A. Investigation on antimicrobial and antioxidant effect of olive leaf extract in karadi sheep meat during storage. *J. Biol. Agric. Healthc.* **2014**, *4*, 46–49.
66. Djenane, D.; Gómez, D.; Yangüela, J.; Roncalés, P.; Ariño, A. Olive leaves extract from algerian oleaster (*Olea europaea* var. sylvestris) on microbiological safety and shelf-life stability of raw halal minced beef during display. *Foods* **2019**, *8*, 10. [CrossRef] [PubMed]
67. Saleh, E.; Morshdy, A.E.; El-Manakhly, E.; Al-Rashed, S.; Hetta, H.F.; Jeandet, P.; Yahia, R.; El-Saber Batiha, G.; Ali, E. Effects of olive leaf extracts as natural preservative on retailed poultry meat quality. *Foods* **2020**, *9*, 1017. [CrossRef]
68. Elama, C.; Tarawa, M.; Al-Rimawi, F. Oleuropein from olive leaf extract as natural antioxidant of frozen hamburger. *J. Food Sci. Eng.* **2017**, *7*, 406–412.
69. Hawashin, M.D.; Al-Juhaimi, F.; Ahmed, I.A.M.; Ghafoor, K.; Babiker, E.E. Physicochemical, microbiological and sensory evaluation of beef patties incorporated with destoned olive cake powder. *Meat Sci.* **2016**, *122*, 32–39. [CrossRef]
70. Nieto, G.; Martínez, L.; Castillo, J.; Ros, G. Hydroxytyrosol extracts, olive oil and walnuts as functional components in chicken sausages. *J. Sci. Food Agric.* **2017**, *97*, 3761–3771. [CrossRef]
71. Lampila, L.E. Major microbial hazards associated with packaged seafood. In *Advances in Meat, Poultry and Seafood Packaging*; Kerry, J.P., Ed.; Woodhead Publishing: New Delhi, India, 2012; pp. 59–85.
72. Viji, P.; Binsi, P.K.; Visnuvinayagam, S.; Bindu, J.; Ravishankar, C.N.; Gopal, T.K.S. Efficacy of mint (*Mentha arvensis*) leaf and citrus (*Citrus aurantium*) peel extracts as natural preservatives for shelf life extension of chill stored Indian mackerel. *J. Food. Sci. Technol.* **2015**, *52*, 6278–6289. [CrossRef]
73. Özen, B.Ö.; Soyer, A. Effect of plant extracts on lipid and protein oxidation of mackerel (*Scomber scombrus*) mince during frozen storage. *J. Food Sci. Technol.* **2018**, *55*, 120–127. [CrossRef]
74. Özen, B.Ö.; Eren, M.; Pala, A.; Özmen, I.; Soyer, A. Effect of plant extracts on lipid oxidation during frozen storage of minced fish muscle. *Int. J. Food Sci. Technol.* **2011**, *46*, 724–731. [CrossRef]
75. Hasani, S.; Alizadeh, E. antioxidant effects of grape pomace on the quality of silver carp (*Hypophthalmichthys molitrix*) fillets during refrigerated storage. *Int. J. Food Prop.* **2015**, *18*, 1223–1230. [CrossRef]
76. Yerlikaya, P.; Ucak, I.; Gumus, B.; Gokoglu, N. Citrus peel extract incorporated ice cubes to protect the quality of common Pandora. *J. Food Sci. Technol.* **2015**, *52*, 8350–8356. [CrossRef] [PubMed]
77. Ali, M.; Imran, M.; Nadeem, M.; Khan, M.K.; Sohaib, M.; Suleria, H.A.R.; Bashir, R. Oxidative stability and Sensoric acceptability of functional fish meat product supplemented with plant−based polyphenolic optimal extracts. *Lipids Health Dis.* **2019**, *18*, 35. [CrossRef] [PubMed]
78. Abdel-Wahab, M.; El-Sohaimy, S.A.; Ibrahim, H.A.; El-Makarem, H.S.A. Evaluation the efficacy of clove, sage and kiwifruit peel extracts as natural preservatives for fish fingers. *Ann. Agric. Sci.* **2020**, *65*, 98–106. [CrossRef]
79. Miraglia, D.; Castrica, M.; Menchetti, L.; Esposto, S.; Branciari, R.; Ranucci, D.; Urbani, S.; Sordini, B.; Veneziani, G.; Servili, M. Effect of an Olive Vegetation Water Phenolic Extract on the Physico-Chemical, Microbiological and Sensory Traits of Shrimp (*Parapenaeus longirostris*) during the Shelf-Life. *Foods* **2020**, *9*, 1647. [CrossRef]
80. Panza, O.; Lacivita, V.; Palermo, C.; Conte, A.; Del Nobile, M.A. Food by-products to extend shelf life: The case of cod sticks breaded with dried olive paste. *Foods* **2020**, *9*, 1902. [CrossRef]
81. Panza, O.; Conte, A.; Del Nobile, M.A. pomegranate by-products as natural preservative to prolong the shelf life of breaded cod stick. *Molecules* **2021**, *26*, 2385. [CrossRef]

82. Cauvain, S.P. Bread and other bakery products. In *The Stability and Shelf Life of Food*; Subramaniam, P., Ed.; Woodhead Publishing: New Delhi, India, 2016; pp. 431–459.
83. Ismail, T.; Akhtar, S.; Riaz, M.; Hameed, A.; Afzal, K.; Sheikh, A.S. Oxidative and microbial stability of pomegranate peel extracts and bagasse supplemented cookies. *J. Food Qual.* **2016**, *39*, 658–668. [CrossRef]
84. Mehta, D.; Prasad, P.; Sangwan, R.S.; Yadav, S.K. Tomato processing byproduct valorization in bread and muffin: Improvement in physicochemical properties and shelf life stability. *J. Food Sci. Technol.* **2018**, *55*, 2560–2568. [CrossRef]
85. Difonzo, G.; Pasqualone, A.; Silletti, R.; Cosmai, L.; Summo, C.; Paradiso, V.M.; Caponio, F. Use of olive leaf extract to reduce lipid oxidation of baked snacks. *Food Res. Int.* **2018**, *108*, 48–56. [CrossRef] [PubMed]

Review

Fresh Fish Degradation and Advances in Preservation Using Physical Emerging Technologies

Jéssica Tavares [1], Ana Martins [1], Liliana G. Fidalgo [1,2], Vasco Lima [1], Renata A. Amaral [1], Carlos A. Pinto [1], Ana M. Silva [3] and Jorge A. Saraiva [1,*]

1 LAQV-REQUIMTE, Department of Chemistry, University of Aveiro, 3810-193 Aveiro, Portugal; jessica.tavares19@ua.pt (J.T.); ana.patricia.martins@ua.pt (A.M.); liliana.fidalgo@ipbeja.pt (L.G.F.); vasco.lima@ua.pt (V.L.); renata.amaral@ua.pt (R.A.A.); carlospinto@ua.pt (C.A.P.)
2 Department of Applied Technologies and Sciences, School of Agriculture, Polytechnic Institute of Beja, 7800-295 Beja, Portugal
3 SONAE, Lugar do Espido, Via Norte, 4471-909 Maia, Portugal; amsilva@sonaemc.com
* Correspondence: jorgesaraiva@ua.pt

Abstract: Fresh fish is a highly perishable food characterized by a short shelf-life, and for this reason, it must be properly handled and stored to slow down its deterioration and to ensure microbial safety and marketable shelf-life. Modern consumers seek fresh-like, minimally processed foods due to the raising concerns regarding the use of preservatives in foods, as is the case of fresh fish. Given this, emergent preservation techniques are being evaluated as a complement or even replacement of conventional preservation methodologies, to assure food safety and extend shelf-life without compromising food safety. This paper reviews the main mechanisms responsible for fish spoilage and the use of conventional physical methodologies to preserve fresh fish, encompassing the main effects of each methodology on microbiological and chemical quality aspects of this highly perishable food. In this sense, conventional storage procedures (refrigeration and freezing) are counterpointed with more recent cold-based storage methodologies, namely chilling and superchilling. In addition, the use of novel food packaging methodologies (edible films and coatings) is also presented and discussed, along with a new storage methodology, hyperbaric storage, that states storage pressure control to hurdle microbial development and slow down organoleptic decay at subzero, refrigeration, and room temperatures.

Keywords: fresh fish; spoilage; shelf-life; chilling/refrigeration; freezing; edible coatings; hyperbaric storage

Citation: Tavares, J.; Martins, A.; Fidalgo, L.G.; Lima, V.; Amaral, R.A.; Pinto, C.A.; Silva, A.M.; Saraiva, J.A. Fresh Fish Degradation and Advances in Preservation Using Physical Emerging Technologies. *Foods* **2021**, *10*, 780. https://doi.org/10.3390/foods10040780

Academic Editor: Luis Guerrero Asorey

Received: 1 March 2021
Accepted: 30 March 2021
Published: 5 April 2021

1. Introduction

Fish is a highly demanded and nutritious food product, yet perishability remains the biggest challenge for its preservation [1]. This food must be stored refrigerated or frozen, and, even under those conditions, it has a very short shelf-life, particularly for refrigeration (5–7 days and 9–12 months under refrigeration and frozen conditions, respectively) [2]. The deterioration of fresh fish during storage is attributed to different damage mechanisms, like microbiological spoilage, autolytic degradation, and lipid oxidation [3].

Fish products contain important nutritional and digestive proteins, including essential amino acids, lipid soluble vitamins, micronutrients, and highly unsaturated fatty acids. The muscle is mostly composed of water (75–85%), and it has a high water activity (0.98–0.99) [4]. Protein represents 20–22% of the muscle [5], while many types of lipids with different chemical composition, such as neutral/non-polar (triglycerides, diglycerides, etc.) and polar (free fatty acids, phospholipids, etc.) lipids, are also present [6]. Fish can be divided in four basic groups regarding its fat content: lean (<2% fat), low-fat (2–4% fat), medium-fat (4–8% fat), and high-fat (>8% fat) [7].

1.1. Fish Spoilage

After fish are caught, spoilage starts rapidly, and rigor mortis is responsible for changes in the fish after its death. A breakdown of various components and the formation of new compounds are responsible for the alterations in odor, flavor, and texture that happen throughout the spoilage process, and deterioration occurs very quickly due to various mechanisms triggered by the metabolic activity of microorganisms, endogenous enzymatic activity (autolysis), and by the chemical oxidation of lipids [1,8].

1.1.1. Autolytic Enzymatic Spoilage

Initially, the main autolytic changes happening are the enzymatic degradation of adenosine 5′-triphosphate (ATP) and its related products, followed by the action of proteolytic enzymes [9], as reported in Table 1. The concentrations of ATP and its breakdown products (adenosine 5′-diphosphate (ADP), adenosine 5′-monophosphate (AMP), inosine 5′-monophosphate (IMP), inosine (INO), and hypoxanthine (Hx)) are one of the most effective and reliable indicators of fish freshness (K-value), varying according to the fish species, muscle types, and storage conditions [1,9]. The K-value, which increases with spoilage, is calculated according to the following ratio (Equation (1)):

$$\mathrm{K-value}\ (\%) = \frac{100 \times (\mathrm{INO} + \mathrm{Hx})}{\mathrm{ATP} + \mathrm{ADP} + \mathrm{AMP} + \mathrm{IMP} + \mathrm{Hx}} \quad (1)$$

Table 1. Enzyme action in chilled fish. Adapted from [10].

Enzyme(s)	Substrate(s)	Effect
Glycolytic enzymes	Glycogen	Lactic acid production resulting in pH drop
Nucleotide breakdown enzymes	ATP, ADP, AMP, IMP	Gradual production of Hx
Cathepsins	Proteins, peptides	Softening of tissue
Chymotrypsin, trypsin, carboxypeptidases	Proteins, peptides	Belly-bursting
Calpain	Myofibrillar proteins	Softening of tissue
Collagenases	Connective tissue	Softening and gaping of tissue
TMAO demethylase	TMAO	Formaldehyde production

ATP: Adenosine triphosphatase; ADP: adenosine diphosphate; AMP: adenosine monophosphate; IMP: inosine monophosphate; Hx: hypoxanthine; TMAO: trimethylamine oxide.

High autolytic activity of the major muscle endogenous proteases causes the hydrolysis of key myofibrillar proteins, contributing to the weakening of the myofibril structure during post-mortem storage. The main proteolytic systems in place are the cytoplasmic calpains (at neutral pH) and the lysosomal cathepsins (at acid pH), such as cathepsins B, L, H, and D [11].

Trimethylamine (TMA) and its N-oxide compounds are usually used as indices for freshness in fishery products. The pathway of the production of formaldehyde and ammonia from TMA and its N-oxide is shown in Figure 1, which are associated with the formation of undesirable odors, and occur in fish by the action of several enzymes, such as trimethylamine N-oxide reductase (TMAO reductase), trimethylamine dehydrogenase (TMA dehydrogenase), dimethylamine dehydrogenase (DMA dehydrogenase), and amine dehydrogenase [5]. Total volatile base nitrogen (TVB-N) and trimethylamine-nitrogen (TMA-N) are quality indicators traditionally used for fish products [12]. TVB-N includes the measurement of volatile basic nitrogenous compounds associated with seafood spoilage, like TMA produced by bacteria, DMA derived from autolytic enzymes action, ammonia produced by the deamination of amino acids, and others [13]. A value of 35 mg N/100 g is proposed as the upper limit for the spoilage initiation [13]. However, some studies present lower limits, depending on the results obtained and the studied fish species [13–15]. TMA-N typically has a fishy odor, and it is produced by the decomposition of TMA N-oxide (major constituent of non-protein nitrogen fraction) caused by bacterial spoilage and enzymatic activity. The upper limit of acceptability is typically around 10–15 mg TMA-N/100 g, however, like TVB-N, lower limits are suggested by other authors [12–14].

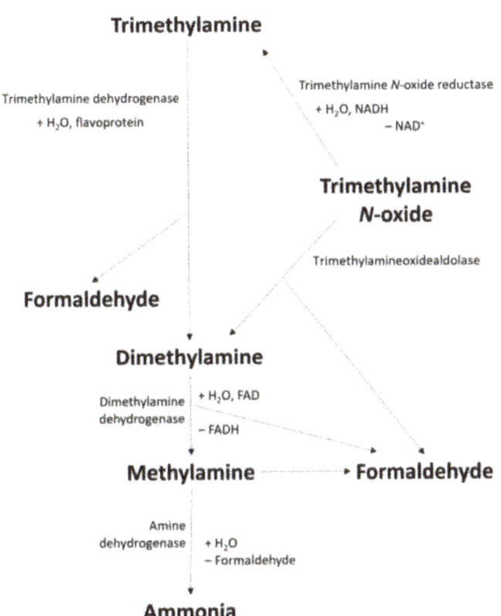

Figure 1. Degradation of trimethylamine and its N-oxide compounds [5].

1.1.2. Lipid Hydrolysis and Oxidative Spoilage

Fish quality can also be affected during storage at different temperatures by lipid oxidation through odors and lipid peroxide formation or by taste, texture, consistency, and nutritional value losses. Transition metals are primary activators of molecular oxygen, leading to oxidation, which consists of oxygen reacting with the double bonds of fatty acids, mostly of polyunsaturated fatty acids (PUFAs), that are highly susceptible to oxidation [16]. Lipid oxidation can occur either enzymatically or non-enzymatically in fish. In the enzymatic hydrolysis (lipolysis) process, glycerides are split by lipases, forming free fatty acids that are responsible for the common off-flavor (rancidity) and from the denaturation of sarcoplasmic and myofibrillar proteins [17]. Main lipolytic enzymes include triacyl lipase, phospholipase A2, and phospholipase B [3], and they can either be endogenous or derived from psychrotrophic microorganisms [17]. Furthermore, the presence of pro-oxidant enzymes, like lipoxygenases and peroxidases, facilitates lipid oxidation [18].

Non-enzymatic oxidation is triggered by the catalysis of hematin compounds, such as hemoglobin, myoglobin, and cytochrome, generating hydroperoxides. The peroxides are unstable and susceptible to hydrolysis, forming volatile compounds (like aldehydes, ketones, and alcohols), which causes off-flavors [19].

Lipid oxidation is relevant for fish quality due to the development of off-odors, especially in fatty fishes. Normally, the degree of lipid oxidation is given by a thiobarbituric acid (TBA) value that measures the malondialdehyde (MDA) content that is formed by the reaction with hydroperoxides (initial products of lipid oxidation). TBA values of 2–4 mg MDA/kg are within quality limits [13]. Nevertheless, this value might not reflect the real rate of lipid oxidation, because MDA can interact with other components of the fish body and produce secondary metabolites that include reactions with carbohydrates, furfural, alkenals, alkadienals, and other aldehydes and ketones [13].

So, in fish, lipid oxidation consists of a complex chain of reactions, with three distinct phases: primary (formation of hydroperoxides), secondary (e.g., hexanal and malondialdehyde formation), and tertiary/interaction compounds (new compounds are formed by the breakdown of secondary oxidation products or through the reaction with other molecules, mostly nucleophilic type) [20].

1.1.3. Microbial Spoilage

Microbial growth is the first mechanism deteriorating fish, being the spoilage factor that most affects the quality of fresh or lightly preserved fish [9]. Initially, the fish muscles are sterile, but after death, they are contaminated by the microbial population present at the fish skin [21]. The high water activity, low acidity (pH > 6), and high amount of non-protein nitrogenous compounds typical of fish results in the fast growth of microorganisms, leading to undesirable changes in appearance, texture, flavor, and odor, reducing its quality. Spoilage created by microorganisms generates volatile amines, biogenic amines, organic acids, sulfides, alcohols, aldehydes, and ketones, which have unpleasant and unacceptable off-flavors [22]. The main compounds formed during microbiological spoilage are listed in Table 2. Biogenic amines, such as histamine, cadaverine, tyramine, and putrescine, are produced by the decarboxylation of specific free amino acids by microorganisms during storage, and are used to monitor fish safety and quality [23,24].

Table 2. Spoilage compounds produced by microorganisms during the storage of fresh fish. Adapted from [25,26].

Spoilage Bacteria	Spoilage Compound(s) Produced
Shewanella putrefaciens	TMA, H_2S, CH_3SH, $(CH_3)_2S$, Hx, and acids
Pseudomonas spp.	CH_3SH, $(CH_3)_2S$, ketones, esters, aldehydes, NH_3, and Hx
Photobacterium phosphoreum	TMA and Hx
Vibrionaceae	TMA and H_2S
Enterobacteriaceae	TMA, H_2S, ketones, esters, aldehydes, NH_3, Hx, and acids
Lactic acid bacteria	H_2S, ketones, esters, aldehydes, NH_3, and acids
Yeast	Ketones, esters, aldehydes, NH_3, and acids
Aerobic spoilers	NH_3, acetic, butyric, and propionic acids
Anaerobic rods	Ketones, esters, aldehydes, and NH_3

TMA: trimethylamine; H_2S: hydrogen sulphide; CH_3SH: methylmercarptan; $(CH_3)_2S$: dimethylsulphide; Hx: hypoxanthine; NH_3: ammonia.

The ability to produce histamine is known in different species of bacteria that have histidine decarboxylase [27]. Being extremely stable, histamine cannot be easily removed or destroyed by cooking, retorting, or freezing [28], and, among amines, it is toxicologically relevant, causing scombroid fish poisoning and food intolerance [29].

It should also be noted that the existing physicochemical conditions and the interactions between the microorganisms will impose a selection of the organisms capable of growing under such conditions. The initial microflora of the fish is dependent on different factors, such as the environment where the fish lives, the fishing season, water temperature, the method of capture, the handling on the ship, or the technological and sale process [21], but, regardless of the variety of microflora present in the fresh fish and the diverse parameters used for preserving, the species growing are consistent in the different products. From the different microbial species that can develop on fish, only one or a few will produce the off-odors and off-flavors, named specific spoilage organisms [30].

2. Chilling, Superchilling, and Freezing

The preservation of food products without using preservatives or additives has been increasingly demanded among consumers, and has brought additional challenges, especially to highly perishable foods, such as meat or fish. Low temperatures during the capture, transportation, and storage of the fish are of major importance worldwide. Chilling, superchilling, and freezing techniques allow for the preservation of fish for longer periods without major changes in quality, and assure economic benefits for the fish companies [31]. Therefore, chilling is one of the most used methods for fish preservation, along with freezing and, recently, superchilling.

2.1. Chilling

Chilling is the process of cooling fish or fish products to a temperature approaching that of melting ice, using, for example, ice. Chilling promotes an increase of shelf-life by slowing physical and chemical reactions and the action of deteriorative microorganisms and enzymes [31,32].

Chilled fish can keep a high organoleptic quality, being highly attractive for consumers, however, it is susceptible to microbial safety problems due to the temperature range at which it is kept, since psychrotrophic pathogens can grow and proliferate without an obvious sensorial impact [33].

Usually done with ice, chilling can maintain the fish at temperatures close to 0 °C and extend the shelf-life up to 30 days (in fatty fish, this can be up to 40 days), depending on several factors, such as the water temperature (temperate or tropical waters) and the type of species (marine or freshwater species) [31]. The shelf-life of different fish species stored in ice in shown in Table 3. However, temperatures close to 0 °C are not easily possible at retail and consumer houses, and, therefore, refrigeration (storage above 0 °C and up to 5 °C), the most usual storage process for fresh fish, results in a much shorter shelf-life [34].

Table 3. Shelf-life of different fish species stored in ice. Adapted from [31,35].

Fish Species	Temperate Waters (Days)	Tropical Waters (Days)
Marine Species	2–24	6–35
Cod (*Gadus morhua*)	9–15	na
Hake (*Merluccius merluccius*)	7–15	na
Catfish	Na	16–19
Batfih (*Ogcocephalus darwini*)	Na	21–24
Halibut (*Hippoglossus stenolepis*)	21–24	na
Sardine (*Sardina pilchardus*)	3–8	9–16
Freshwater species	9–17	6–40
Catfish (*Silurus glanis*)	12–13	15–27
Trout (*Oncorhynchus mykiss*)	9–11	16–24
Perch (*Perca* spp.)	8–17	13–32
Tilapia (*Oreochromis niloticus*)	Na	10–27
Corvina (*Argyrosomus regius*)	Na	30

na—not applicable.

2.2. Superchilling

Superchilling, also known as partial freezing or deep chilling, is characterized by low temperatures (between conventional chilling and freezing), in which a decrease of 1–2 °C occurs below the initial freezing point of the food product [32,36,37]. Most foods have a freezing point that varies from −0.5 to −2.4 °C and, specifically for fishery products, this parameter is between −0.8 and −1.4 °C [38,39].

The superchilling process results in the conversion of a small fraction of water (≈5–30%) into ice, forming a thin layer of ice (≈1–3 mm) on the surface of the food and an internal ice reservoir [40,41]. Thus, the combined effect of low temperature and internal/external ice on food produce slows deteriorative processes (such as microbial activity) and, for short periods, ice may not be necessary during transport or storage [32,36,40,42].

Ideally, in superchilling, a small amount of water content is transformed into ice and therefore, there is less freeze protein and structural damage (detachments and breaks of myofibers) by ice crystals compared to frozen storage. The shelf-life of superchilled food can be one and a half to four times longer when compared to the chilling process due to the reduction of microbial and enzymatic activity [32,37,39,41]. For example, according to [43], the shelf-life of fresh cod loins (*Gadus morhua*) has been extended from nine days in ice-chilled storage to 16 or 17 days in superchilled storage, which means that the shelf-life has increased by about 1.9 times. At superchilling temperatures, biochemical/physicochemical reactions are affected (at a higher rate or even accelerated over time), and there are some negative effects on quality parameters, for instance, changes in muscle texture [32,36,39]. Therefore, superchilled technology can be combined with other preservation methods, such as modified atmosphere packaging (MAP) and vacuum packaging (VP) as combined methodologies, to minimize possible detrimental effects [36]. The synergetic effect of MAP and superchilled technologies was described by [44], in which the extension of the shelf-life of fresh Atlantic salmon fillets (*Salmo salar*) was investigated. It was concluded that the salmon samples packaged in a MAP atmosphere (90% CO_2 and a gas-to-product volume ratio of 2.5) increased from 11 to 22 days in terms of shelf-life when compared to the control samples (wrapped and exposed to atmospheric oxygen). Additionally, the sensorial and physicochemical properties (drip loss, pH, total volatiles amines, etc.) were evaluated, and all were within the acceptable limits, therefore, the shelf-life was determined only by microbial growth [44].

Superchilling has raised interest in its application to some food products, namely fishery products, due to the shelf-life extension and quality improvement, in comparison to traditional preservation methods. Table 4 presents the conditions for different superchilled fish species, including data from other preservation technologies and from combination with diverse packaging methods. Therefore, in general, the shelf-life is longer when superchilling technology is combined either with VP or MAP methods and when compared to the shelf-life obtained in each of these individually.

Table 4. The effect of superchilling, chilling, and/or freezing technologies and/or the synergistic effect with packaging (vacuum packaging or MAP: modified atmosphere packaging) on the quality and shelf-life of fish muscle foods.

Species (Scientific Name)	Storage Conditions	Main Results	References
Atlantic Salmon (*Salmo salar*)	Superchilling: −1.0 °C up to 16 days Chilling: +4.0 °C up to 16 days Freezing: −40.0 °C up to 30 days	Superchilled salmon for nine days, without salting, presented the best results when compared with ice-chilled and frozen samples due to the reduction of biochemical quality degradation and a low/less degree of protein denaturation and structural damage. The highest production yield of salted salmon (in the weight of the salmon) was observed for the superchilling method at the ninth day of storage.	[45]
	Superchilling: −1.4 °C and −3.6 °C up to 34 days Chilling: +4.0 °C up to 21 days Freezing: −40.0 °C up to 37 days Packaging: Vacuum	The storage time of vacuum-packed, superchilled salmon fillets can be doubled when compared to ice-chilled samples. The highest drip loss value and degree of protein and myosin denaturation for superchilled salmon stored at −1.4 °C. Better muscle hardness in superchilled samples stored at −3.6 °C and stable activity of cathepsins enzymes (B and B + L) in all salmon samples.	[46]
	Superchilling: −1.5 °C up to 28 days Chilling: +2.0 °C up to 21 days Packaging: MAP-CO_2, N_2 (CO_2 compositions: 25%, 40%, 60%, 75%, and 90% with different gas-to-product ratios).	Extension of shelf life for superchilled product (using 90% CO_2) salmon fillets, from 11 days to 22 days, compared to chilled/control samples.	[44]

Table 4. *Cont.*

Species (Scientific Name)	Storage Conditions	Main Results	References
	Superchilling: −2.0 °C air overwrap up to 24 days Chilling: +4.0 °C air overwrap up to 24 days Packaging: MAP—60% CO_2, 40% N_2	Spoilage of MAP and air-stored salmon fillets after 10 days and seven days, respectively. Good microbial and sensorial quality during 24 days of storage for superchilled MAP salmon, compared to 21 days of superchilled salmon with air overwrap.	[47]
	Superchilling: −1.5 °C up to 28 days Packaging: Vacuum	The liquid loss value of the superchilled salmon fillets was significant after one day of storage and not significant between three and 14 days of storage, and after 21 days, this parameter decreased. The drip loss parameter of the superchilled salmon fillets remained without significant differences between one and 14 days of storage, but increased after 21 days.	[48]
Cod (*Gadus morhua*)	Superchilling: −1.5 °C up to 15 days Chilling: +0.5 °C up to 15 days	The shelf-life of superchilled cod fillets increased by three days when compared to the chilling process, resulting in a total of 15 days of shelf-life. The shelf-life of cod samples stored at a chilled temperature (+0.5 °C) only increased from 12.5 to 14 days.	[49]
	Superchilling: −1.0 °C up to 42 days Chilling: +4.0 °C up to 37 days Freezing: −21.0 °C for 36 days or −40.0 °C for 43 days Packaging: Vacuum	The superchilling technology combined with vacuum packaging prolonged the shelf-life of the cod fillets by several weeks when compared to the traditional chilling method. Drip loss and liquid loss parameters of superchilled cod fillets decreased and increased, respectively, compared to the chilled samples.	[38]
	Superchilling: −0.9 °C up to 21 days Chilling: +1.5 °C up to 21 days Packaging: MAP—50% CO_2, 45% N_2, 5% O_2	The shelf-life of fresh cod loins has been extended from nine days for ice-chilled storage to 16 or 17 days by superchilled storage. In addition, MAP combined with chilling and superchilling methods increased the shelf-life to 14 and 21 days, respectively. The superchilled MAP cod loins after seven days showed significant differences in muscle texture when compared to other storage methods due to the formation of ice crystals, when the storage temperature reached the freezing point of the fish.	[43]
	Superchilling: −2.0 °C or −3.6 °C up to four weeks Chilling: +0.0 °C up to four weeks Packaging: MAP—50% CO_2, 45% N_2, 5% O_2	The effect of brine (2.5 ± 1.0% NaCl) on cod loins was evaluated using the combined superchilling/MAP technology. The synergistic effect of superchilling/MAP extended the shelf-life of unbrined cod loins by 21 days (at −2.0 °C) instead of 14 to 15 days (at −2.0 °C) of the superchilled samples. The brined samples showed a shorter shelf-life compared to unbrined samples, especially in superchilling/MAP cod loins.	[50]

Table 4. *Cont.*

Species (Scientific Name)	Storage Conditions	Main Results	References
Tilapia (*Oreochromis niloticus*) fillets	Superchilling: −1.0 °C Chilling: +1.0 °C Packaging: MAP—50% CO_2, 50% N_2	The MAP tilapia fillets remained good for consumption at the microbiological level, even after 23 days of storage at both temperatures (−1.0 °C and +1.0 °C). However, some detrimental effects were observed in color, drip loss, and texture of samples. The best storage conditions, considering both sensorial evaluation and microbial counts, were 13–15 and 21 days for the chilled and superchilled tilapia fillets, respectively.	[51]

Considering the information presented, it is necessary to understand how the technology influences the degree of superchilling, the growth of ice crystals, and the rate of biochemical reactions (like protein denaturation, enzymatic activity, or lipid oxidation) that indirectly influences quality parameters (like color, texture, flavor, drip/liquid loss, among others) of fish and fishery products [32,36–38].

Superchilling is also a promising and eco-friendlier technology due to an 18% reduction in environmental impact when compared to the conventional cold chain. Additionally, it improves the overall quality of food and extends its shelf-life by reducing microbiological contamination/propagation and, on an industrial scale, promotes higher production yields and reduces labor and transportation costs [32,42].

2.3. Freezing

Of all of the low-temperature preservation methods used, freezing (frozen storage) is the one that can maintain fish and fish products conserved for longer periods, but some quality parameters can be affected. It is typically applied at temperatures between −18 to −40 °C depending on the type of fish stored, and, contrary to what happens with chilling, for frozen storage, most deteriorative and pathogenic microorganisms are unable to proliferate at temperatures below −10 °C [32,52]. At this temperature, approximately 80% of the water is converted to ice, decreasing the water activity, which inhibits microbial activity [53,54].

The shelf-life of the frozen fish depends on several factors, such as the initial quality, storage conditions, and fish species, while the quality depends mainly on the storage temperature and temperature fluctuations [52,55]. Table 5 presents the shelf-life of some fish species stored at different freezing temperatures, according to fat content and fish size and shape. Notwithstanding the advantage over chilling regarding the inhibition of microbial growth, the impact of freezing temperatures in quality parameters is quite important when choosing the preservation technique. Some textural changes take place due to the formation of ice crystals that damage the tissues (mainly related to protein denaturation), which promotes dryness and toughness, and occurs more frequently in lean fish than in fatty or semi-fatty fish species. This can be minimized by fast-freezing processes, leading to smaller ice crystals and lower cell wall rupture and drip loss during the thawing process [52,53].

Table 5. Shelf-life of fish species stored at different freezing temperatures. Adapted from [56].

Type of Fish	Storage Time (Months)		
	−18 °C	−25 °C	−30 °C
Fatty fish (sardines, salmon)	4	8	12
Lean fish (cod, haddock)	8	18	24
Flat fish (flounder, plaice)	9	18	24

Flavor and odor changes also occur in frozen fish due to fatty acid oxidation and development of rancidity [53,57]. Color changes, especially in fatty fish, are directly related to lipid-protein cross links promoting the decrease in protein solubility. These reactions can be minimized by glazing and packaging, and through the exclusion of oxygen and light. In fish like salmon, due to the presence of carotenoids, oxidation occurrence promotes color changes [53,55].

The negative impacts of freezing in the quality parameters of the fish can be attenuated by adjusting and controlling storage temperature, rate of freezing, and fluctuation of temperature during storage through several types of freezing processes. Three basic methods can be used for fish or fish products: air blast freezers, contact or plate freezers, and immersion or spray freezers [56,58].

3. Emergent Preservation Techniques

Several strategies have been evaluated to increase the shelf-life of fresh fish with minimal impact on quality, particularly texture, and to extend shelf-life compared to refrigeration to try to avoid freezing preservation. These strategies rely on the application of additional hurdles prior to conventional storage, such as edible films and coatings, or even on the application of nonthermal preservation methodologies, such as hyperbaric storage, to slow down microbial proliferation, as in refrigeration, and also to possibly reduce microbial loads to more desirable levels and lessen degradation reactions to increase the shelf-life of fresh fish. This section will cover some of the main emergent approaches to preserve fresh fish, encompassing the main effects of each methodology on microbiological and chemical aspects of fresh fish [59].

3.1. Edible Films and Coatings

Edible films and coatings are other innovative strategies applied to food preservation that have been shown to be effective in protecting the sensorial and nutritional properties of food, while improving its safety and prolonging shelf-life by reducing/inhibiting microbial growth during the supply chain [60,61]. In fact, these technologies are similar to active packaging, however, they do not act as a package itself, even though the film/coating is in close contact with the food [62].

Edible films and coatings are defined as a thin layer of edible material, whereby, in the first phase, the film is produced separately (like solid sheets) and placed on the surface or between the food products (as wraps or separation layers, respectively). Meanwhile, edible coatings are formed directly on the surface of the food products by dipping (most used in fish and fishery products), spraying, or a fluidized bed, which are selected according to the characteristics of the food product and the film/coating [41,60,63,64].

These films and coatings are bio-based materials, and they are therefore named biopolymers due to their sustainable and eco-friendly source, as residues from the food industry and undervalued components of proteins (such as corn zein, gelatin, and casein), lipids (like shellac resin, waxes, and triglycerides), polysaccharides (such as starch, chitosan, and carrageenan), or their combinations [41,60,61,64].

In addition to high biodegradability, these biopolymers are edible, or can be washed or disintegrated due to further processing [41,65]. The most common natural polymers applied in fishery products are chitosan, alginate, whey proteins, gelatin, or their combinations [41]. Chitosan belongs to the group of polysaccharides, being one of the most abundant, and it has been investigated to be applied as a film/coating material for fishery products due to its non-toxicity, biocompatibility, biodegradability, biofunctionality, antimicrobial and antifungal properties, film-forming properties, selective gas permeability, and low-fat diffusion. Moreover, other protein and polysaccharide biopolymers have been developed for fish and fish-based products [41,63,66].

Similarly to active packaging films, active compounds, such as antioxidants, antimicrobials, and/or flavorants (namely essential oils, natural extracts from herbs and spices, enzymes, and protein hydrolysates, among others) as edible films and coating materials can be added to improve the safety, quality, and stability of foodstuff due to the low/reduced biological activity of biopolymers against spoilage microorganisms. Additionally, other additives, like plasticizers, and crosslinking agents (to improve or modify the physicochemical properties of films/coating polymers) are also incorporated [41,60,61,64].

Edible films and coatings, both biopolymers and active compounds, must comply with European Union Regulations, namely Commission Regulation (EC) No 450/2009, which establishes which active, intelligent, and article materials can enter in contact with food, and Commission Regulation (EC) No 1333/2008, related to food additives [67–69].

Edible films and coatings composed of biopolymers and enriched with active compounds have raised the interest of the food industry and technologists for application in fish, meat, and derived products in order to prevent lipid/protein/pigment oxidation, off-odors, off-flavors, moisture and color loss, oxygen penetration into the food matrix, and solute transport out of the food, and, therefore, improve preservation, quality, and sensorial properties of the products [64,67,68]. Besides this, these films/coatings add value to food products, as they increase their shelf-life by reducing/inhibiting the growth of spoilage and pathogenic microorganisms [64,67].

However, edible film and coating technologies have some associated concerns for both consumers (food safety) and the food industry, such as the initial investment and production cost, equipment and production process complexity, scale-up process, and associated regulations [41,64]. Table 6 shows some examples of edible films and coatings enriched with active compounds applied to fishery products and their effects on physicochemical, quality, and sensorial properties and the shelf-life of these products. Overall, the application of edible films and coatings enriched with active compounds in fish and fishery products enhanced or maintained its quality and sensorial properties, due essentially to its inhibitory action on the growth of spoilage and pathogenic microorganisms, throughout the storage period and, therefore, led to an extension of shelf-life. Ojagh et al. reported an extension from 12 days to 16 days (refrigerated storage, 4 °C) in rainbow trout (*Oncorhynchus mykiss*) coated with a film of chitosan and cinnamon oil [69]. Meanwhile, the shelf-life of beluga sturgeon (*Huso huso*) fillets covered with whey protein concentrate coating with 1.5% cinnamon essential oil (stored at 4 °C) was extended by eight days [70].

Table 6. Edible film and coating enriched with active compounds applied for fishery products.

Fish Product	Film/Coating Material	Active Compounds	Main Results	Reference
Beluga Sturegeon (*Huso huso*) fillet	Coating composed of 8.0% (*w*/*v*) whey protein concentrate, glycerol, and Tween 80	Cinnamon essential oil (1.5% *v*/*v*)	Shelf-life extension by eight days (compared to uncoated control samples) by reducing lipid oxidation and microbial activity.	[70]
Bluefin tuna (*Thunnus thynnus*) slides	Film composed of 1.0% (*w*/*v*) fish myofibrillar protein, 25.0% (*w*/*w*) glycerol, and 2.0% (*w*/*w*) MTGase	Catechin-Kradon (*Careya sphaerica* Roxb.) extract (0.9% *v*/*v*)	Maintenance of sensorial qualities after eight days in comparison to the four days of the unwrapped control samples, due to the reduced microbial growth, lipid oxidation, and formation of metmyoglobin.	[71]

Table 6. *Cont.*

Fish Product	Film/Coating Material	Active Compounds	Main Results	Reference
Freshwater rainbow trout (*Oncorhynchus mykiss*) fillets	Coating composed of 1.0% (*w*/*w*) carrageenan	Essential lemon oil (1.0% *w*/*w*)	The lemon essential oil incorporated in the carrageenan coating and applied to samples of rainbow trout fillets showed good antimicrobial activity and reduced lipid oxidation during the 15 days of storage at 4 °C.	[72]
Grass carp (*Ctenopharyngodon idellus*) fillets	Coating composed of 2.0% (*w*/*v*) chitosan	Glycerol monolaurate (0.1% and 0.3%)	After 20 days of storage at refrigerated temperature (4 °C), grass carp fillets maintained microbial safety, good quality, and sensorial properties.	[73]
Rainbow trout (*Oncorhynchus mykiss*)	Coating composed of 2.0% (*w*/*v*) chitosan, 1.0% acetic acid, 0.75% glycerol, and 0.2% Tween 80	Cinnamon oil (1.5% *v*/*v*)	The coated rainbow trout during the refrigerated storage (4 °C) for 16 days maintained the overall quality and sensorial properties without significant microbial growth, compared to 12 days of shelf-life for the control samples.	[69]
Rainbow trout (*Oncorhynchus mykiss*) fillets	Film composed of 1.5% (*w*/*w*) quince seed mucilage (QSM), 35.0% (*w*/*w*) glycerol, and 0.1–0.2% (*w*/*v*) Tween 80	Oregano or thyme essential oils (1.0, 1.5, or 2.0% *v*/*v*)	Shelf-life extension of rainbow trout fillets of up to 11 days compared to the control samples.	[74]
Rainbow trout (*Oncorhynchus mykiss*) slices	Film composed of 75.0% fish gelatin and 25.0% sodium alginate, glycerol, and Tween 80	Oregano essential oil (OEO) (1.5% *w*/*v*)	The use of OEO gelatin-alginate film decreased microbial growth during the 15 days of the storage period compared to the control samples.	[75]
Red drum (*Sciaenops ocellatus*) fillets	Coating composed of 1.5% (*w*/*v*) chitosan, 25% glycerol, and 1.0% (*v*/*v*) acetic acid	Grape seed extract (0.2% *w*/*v*) or tea polyphenols (0.2% *w*/*v*)	Considering the results of microbiological, physicochemical (such as pH, TVB-N value, etc.), and sensorial analysis of red drum fillet samples immersed in the active coating, the shelf-life was extended by six to eight days in refrigerated storage (4 °C) when compared to the uncoated control samples.	[76]
Salmon slices *	Film composed of 8.0% (*w*/*v*) catfish-skin gelatin and chitosan, sorbitol, and glycerol	Clove essential oil (7.5% *v/w*)	Samples of salmon slices wrapped in edible films of gelatin-chitosan with clove essential oil showed an inhibitory action on spoilage and pathogenic microorganisms when stored at 2 °C, with a shelf-life of 11 days against nine days of the control samples.	[77]

* The authors did not mention the scientific name of the species.

3.2. Hyperbaric Storage: A Novel Methodology for Fish Preservation

High-pressure processing is used as a promising "nonthermal" technique for food preservation that efficiently inactivates the vegetative microorganisms most commonly related to foodborne diseases. High-pressure processing is carried out with intense pressure in the range of 100–1000 MPa, allowing preservation with minimal effect on food taste and nutritional characteristics [78]. One of the advantages of high-pressure processing is that food products are treated instantly, regardless of their shape and size (isostatic principle). The application of elevated pressures (100–600 MPa) can be used for a variety of food processing and preservation applications, including high-pressure freezing and thawing, blanching, pasteurization, and commercial sterilization [79]. Research about the application of high pressure on fish muscles has been mainly focused on three main areas, including the extension of refrigerated/frozen shelf-life [80,81], pressure-induced texturation (gel-forming) [82], and high-pressure freezing/thawing [83].

3.2.1. Hyperbaric Storage

Recently, a novel food preservation storage methodology based on storage under moderated pressure (hyperbaric storage, HS, from 25 to 100 MPa) has attracted interest due to its high potential energy savings and shelf-life extension. HS opened the possibility to store food products and other biomaterials above atmospheric pressure (AP, 0.1 MPa) as a possible enhancement of conventional refrigeration storage, increasing shelf-life and food quality. This methodology allows the storage of food under pressure at subzero (ST), low (LT), and room (RT) temperatures, HS/ST being particularly important for solid foods, on which freezing/thawing can cause substantial damages to cellular/tissue structures, leading to textural modifications [84].

The possibility to use HS for food preservation or other biomaterials occurred by chance, after the observation of recovered items from the research submersible Alvin (owned by Woods Hole Oceanographic Institution), which sank about 1540 m during ten months, containing two bottles filled with bouillon and a plastic box containing sandwiches and apples. The environmental conditions at a depth of 1500 m are fairly constant, and are estimated to be 3–4 °C and ≈15 MPa. After being recovered, the sandwiches appeared fresh by taste and smell, and apples showed no sign of obvious deterioration. The pH value of the apples was the same as fresh ones, and the tyrosinase activity was about half that of a fresh apple [85].

3.2.2. Hyperbaric Storage at Subzero and Low Temperatures

Two freshly dressed whole fishes (pollock and cod, from the species *Gadus chalcogrammus* and *G. morhua*, respectively) were stored at 24.1 MPa and 1 °C during 12 and 21 days, respectively. According to an expert panel, both types of fish were assessed to have better sensory attributes than fish samples stored at the same temperature and AP. The ratings by the expert panel corresponded to the rating these fish samples would have received if they had been stored for a shorter period. This means that the pollock/cod samples stored for 12/21 days with HS at 1 °C received ratings typical of pollock/cod samples stored at AP at 1 °C for 6.7/8.2 days [86].

Some authors used HS/ST without freezing to extend fish shelf-life, while avoiding the damages caused by freezing. This advantage led to the first studies in HS/ST applied in fresh fish (Table 7). High-pressure application decreases the freezing and melting point of water to a minimum of −22 °C at 209 MPa [87], as pressure acts in opposition to the volume increase that occurs with the formation of type I ice crystals, resulting in tissue damage [88]. Charm et al. [86] showed that HS/ST (−3 °C and 22.8 MPa) for 36 days reduced microbial growth on cod fillets, while AP samples presented higher microbial loads. These samples were evaluated by an expert panel, and the HS/ST samples showed comparable or better quality than those stored at AP and −3 or −20 °C. These results suggested that HS/ST is a non-freezing storage method that improves the preservation of cod fillets compared to conventional methods of freezing or refrigeration, because HS/ST inhibits microbial growth and enzymatic activity and prevents damage caused by freezing/thawing. Furthermore, enzymatic degradation of nucleic acid-related substances (ATP, ADP, AMP, and IMP) from carp and chicken muscles stored under HS/ST for 50 days (−8 °C and 110 MPa, or −15 °C and 170 MPa) was slightly slower than for storage at −8 °C (AP), while the enzymatic activity was significantly reduced only by freezing at −18 °C (AP) [89].

Table 7. Main results of the application of hyperbaric storage (HS) on fresh fish.

Samples	Storage Conditions	Main Results	References
Cod fish fillets (*Gadus morhua*)	22.8 MPa, at −3 °C for 36 days	An expert panel rated these samples with a similar or better quality than samples stored at −3 °C and at atmospheric pressure.	[86]
Dressed pollock (*Pollachius pollachius*)	24.1 MPa, at 1 °C for 12 days	Microbial load with no changes during storage time.	[86]
Dressed cod fish (*Gadus morhua*)	24.1 MPa, at 1 °C for 21 days		
Carp *	110 MPa/−8 °C and 170 MPa/−15 °C for 50 days	Enzymatic activity associated to nucleic acid degradation was slower than in samples stored at −8 °C and at atmospheric pressure.	[89]
Tilapia fillets (*Oreochromis niloticus*)	100–200 MPa, at 25 °C for 12 h	Total plate counts remained stable at 100 MPa and showed a reduction of about two log units at 200 MPa.	[90]
Cape hake loins (*Merluccius capensis*)	50 MPa, at 5 °C for seven days	Microbial counts and total volatile basic nitrogen content remained unaltered during storage. Drip losses, shear resistance, and whiteness increased, but, after cooking, these differences almost disappeared.	[91]
Atlantic salmon (*Salmo salar*)	50–75 MPa, at 25–37 °C for 10 days	Initial microbial counts were reduced at 75 MPa, while no effect was observed at 50 MPa, and there was an inhibition at 60 MPa. No effect on color parameters. Increase of lipid oxidation state compared to refrigeration.	[92]
Atlantic mackerel (*Scomber scombrus*, L.) fillets	50 MPa, at 5 °C for 12 days	Initial microbial counts were maintained or reduced. No significant lipid degradation was observed, and better fish quality indicators were observed (pH, TVB-N, drip loss, water holding capacity, and firmness after cooking) than under refrigeration.	[93]
Atlantic salmon (*Salmo salar*)	40–60 MPa at 5–15 °C for 10 days	Microbial growth was slowed down with inactivation at 60 MPa. Low values of volatile base nitrogen at 60 MPa up to 15 days with stable trimethylamine-nitrogen. Increase of formaldehyde and dimethylamine-nitrogen content.	[94]
	50–75 MPa at 10–37 °C for 10 days	Initial activities of acid phosphatase, cathepsin B and D, and calpains decreased by increasing the storage temperature. A pronounced increase of myofibrillar fragmentation index at 75 MPa and 25 or 37 °C after 10 days.	[95]
	60 MPa at 10 °C for 30 days	No variations in drip loss, water holding capacity, or myofibrillar fragmentation index. Low changes in muscle fibers, visible by scanning electron micrographs, and a decrease of the resilience property. No effect on fatty acids content, with a lower polyene index compared to refrigeration. Retention of fresh-like volatile compounds.	[96]

Table 7. *Cont.*

Samples	Storage Conditions	Main Results	References
	75 MPa at 25 °C for 30 days	Only resilience (textural property) was affected, decreasing after 30 days. Slower myofibrillar fragmentation index and no effect on fatty acids content, with a lower polyene index, compared to refrigeration. Retention of fresh-like volatile compounds.	[97]
	60 MPa/10 °C and 75 MPa/25 °C for 30 days	Inhibition and inactivation the spoilage microorganisms. Reduction of surrogate pathogenic microorganism (*Bacillus subtilis* endospores, *Escherichia coli*, and *Listeria innocua*) counts.	[98]

* The authors did not mention the scientific name of the species.

Recently, several studies were published regarding the use of HS/LT to preserve fresh fish. Cape hake loins, *Merluccius* spp. [91], and Atlantic mackerel, *S. scombrus*, fillets [93] were both stored at 50 MPa/5 °C for 7and 12 days, respectively, and showed almost no changes in microbial load or TVB-N content compared to the initial samples. For HS samples, an increase was found during storage for water content, water holding capacity, shear resistance, and whiteness. After cooking, HS samples presented a weight lost less than half of the control samples, with no differences in whiteness and only moderate differences by sensorial analysis [91,93].

The quality of Atlantic salmon (*S. salar*) was evaluated by HS/LT (40–60 MPa, 5–15 °C) during 10 days, and a slowdown of spoilage microbial growth was observed, while an additional longer storage experiment (50 days) at 60 MPa/10 °C revealed microbial inactivation (Fidalgo et al., 2019) [94]. Furthermore, the established limit of total volatile base nitrogen was surpassed at 60 MPa/10 °C after 30 days (contrarily to six days at AP/10 °C), but with stable TMA-N content in the former. Formaldehyde and dimethylamine-nitrogen content increased after six days of HS/LT, but only the former progressively increased until the tenth day, indicating a possible formation by the action of enzymatic activity, but also by other chemical reactions. Additionally, HS/LT slightly increased secondary product content from the lipid oxidation, although to a lower extent compared to AP (at the different storage temperatures). This condition of 60 MPa/10 °C also showed no variations in drip loss, water holding capacity, or myofibrillar fragmentation index in Atlantic salmon, with low changes in muscle fibers, visible by scanning electron micrographs [96]. Furthermore, a decrease of resilience (a texture property) and a retention of fresh-like alcohols and aldehydes (not detected in AP samples after 15 days) were observed in these salmon samples stored at 60 MPa/10 °C for 30 days [96].

Proteolytic enzymes and muscle proteins of Atlantic salmon (*S. salar*) were studied under HS [95]. Generally, activities of acid phosphatase, cathepsin B and D, and calpains decreased when compared to fresh salmon, with a more pronounced effect of storage temperature of 37 °C, regardless of the pressure condition. However, activity recovery was observed for some enzymes, as the case of cathepsins B and D, and calpains, which showed an increase of residual activity for samples stored at 60 MPa/10 °C and 75 MPa/25 °C after 50 and 25 days, respectively. A pronounced increase of the myofibrillar fragmentation index was observed at 75 MPa (25/37 °C) after 10 days. Otherwise, at 60 MPa/10 °C, a decrease of myofibrillar fragmentation index values was observed after 50 days of storage. For sarcoplasmic proteins, no effect was observed at 60 MPa/10 °C during 30 days of storage, with a slight increase after 50 days. At 75 MPa/25 °C, a decrease of sarcoplasmic protein content was obtained after 10 days, with no further changes during the 25 days of storage [95].

Spoilage and inoculated surrogate pathogenic (*Bacillus subtilis* endospores, *Escherichia coli*, and *Listeria innocua*) microorganisms were monitored during HS using Atlantic salmon. HS/LT inhibited and inactivated the spoilage microorganisms, and *B. subtilis* endospores reached counts below the detection limit after 30 days, verifying a similar reduction for *E. coli* and *L. innocua* counts [98].

3.2.3. Hyperbaric Storage at Room Temperatures

Similarly, the concept of HS/RT arose as an opportunity to preserve food products, namely fresh fish products, with the published works demonstrating a great opportunity to preserve fresh fish. Tilapia fillets (*O. niloticus*) stored under HS/RT (100 MPa/25 °C) over 12 h revealed almost no changes in the microbial load of mesophiles (4.7 log CFU/g) and psychrophiles (4.59 log CFU/g), and a decrease to about 2.0 log CFU/g when stored at 200 MPa/25 °C for 12 h. Tilapia fillets stored at 200 MPa/25 °C for 12 h showed a lower K-value (40%) than samples stored at AP (92%). This result is significant, because a K-value above 60% indicates putrefaction, and only the HS/RT tilapia fillets were below this limit [90].

Recently, HS at 75 MPa caused a reduction in the initial microbial counts of Atlantic salmon, leading to an increase of the microbial shelf-life of at least 25 days, compared to three days of refrigeration, while, at 60 MPa, a microbial growth slowdown was observed, increasing the microbial shelf-life to at least six days. Additionally, besides the maintenance of muscle color during the 25 days, an enhancement of primary and secondary lipid oxidation products was observed, but to a lower extent compared to AP samples [92]. Later, vacuum-packaged fresh Atlantic salmon loins were studied for 30 days under HS/RT conditions, verifying the retention of important physicochemical properties for at least 15 days, such as fatty acids (n-3 polyunsaturated fatty acids) and fresh-like volatile compounds, or lower lipid oxidation and myofibrillar fragmentation index, while refrigeration after five days showed already volatile spoilage-like compounds due to microbial activity [97].

In addition, spoilage and inoculated surrogate pathogenic (*B. subtilis* endospores, *E. coli*, and *L. innocua*) microorganisms were also monitored during HS/RT using Atlantic salmon. HS/RT inhibited and inactivated the spoilage, and inoculated surrogate pathogenic (*B. subtilis* endospores, *E. coli*, and *L. innocua*) microorganisms reached counts below the detection limit after 30 days, showing that, besides shelf-life extension (due to microbial growth inhibition), it also increased microbial safety (by microbial inactivation) of vacuum-packaged Atlantic salmon [98].

This concept of HS/RT arose as an opportunity to preserve food products, providing an opportunity to reduce energy consumption, carbon footprint, and its associated costs (Figure 2).

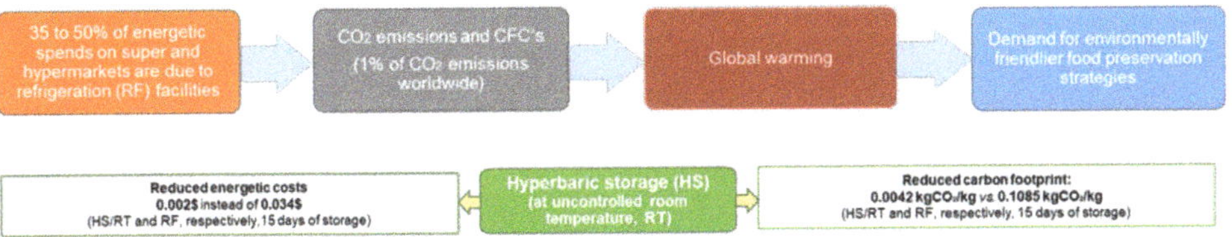

Figure 2. Schematic representation of the energetic costs and environmental impact of hyperbaric storage at uncontrolled room temperature compared to conventional refrigeration [99,100].

Consequently, this method has attracted the attention of many researchers during the last few years, and some studies have been made recently to assess the feasibility of this technology for food preservation compared to refrigeration [101]. HS/RT has been shown to inhibit microbial growth at 50–100 MPa, to inactivate microorganisms at higher pressures (100–220 MPa), and to attenuate some of the physicochemical changes that occur during storage at AP, thereby yielding similar or better products than those obtained with refrigerated storage. HS/RT requires energy only during the short compression and decompression phases, and no additional energy is required to maintain the product while stored under pressure for prolonged periods. Energy cost savings with HS/RT were

101investigated by [99], who concluded that energetic costs of HS/RT were lower than refrigerated storage (Figure 3). Additionally, the carbon footprint associated to HS/RT is also lower than refrigeration. With regard to refrigeration, the two main sources of CO_2 production are from energy utilization and the leakage of cooling gas, while, for HS/RT, the CO_2 produced by energy consumption is negligible, and the main source of CO_2 emissions are attributable to the production of construction materials used for the hyperbaric chamber, thereby demonstrating that HS/RT generates considerably less CO_2 than conventional refrigeration processes [99].

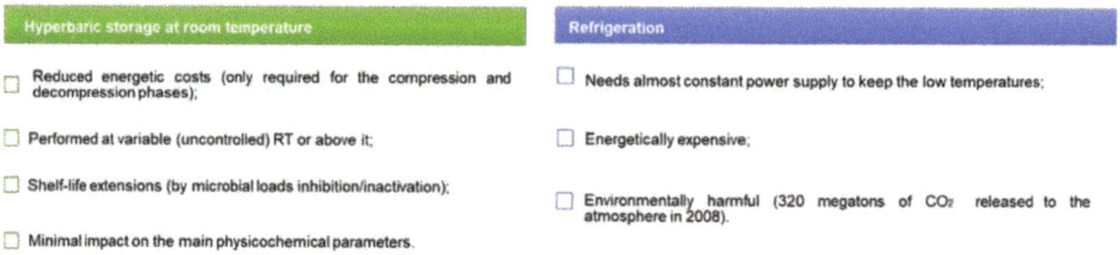

Hyperbaric storage at room temperature	Refrigeration
☐ Reduced energetic costs (only required for the compression and decompression phases);	☐ Needs almost constant power supply to keep the low temperatures;
☐ Performed at variable (uncontrolled) RT or above it;	☐ Energetically expensive;
☐ Shelf-life extensions (by microbial loads inhibition/inactivation);	☐ Environmentally harmful (320 megatons of CO_2 released to the atmosphere in 2008).
☐ Minimal impact on the main physicochemical parameters.	

Figure 3. Schematic representation of the advantages of hyperbaric storage at room temperature compared to conventional refrigeration [102].

4. Conclusions

As stated, fish is a highly perishable food characterized by a short shelf-life. Refrigeration is probably one of the most used methods for fish preservation, along with freezing, and, more recently, superchilling. However, several deteriorative fish quality changes occur during refrigerated storage, particularly in texture, color, and flavor, limiting shelf-life. Frozen storage can avoid these changes (except for texture), but freezing/thawing largely alters the fish fresh-like characteristics. Emerging food packaging techniques, such as the use of edible films and coatings, also meet consumer demands due to their biodegradability and sustainability, while improving the safety and extending the shelf-life of fish and fishery products. Other emergent technologies are arising, as in the case of hyperbaric storage. This methodology uses different pressure and temperature conditions applied at subzero, low, and room temperatures, and has shown the possibility to increase fish shelf-life by microbial inhibition/inactivation, maintaining textural, sensorial, and nutritional characteristics when compared to conventional methods of storage, with the additional advantage of potentially high energy savings, especially when performed at naturally variable/uncontrolled room temperatures. However, currently available high pressure equipment was designed to operate at very high pressure (up to 600 MPa for short minutes), and not to perform hyperbaric storage (up to a maximum of 200 MPa, but for weeks/months), and so specific pressure requirements for hyperbaric storage are of interest to be built.

Author Contributions: Writing, original draft, J.T., A.M., L.G.F., and V.L.; Writing, review and editing, J.T., L.G.F., V.L., R.A.A., and C.A.P.; Validation, A.M.S. and J.A.S.; Resources, A.M.S. and J.A.S.; Project administration, A.M.S. and J.A.S. All authors have read and agreed to the published version of the manuscript.

Funding: This research was funded by project VALORMAR: Valorização Integral dos Recursos Marinhos: Potencial, Inovação Tecnológica e Novas Aplicações (POCI-01-0247-FEDER-024517).

Data Availability Statement: The data that support the findings of this study are available from the corresponding author, J.A.S., upon reasonable request.

Acknowledgments: Thanks are due to the University of Aveiro and FCT/MCT for the financial support for the Associate Laboratory LAQV-REQUIMTE (FCT UIDB/50006/2020) through national founds and, where applicable, co-financed by the FEDER, within the PT2020 Partnership Agreement, and for financing the PhD grant of R.A.A. (SFRH/BD/146009/2019), V.L. (SFRH/BD/146080/2019), A.M. (SFRH/BD/146369/2019), and C.A.P. (SFRH/BD/137036/2018). The authors gratefully acknowledge the financial support of Fundo Europeu de Desenvolvimento Regional (FEDER) através do Programa Operacional Competitividade e Internacionalização (POCI) through the research project VALORMAR: Valorização Integral dos Recursos Marinhos: Potencial, Inovação Tecnológica e Novas Aplicações (POCI-01-0247-FEDER-024517).

Conflicts of Interest: The authors have no conflict of interest to declare. There are no relevant financial or non-financial competing interests to report.

References

1. Prabhakar, P.K.; Vatsa, S.; Srivastav, P.P.; Pathak, S.S. A Comprehensive Review on Freshness of Fish and Assessment: Analytical Methods and Recent Innovations. *Food Res. Int.* **2020**, *133*, 109157. [CrossRef]
2. Ježek, F.; Buchtová, H. Physical and Chemical Changes in Fresh Chilled Muscle Tissue of Common Carp (*Cyprinus Carpio* L.) Packed in a Modified Atmosphere. *Acta Vet. Brno* **2007**, *76*, S83–S92. [CrossRef]
3. Ghaly, A.; Dave, D.; Budge, S.; Brooks, M. Fish Spoilage Mechanisms and Preservation Techniques: Review. *Am. J. Appl. Sci.* **2010**, *7*, 859–877. [CrossRef]
4. Simopoulos, A. Nutritional aspects of fish. In *Producer to Consumer, Integrated Approach to Quality*; Luten, J., Börrensem, T., Oehlenschläger, J., Eds.; Elsevier Science: London, UK, 1997; pp. 587–607.
5. Simpson, B.K.; Nollet, L.M.L.; Toldrá, F.; Benjakul, S.; Paliyath, G.; Hui, Y.H. *Food Biochemistry and Food Processing*, 2nd ed.; Wiley-Blackwell: Hoboken, NJ, USA, 2012; pp. 13–288. ISBN 0813803780.
6. Manirakiza, P.; Covaci, A.; Schepens, P. Comparative Study on Total Lipid Determination Using Soxhlet, Roese-Gottlieb, Bligh & Dyer, and Modified Bligh & Dyer Extraction Methods. *J. Food Compos. Anal.* **2001**, *14*, 93–100. [CrossRef]
7. Kolakowska, A.; Olley, J.; Dunstan, G.A. Fish lipids. In *Chemical and Functional Properties of Food Lipids*; Zdzislaw, Z., Sikorski, E., Kolakowska, A., Eds.; CRC Press Inc.: Boca Raton, FL, USA, 2010; pp. 221–265.
8. Gram, L.; Huss, H.H. Microbiological Spoilage of Fish and Fish Products. *Int. J. Food Microbiol.* **1996**, *33*, 121–137. [CrossRef]
9. Boziaris, I.S. Introduction to seafood processing-assuring quality and safety of seafood. In *Seafood Processing*; John Wiley & Sons, Ltd.: Chichester, UK, 2013; pp. 1–8.
10. FAO. *Post-Harvest Changes in Fish. FAO—Fisheries and Aquaculture Department, Food and Agriculture Organization*; FAO: Rome, Italy, 2005.
11. Stagg, N.J.; Amato, P.M.; Giesbrecht, F.; Lanier, T.C. Autolytic Degradation of Skipjack Tuna during Heating as Affected by Initial Quality and Processing Conditions. *J. Food Sci.* **2012**, *77*, C149–C155. [CrossRef]
12. Li, X.; Li, J.; Zhu, J.; Wang, Y.; Fu, L.; Xuan, W. Postmortem Changes in Yellow Grouper (Epinephelus Awoara) Fillets Stored under Vacuum Packaging at 0 °C. *Food Chem.* **2011**, *126*, 896–901. [CrossRef]
13. Kostaki, M.; Giatrakou, V.; Savvaidis, I.N.; Kontominas, M.G. Combined Effect of MAP and Thyme Essential Oil on the Microbiological, Chemical and Sensory Attributes of Organically Aquacultured Sea Bass (Dicentrarchus Labrax) Fillets. *Food Microbiol.* **2009**, *26*, 475–482. [CrossRef] [PubMed]
14. Kykkidou, S.; Giatrakou, V.; Papavergou, A.; Kontominas, M.G.; Savvaidis, I.N. Effect of Thyme Essential Oil and Packaging Treatments on Fresh Mediterranean Swordfish Fillets during Storage at 4 °C. *Food Chem.* **2009**, *115*, 169–175. [CrossRef]
15. Genç, I.Y.; Esteves, E.; Aníbal, J.; Diler, A. Effects of Chilled Storage on Quality of Vacuum Packed Meagre Fillets. *J. Food Eng.* **2013**, *115*, 486–494. [CrossRef]
16. Hultin, H.O. Oxidation of lipids in seafoods. In *Seafoods Chemistry, Processing Technology and Quality*; Shahidi, F., Botta, J.R., Eds.; Blackie Academic and Professional: London, UK, 1994; pp. 49–74.
17. Huis In't Veld, J.H.J. Microbial and Biochemical Spoilage of Foods: An Overview. *Int. J. Food Microbiol.* **1996**, *33*, 1–18. [CrossRef]
18. Erickson, M. Lipid oxidation: Flavor and nutritional quality deterioration in frozen foods. In *Quality in Frozen Food*; Erickson, M., Hung, Y.-C., Eds.; Chapman and Hall: New York, NY, USA, 1997; pp. 141–173.
19. Khayat, A.; Schwall, D. Lipid Oxidation in Seafood. *Food Technol.* **1983**, *37*, 130–140.
20. Barriuso, B.; Astiasarán, I.; Ansorena, D. A Review of Analytical Methods Measuring Lipid Oxidation Status in Foods: A Challenging Task. *Eur. Food Res. Technol.* **2013**, *236*, 1–15. [CrossRef]
21. Comi, G. Chapter 8—Spoilage of meat and fish. In *Woodhead Publishing Series in Food Science, Technology and Nutrition*; Bender, D.A., Purchase, R., Ruthven, A.D., Waldron, K.W., Johnson, I.T., Fenwick, G.R., Kress-Rogers, E., Bergenstahl, B., Monnier, V., Baynes, J., et al., Eds.; Woodhead Publishing: Sawston, UK, 2017; pp. 179–210. ISBN 978-0-08-100502-6.
22. Sperber, W.H.; Doyle, M.P. (Eds.) *Compendium of the Microbiological Spoilage of Foods and Beverages*; Springer: Griffith, GA, USA, 2010; ISBN 978-1-4419-0826-1.
23. Biji, K.B.; Ravishankar, C.N.; Venkateswarlu, R.; Mohan, C.O.; Gopal, T.K.S. Biogenic Amines in Seafood: A Review. *J. Food Sci. Technol.* **2016**, *53*, 2210–2218. [CrossRef]
24. Silbande, A.; Adenet, S.; Chopin, C.; Cornet, J.; Smith-Ravin, J.; Rochefort, K.; Leroi, F. Effect of Vacuum and Modified Atmosphere Packaging on the Microbiological, Chemical and Sensory Properties of Tropical Red Drum (Sciaenops Ocellatus) Fillets Stored at 4 °C. *Int. J. Food Microbiol.* **2018**, *266*, 31–41. [CrossRef]
25. Church, N. MAP Fish and Crustaceans Sensory Enhancement. *Food Sci. Technol. Today* **1998**, *12*, 73–83.
26. Gram, L.; Huss, H.H. Fresh and processed fish and shellfish. In *The Microbiological Safety and Quality of Foods*; Lund, B.M., Baird-Parker, A.C., Gould, G.W., Eds.; Chapman and Hall: London, UK, 2000; pp. 472–506.

27. Kim, S.-H.; Ben-Gigirey, B.; Barros-Velázquez, J.; Price, R.J.; An, H. Histamine and Biogenic Amine Production by Morganella Morganii Isolated from Temperature-Abused Albacore. *J. Food Prot.* **2000**, *63*, 244–251. [CrossRef] [PubMed]
28. Tahmouzi, S.; Ghasemlou, M.; Aliabadi, F.S.; Shahraz, F.; Hosseini, H.; Khaksar, R. Histamine Formation and Bacteriological Quality in Skipjack Tuna (Katsuwonus Pelamis): Effect of Defrosting Temperature. *J. Food Process. Preserv.* **2013**, *37*, 306–313. [CrossRef]
29. Hungerford, J.M. Scombroid Poisoning: A Review. *Toxicon* **2010**, *56*, 231–243. [CrossRef]
30. Leisner, J.J.; Gram, L. Fish: Spoilage of Fish. In *Encyclopedia of Food Microbiology*; Academic press: Cambridge, MA, USA, 2014; pp. 932–937.
31. Shawyer, M.; Pizzali, M. (Eds.) The use of ice on small fishing vessels. FAO Fisheries Technical Paper 436. In *Food and Agriculture Organization of the United Nation*; FAO: Rome, Italy, 2003; pp. 1–34. ISBN 92-5-105010-4.
32. Kaale, L.D.; Eikevik, T.M.; Rustad, T.; Kolsaker, K. Superchilling of Food: A Review. *J. Food Eng.* **2011**, *107*, 141–146. [CrossRef]
33. Brackett, R.E. Microbiological Safety of Chilled Foods: Current Issues. *Trends Food Sci. Technol.* **1992**, *3*, 81–85. [CrossRef]
34. Berk, Z. (Ed.) Chapter 19—Refrigeration—Chilling and freezing. In *Food Process Engineering and Technology*, 3rd ed.; Food Science and Technology; Academic Press: Cambridge, MA, USA, 2018; pp. 439–461. ISBN 978-0-12-812018-7.
35. Graham, J.; Johnston, W.; Nicholson, F. *Ice in Fisheries: Food and Agriculture Organization*; FAO: Rome, Italy, 1993.
36. Banerjee, R.; Maheswarappa, N.B. Superchilling of Muscle Foods: Potential Alternative for Chilling and Freezing. *Crit. Rev. Food Sci. Nutr.* **2019**, *59*, 1256–1263. [CrossRef] [PubMed]
37. Magnussen, O.M.; Haugland, A.; Hemmingsen, A.K.T.; Johansen, S.; Nordtvedt, T.S. Advances in Superchilling of Food—Process Characteristics and Product Quality. *Trends Food Sci. Technol.* **2008**, *19*, 418–424. [CrossRef]
38. Duun, A.S.; Rustad, T. Quality Changes during Superchilled Storage of Cod (Gadus Morhua) Fillets. *Food Chem.* **2007**, *105*, 1067–1075. [CrossRef]
39. Erikson, U.; Misimi, E.; Gallart-Jornet, L. Superchilling of Rested Atlantic Salmon: Different Chilling Strategies and Effects on Fish and Fillet Quality. *Food Chem.* **2011**, *127*, 1427–1437. [CrossRef]
40. Eliasson, S.; Arason, S.; Margeirsson, B.; Bergsson, A.B.; Palsson, O.P. The Effects of Superchilling on Shelf-Life and Quality Indicators of Whole Atlantic Cod and Fillets. *LWT* **2019**, *100*, 426–434. [CrossRef]
41. Yu, D.; Wu, L.; Regenstein, J.M.; Jiang, Q.; Yang, F.; Xu, Y.; Xia, W. Recent Advances in Quality Retention of Non-Frozen Fish and Fishery Products: A Review. *Crit. Rev. Food Sci. Nutr.* **2020**, *60*, 1747–1759. [CrossRef]
42. Hoang, H.M.; Brown, T.; Indergard, E.; Leducq, D.; Alvarez, G. Life Cycle Assessment of Salmon Cold Chains: Comparison between Chilling and Superchilling Technologies. *J. Clean. Prod.* **2016**, *126*, 363–372. [CrossRef]
43. Wang, T.; Sveinsdóttir, K.; Magnússon, H.; Martinsdóttir, E. Combined Application of Modified Atmosphere Packaging and Superchilled Storage to Extend the Shelf Life of Fresh Cod (Gadus Morhua) Loins. *J. Food Sci.* **2008**, *73*, S11–S19. [CrossRef]
44. Fernández, K.; Aspe, E.; Roeckel, M. Shelf-Life Extension on Fillets of Atlantic Salmon (Salmo Salar) Using Natural Additives, Superchilling and Modified Atmosphere Packaging. *Food Control* **2009**, *20*, 1036–1042. [CrossRef]
45. Gallart-Jornet, L.; Rustad, T.; Barat, J.M.; Fito, P.; Escriche, I. Effect of Superchilled Storage on the Freshness and Salting Behaviour of Atlantic Salmon (Salmo Salar) Fillets. *Food Chem.* **2007**, *103*, 1268–1281. [CrossRef]
46. Duun, A.S.; Rustad, T. Quality of Superchilled Vacuum Packed Atlantic Salmon (Salmo Salar) Fillets Stored at −1.4 and −3.6 °C. *Food Chem.* **2008**, *106*, 122–131. [CrossRef]
47. Stevik, A.M.; Duun, A.S.; Rustad, T.; O'Farrell, M.; Schulerud, H.; Ottestad, S. Ice Fraction Assessment by Near-Infrared Spectroscopy Enhancing Automated Superchilling Process Lines. *J. Food Eng.* **2010**, *100*, 169–177. [CrossRef]
48. Kaale, L.D.; Eikevik, T.M.; Rustad, T.; Nordtvedt, T.S. Changes in Water Holding Capacity and Drip Loss of Atlantic Salmon (Salmo Salar) Muscle during Superchilled Storage. *LWT Food Sci. Technol.* **2014**, *55*, 528–535. [CrossRef]
49. Olafsdottir, G.; Lauzon, H.L.; Martinsdóttir, E.; Oehlenschläger, J.; Kristbergsson, K. Evaluation of Shelf Life of Superchilled Cod (Gadus Morhua) Fillets and the Influence of Temperature Fluctuations during Storage on Microbial and Chemical Quality Indicators. *J. Food Sci.* **2006**, *71*. [CrossRef]
50. Lauzon, H.L.; Magnússon, H.; Sveinsdóttir, K.; Gudjónsdóttir, M.; Martinsdóttir, E. Effect of Brining, Modified Atmosphere Packaging, and Superchilling on the Shelf Life of Cod (Gadus Morhua) Loins. *J. Food Sci.* **2009**, *74*. [CrossRef] [PubMed]
51. Cyprian, O.; Lauzon, H.L.; Jóhannsson, R.; Sveinsdóttir, K.; Arason, S.; Martinsdóttir, E. Shelf Life of Air and Modified Atmosphere-packaged Fresh Tilapia (*Oreochromis Niloticus*) Fillets Stored under Chilled and Superchilled Conditions. *Food Sci. Nutr.* **2013**, *1*, 130–140. [CrossRef]
52. Gökoğlu, N.; Yerlikaya, P. Freezing and frozen storage of fish. In *Seafood Chilling, Refrigeration and Freezing*; John Wiley & Sons, Ltd.: Chichester, UK, 2015; pp. 186–207.
53. Hall, G.M. Freezing and chilling of fish and fish products. In *Fish Processing: Sustainability and New Opportunities*; Wiley-Blackwell: Oxford, UK, 2010; pp. 77–97. ISBN 9781405190473.
54. Tolstorebrov, I.; Eikevik, T.M.; Bantle, M. Effect of Low and Ultra-Low Temperature Applications during Freezing and Frozen Storage on Quality Parameters for Fish. *Int. J. Refrig.* **2016**, *63*, 37–47. [CrossRef]
55. Burgaard, M.G.; Jørgensen, B.M. Effect of Frozen Storage Temperature on Quality-Related Changes in Rainbow Trout (Oncorhynchus Mykiss). *J. Aquat. Food Prod. Technol.* **2011**, *20*, 53–63. [CrossRef]
56. Johnston, W.A.; Nicholson, F.J.; Roger, A. *Freezing and Refrigerated Storage in Fisheries*; Johnston, W.A., Ed.; FAO Fisheries Technical Paper-T340; Food & Agriculture Organization: Rome, Italy, 1994.

57. Jessen, F.; Nielsen, J.; Larsen, E. Chilling and freezing of fish. In *Seafood Processing*; John Wiley & Sons, Ltd.: Chichester, UK, 2013; pp. 33–59.
58. Muthukumarappan, K.; Marella, C.; Sunkesula, V. Food freezing technology. In *Handbook of Farm, Dairy and Food Machinery Engineering*; Academic press: Cambridge, MA, USA, 2019; pp. 389–415.
59. Tsironi, T.; Houhoula, D.; Taoukis, P. Hurdle Technology for Fish Preservation. *Aquac. Fish.* **2020**, *5*, 65–71. [CrossRef]
60. Valdés, A.; Ramos, M.; Beltrán, A.; Jiménez, A.; Garrigós, M.C. State of the Art of Antimicrobial Edible Coatings for Food Packaging Applications. *Coatings* **2017**, *7*, 56. [CrossRef]
61. Jancikova, S.; Dordevic, D.; Jamroz, E.; Behalova, H.; Tremlova, B. Chemical and Physical Characteristics of Edible Films, Based on κ- and ι-Carrageenans with the Addition of Lapacho Tea Extract. *Foods* **2020**, *9*, 357. [CrossRef] [PubMed]
62. Gómez-Estaca, J.; López-de-Dicastillo, C.; Hernández-Muñoz, P.; Catalá, R.; Gavara, R. Advances in Antioxidant Active Food Packaging. *Trends Food Sci. Technol.* **2014**, *35*, 42–51. [CrossRef]
63. Tsironi, T.N.; Taoukis, P.S. Current Practice and Innovations in Fish Packaging. *J. Aquat. Food Prod. Technol.* **2018**, *27*, 1024–1047. [CrossRef]
64. Umaraw, P.; Munekata, P.E.S.; Verma, A.K.; Barba, F.J.; Singh, V.P.; Kumar, P.; Lorenzo, J.M. Edible Films/Coating with Tailored Properties for Active Packaging of Meat, Fish and Derived Products. *Trends Food Sci. Technol.* **2020**, *98*, 10–24. [CrossRef]
65. Ganiari, S.; Choulitoudi, E.; Oreopoulou, V. Edible and Active Films and Coatings as Carriers of Natural Antioxidants for Lipid Food. *Trends Food Sci. Technol.* **2017**, *68*, 70–82. [CrossRef]
66. Galus, S.; Arik Kibar, E.A.; Gniewosz, M.; Kraśniewska, K. Novel Materials in the Preparation of Edible Films and Coatings—A Review. *Coatings* **2020**, *10*, 674. [CrossRef]
67. Lee, K.T. Quality and Safety Aspects of Meat Products as Affected by Various Physical Manipulations of Packaging Materials. *Meat Sci.* **2010**, *86*, 138–150. [CrossRef]
68. Alishahi, A.; Aïder, M. Applications of Chitosan in the Seafood Industry and Aquaculture: A Review. *Food Bioprocess Technol.* **2012**, *5*, 817–830. [CrossRef]
69. Ojagh, S.M.; Rezaei, M.; Razavi, S.H.; Hosseini, S.M.H. Effect of Chitosan Coatings Enriched with Cinnamon Oil on the Quality of Refrigerated Rainbow Trout. *Food Chem.* **2010**, *120*, 193–198. [CrossRef]
70. Bahram, S.; Rezaie, M.; Soltani, M.; Kamali, A.; Abdollahi, M.; Khezri Ahmadabad, M.; Nemati, M. Effect of Whey Protein Concentrate Coating Cinamon Oil on Quality and Shelf Life of Refrigerated Beluga Sturegeon (Huso Huso). *J. Food Qual.* **2016**, *39*, 743–749. [CrossRef]
71. Kaewprachu, P.; Osako, K.; Benjakul, S.; Suthiluk, P.; Rawdkuen, S. Shelf Life Extension for Bluefin Tuna Slices (Thunnus Thynnus) Wrapped with Myofibrillar Protein Film Incorporated with Catechin-Kradon Extract. *Food Control* **2017**, *79*, 333–343. [CrossRef]
72. Volpe, M.G.; Siano, F.; Paolucci, M.; Sacco, A.; Sorrentino, A.; Malinconico, M.; Varricchio, E. Active Edible Coating Effectiveness in Shelf-Life Enhancement of Trout (Oncorhynchusmykiss) Fillets. *LWT Food Sci. Technol.* **2015**, *60*, 615–622. [CrossRef]
73. Yu, D.; Jiang, Q.; Xu, Y.; Xia, W. The Shelf Life Extension of Refrigerated Grass Carp (Ctenopharyngodon Idellus) Fillets by Chitosan Coating Combined with Glycerol Monolaurate. *Int. J. Biol. Macromol.* **2017**, *101*, 448–454. [CrossRef]
74. Jouki, M.; Yazdia, F.T.; Mortazavia, S.A.; Koocheki, A.; Khazaei, N. Effect of Quince Seed Mucilage Edible Films Incorporated with Oregano or Thyme Essential Oil on Shelf Life Extension of Refrigerated Rainbow Trout Fillets. *Int. J. Food Microbiol.* **2014**, *174*, 88–97. [CrossRef]
75. Kazemi, S.M.; Rezaei, M. Antimicrobial Effectiveness of Gelatin-Alginate Film Containing Oregano Essential Oil for Fish Preservation. *J. Food Saf.* **2015**, *35*, 482–490. [CrossRef]
76. Li, T.; Li, J.; Hu, W.; Li, X. Quality Enhancement in Refrigerated Red Drum (Sciaenops Ocellatus) Fillets Using Chitosan Coatings Containing Natural Preservatives. *Food Chem.* **2013**, *138*, 821–826. [CrossRef]
77. Gómez-Estaca, J.; López De Lacey, A.; Gómez-Guillén, M.C.; López-Caballero, M.E.; Montero, P. Antimicrobial Activity of Composite Edible Films Based on Fish Gelatin and Chitosan Incorporated with Clove Essential Oil. *J. Aquat. Food Prod. Technol.* **2009**, *18*, 46–52. [CrossRef]
78. Yordanov, D.G.; Angelova, G.V. High Pressure Processing for Foods Preserving. *Biotechnol. Biotechnol. Equip.* **2010**, *24*, 1940–1945. [CrossRef]
79. Balasubramaniam, V.M.; Martínez-Monteagudo, S.I.; Gupta, R. Principles and Application of High Pressure-Based Technologies in the Food Industry. *Annu. Rev. Food Sci. Technol.* **2015**, *6*, 435–462. [CrossRef]
80. Fidalgo, L.G.; Saraiva, J.A.; Aubourg, S.P.; Vázquez, M.; Torres, J.A. Effect of High-Pressure Pre-Treatments on Enzymatic Activities of Atlantic Mackerel (*Scomber Scombrus*) during Frozen Storage. *Innov. Food Sci. Emerg. Technol.* **2014**, *23*, 18–24. [CrossRef]
81. Yang, H.; Gao, Y.; Zhang, H.; Qi, X.; Pan, D. Effects of High Hydrostatic Pressure Treatment on the Qualities of Cultured Large Yellow Croaker (*Pseudosciaena Crocea*) during Cold Storage. *J. Food Process. Preserv.* **2014**. [CrossRef]
82. Moreno, H.M.; Bargiela, V.; Tovar, C.A.; Cando, D.; Borderias, A.J.; Herranz, B. High Pressure Applied to Frozen Flying Fish (*Parexocoetus Brachyterus*) Surimi: Effect on Physicochemical and Rheological Properties of Gels. *Food Hydrocoll.* **2015**, *48*, 127–134. [CrossRef]
83. Tironi, V.; de Lamballerie, M.; Le-Bail, A. Quality Changes during the Frozen Storage of Sea Bass (*Dicentrarchus Labrax*) Muscle after Pressure Shift Freezing and Pressure Assisted Thawing. *Innov. Food Sci. Emerg. Technol.* **2010**, *11*, 565–573. [CrossRef]

84. Nakazawa, N.; Okazaki, E. Recent Research on Factors Influencing the Quality of Frozen Seafood. *Fish Sci.* **2020**, *86*, 231–244. [CrossRef]
85. Jannasch, H.W.; Eimhjellen, K.; Wirsen, C.O.; Farmanfarmaian, A. Microbial Degradation of Organic Matter in the Deep Sea. *Science* **1971**, *171*, 672–675. [CrossRef]
86. Charm, S.E.; Longmaid, H.E., III; Carver, J. A Simple System for Extending Refrigerated, Nonfrozen Preservation of Biological Material Using Pressure. *Cryobiology* **1977**, *14*, 625–636. [CrossRef]
87. Wagner, W.; Saul, A.; Pruss, A. International Equations for the Pressure along the Melting and along the Sublimation Curve of Ordinary Water Substance. *J. Phys. Chem. Ref. Data* **1994**, *23*, 515. [CrossRef]
88. Kalichevsky, M.T.; Knorr, D.; Lillford, P.J. Potential Food Applications of High-Pressure Effects on Ice-Water Transitions. *Trends Food Sci. Technol.* **1995**, *6*, 253–259. [CrossRef]
89. Ooide, A.; Kameyama, Y.; Iwata, N.; Uchio, R.; Karino, S.; Kanyama, N. Non-freezing preservation of fresh foods under subzero temperature. In *High Pressure Bioscience*; Hayashi, R.S.K., Shimada, S., Suzuki, A., Eds.; San-Ei Shuppan Co.: Kyoto, Japan, 1994; pp. 344–351.
90. Ko, W.-C.; Hsu, K.-C. Changes in K Value and Microorganisms of Tilapia Fillet during Storage at High-Pressure, Normal Temperature. *J. Food Prot.* **2001**, *64*, 94–98. [CrossRef]
91. Otero, L.; Pérez-Mateos, M.; López-Caballero, M.E. Hyperbaric Cold Storage versus Conventional Refrigeration for Extending the Shelf-Life of Hake Loins. *Innov. Food Sci. Emerg. Technol.* **2017**, *41*, 19–25. [CrossRef]
92. Fidalgo, L.G.; Lemos, Á.T.; Delgadillo, I.; Saraiva, J.A. Microbial and Physicochemical Evolution during Hyperbaric Storage at Room Temperature of Fresh Atlantic Salmon (*Salmo Salar*). *Innov. Food Sci. Emerg. Technol.* **2018**, *45*, 264–272. [CrossRef]
93. Otero, L.; Pérez-Mateos, M.; Holgado, F.; Márquez-Ruiz, G.; López-Caballero, M.E. Hyperbaric Cold Storage: Pressure as an Effective Tool for Extending the Shelf-Life of Refrigerated Mackerel (*Scomber Scombrus*, L.). *Innov. Food Sci. Emerg. Technol.* **2019**, *51*, 41–50. [CrossRef]
94. Fidalgo, L.G.; Castro, R.; Trigo, M.; Aubourg, S.P.; Delgadillo, I.; Saraiva, J.A. Quality of Fresh Atlantic Salmon (*Salmo Salar*) under Hyperbaric Storage at Low Temperature by Evaluation of Microbial and Physicochemical Quality Indicators. *Food Bioprocess Technol.* **2019**, *12*, 1895–1906. [CrossRef]
95. Fidalgo, L.G.; Delgadillo, I.; Saraiva, J.A. Autolytic Changes Involving Proteolytic Enzymes on Atlantic Salmon (Salmo Salar) Preserved by Hyperbaric Storage. *LWT* **2020**, *118*, 108755. [CrossRef]
96. Fidalgo, L.G.; Simões, M.M.Q.; Casal, S.; Lopes-da-Silva, J.A.; Carta, A.M.S.; Delgadillo, I.; Saraiva, J.A. Physicochemical Parameters, Lipids Stability, and Volatiles Profile of Vacuum-Packaged Fresh Atlantic Salmon (Salmo Salar) Loins Preserved by Hyperbaric Storage at 10 °C. *Food Res. Int.* **2020**, *127*, 108740. [CrossRef] [PubMed]
97. Fidalgo, L.G.; Simões, M.M.Q.; Casal, S.; Lopes-da-Silva, J.A.; Delgadillo, I.; Saraiva, J.A. Enhanced Preservation of Vacuum-Packaged Atlantic Salmon by Hyperbaric Storage at Room Temperature versus Refrigeration. *Sci. Rep.* **2021**, *11*, 1–13. [CrossRef] [PubMed]
98. Fidalgo, L.G.; Pint Fernandeso, C.A.; Delgadillo, I.; Saraiva, J.A. Hyperbaric Storage of Vacuum-Packaged Fresh Atlantic Salmon (Salmo Salar) Loins by Evaluation of Spoilage Microbiota and Inoculated Surrogate-Pathogenic Microorganisms. *Food Eng. Rev.* **2021**. [CrossRef]
99. Bermejo-Prada, A.; Colmant, A.; Otero, L.; Guignon, B. Industrial Viability of the Hyperbaric Method to Store Perishable Foods at Room Temperature. *J. Food Eng.* **2017**, *193*, 76–85. [CrossRef]
100. Segovia-Bravo, K.A.; Guignon, B.; Bermejo-Prada, A.; Sanz, P.D.; Otero, L. Hyperbaric Storage at Room Temperature for Food Preservation: A Study in Strawberry Juice. *Innov. Food Sci. Emerg. Technol.* **2012**, *15*, 14–22. [CrossRef]
101. Fernandes, P.A.R.; Moreira, S.A.; Fidalgo, L.G.; Santos, M.D.; Queirós, R.P.; Delgadillo, I.; Saraiva, J.A. Food Preservation under Pressure (Hyperbaric Storage) as a Possible Improvement/Alternative to Refrigeration. *Food Eng. Rev.* **2015**, *7*, 1–10. [CrossRef]
102. Gilbert, N. One-Third of Our Greenhouse Gas Emissions Come from Agriculture. *Nature* **2012**, 1–4. [CrossRef]

MDPI
St. Alban-Anlage 66
4052 Basel
Switzerland
Tel. +41 61 683 77 34
Fax +41 61 302 89 18
www.mdpi.com

Foods Editorial Office
E-mail: foods@mdpi.com
www.mdpi.com/journal/foods

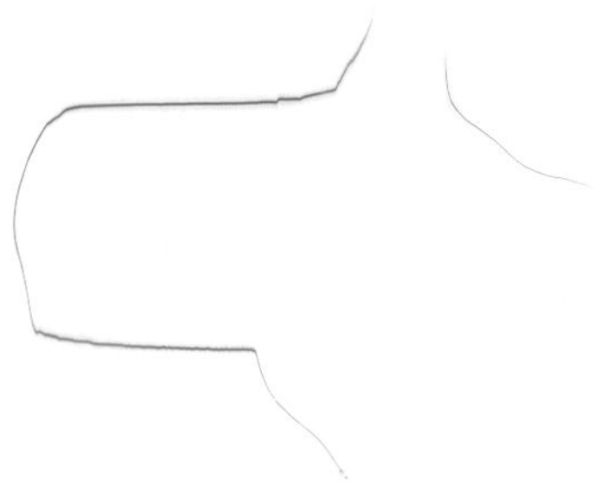

www.ingramcontent.com/pod-product-compliance
Lightning Source LLC
LaVergne TN
LVHW071253170726
843515LV00006BA/1395

* 9 7 8 3 0 3 6 5 3 9 1 3 3 *